Riti Thapar Kapoor and Mohd Rafatullah (Eds.)
Bioremediation Technologies

Also of interest

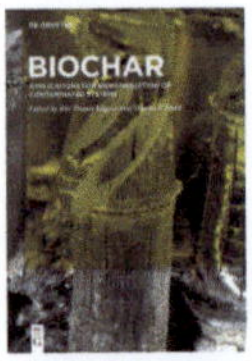

BioChar.
Applications for Bioremediation of Contaminated Systems
Edited by Riti Thapar Kapoor, Maulin P. Shah, 2022
ISBN 978-3-11-073858-2, e-ISBN 978-3-11-073400-3

Green Pulp and Paper Industry.
Biotechnology for Ecofriendly Processing
Edited by Amit Kumar, Puneet Pathak, Dharm Dutt, 2021
ISBN 978-3-11-059184-2, e-ISBN 978-3-11-059241-2

Industrial Waste.
Characterization, Modification and Applications of Residues
Edited by Herbert Pöllmann, 2021
ISBN 978-3-11-067486-6, e-ISBN 978-3-11-067494-1

Bioelectrochemistry.
Design and Applications of Biomaterials
Edited by Serge Cosnier, 2019
ISBN 978-3-11-056898-1, e-ISBN 978-3-11-057052-6

Photosynthesis.
Biotechnological Applications with Micro-Algae
Edited by Matthias Rögner, 2021
ISBN 978-3-11-071691-7, e-ISBN 978-3-11-071697-9

Aquatic Chemistry.
For Water and Wastewater Treatment Applications
Ori Lahav, Liat Birnhack, 2019
ISBN 978-3-11-060392-7, e-ISBN 978-3-11-060395-8

Bioremediation Technologies

For Wastewater and Sustainable Circular Bioeconomy

Edited by
Riti Thapar Kapoor and Mohd Rafatullah

DE GRUYTER

Editors
Dr. Riti Thapar Kapoor
Amity Institute of Biotechnology
Amity University Uttar Pradesh
Sector 125
Noida 201313
Uttar Pradesh
India
rititkapoor@gmail.com

Dr. Mohd Rafatullah
Division of Environmental Technology
School of Industrial Technology
Universiti Sains Malaysia
11800 Penang
Malaysia
mrafatullah@usm.my

ISBN 978-3-11-101665-8
e-ISBN (PDF) 978-3-11-101682-5
e-ISBN (EPUB) 978-3-11-101701-3

Library of Congress Control Number: 2023931061

Bibliographic information published by the Deutsche Nationalbibliothek
The Deutsche Nationalbibliothek lists this publication in the Deutsche Nationalbibliografie; detailed bibliographic data are available on the Internet at http://dnb.dnb.de.

Cover image: adventtr/iStock/Getty Images Plus
Typesetting: Integra Software Services Pvt. Ltd.
Printing and binding: CPI books GmbH, Leck

www.degruyter.com

Contents

Editors Biography

Dr. Riti Thapar Kapoor is Associate Professor in Amity Institute of Biotechnology, Amity University Uttar Pradesh, Noida, India. Dr. Kapoor received her Ph.D from University of Allahabad and worked as post - doctoral fellow at Banaras Hindu University, Varanasi, India. Dr. Kapoor has fourteen years of teaching and research experience and her area of specialization is environmental biotechnology, bioremediation, wastewater treatment and abiotic stresses. Dr. Kapoor has published 7 books and over 105 research papers in various journals of national and international repute. Dr. Kapoor has visited 8 countries for participation in different academic programmes. Dr. Kapoor has received prestigious travel award from Bill & Melinda Gates Research Foundation under CGIAR project for participation in International training programme held at International Rice Research Institute (IRRI), Manila, Philippines in 2010. She is also recipient of DST travel grant for participation in International Conference held at Sri Lanka in 2013. Dr. Kapoor has been awarded with Teacher's Research Fellowship from Indian Academy of Sciences, Bengaluru in 2019. She has supervised and mentored a number of research projects granted by different government funding agencies such as DAE, DST and UPCST etc. She has successfully supervised three research students for Ph.D. degree besides several (more than 70) masters and graduate students for their dissertation thesis.

Dr. Riti Thapar Kapoor, Associate Professor, Amity Institute of Biotechnology, Amity University Uttar Pradesh, Sector - 125, Noida - 201 313, India.
E.mail: rititkapoor@gmail.com

Dr. Mohd Rafatullah I am presently working as an Associate Professor of Environmental Technology in the School of Industrial Technology, Universiti Sains Malaysia (USM), Malaysia. I joined this School in the year 2008 as a Post Doctoral Fellow. I have completed my education; Ph. D. in Environmental Chemistry, Master of Science in Analytical Chemistry and Bachelor of Science in Chemistry from Aligarh Muslim University (AMU), India. My research interest is in the areas of environmental water pollutants and their safe removal; preparation of various nanomaterials to protect the environment; water and wastewater treatment; adsorption and ion exchange; microbial fuel cells; advance oxidation process; activated carbons and their electrochemical properties. My contribution was recognized by Guest Editors and Member of Editorial Board of various scientific journals. Listed in Word's top 2% Scientist by Stanford University, listed in Top 1% peer reviewer, in Chemistry, Environmental Science and cross-field on Publons global reviewer, Web of Science. Life time Fellow member of International Society of Sustainable Developments and member of various professional international societies. I have published several reviews articles and regular research papers in the journals of international repute and presented my research work in various national and international conferences. I have also attended many workshops and seminars of environmental chemistry. Based on my performance and contribution in research (total citations: > 11900 and h index: 47 @ Scopus)

Dr. Mohd Rafatullah, Associate Professor, Division of Environmental Technology
School of Industrial Technology, Universiti Sains Malaysia
11800 Penang, Malaysia
E.mail: mrafatullah@usm.my

https://doi.org/10.1515/9783111016825-203

List of authors

A. I. Abd-Elhamid
Composites and Nanostructured Materials
Research Department,
Advanced Technology and New Materials
Research Institute (ATNMRI),
City of Scientific Research and Technological
Applications (SRTA-City),
New Borg Al-Arab,
Alexandria 21934,
Egypt

AbdElAziz A. Nayl
Department of Chemistry,
College of Science,
Jouf University,
Sakaka 72341,
Saudi Arabia

H. F. Aly
Hot Laboratories Center,
Egyptian Atomic Energy Authority,
Cairo 13759,
Egypt

Arniza Ghazali
Division of Bioresource Technology,
School of Industrial Technology,
Universiti Sains Malaysia,
11800 USM,
Penang,
Malaysia

AnandKumar J.
National Institute of Technology Raipur
Raipur, Chhattisgarh 492010,
India

Jyoti Kant Choudhari
Maulana Azad National Institute
of Technology Bhopal,
Bhopal 462006,
Madhya Pradesh,
India

Mukesh Kumar Verma
National Institute of Technology Raipur,
Raipur, Chhattisgarh 492010,
India

Jyotsna Choubey
Chhattisgarh Swami Vivekanand
Technical University,
Bhilai 491107,
Chhattisgarh,
India

Biju Prava Sahariah
Chhattisgarh Swami Vivekanand
Technical University,
Bhilai 491107,
Chhattisgarh,
India

Harlina Ahmad
Environmental Technology Division,
School of Industrial Technology,
Universiti Sains Malaysia,
Penang 11800,
Malaysia

Mardiana Idayu Ahmad
Environmental Technology Division,
School of Industrial Technology,
Universiti Sains Malaysia,
Penang 11800,
Malaysia

Rekah Nadarajah
Environmental Technology Division,
School of Industrial Technology,
Universiti Sains Malaysia,
Penang 11800,
Malaysia

Nishalini Ratha Pukallenthy
Environmental Technology Division,
School of Industrial Technology,
Universiti Sains Malaysia,
Penang 11800,
Malaysia

https://doi.org/10.1515/9783111016825-204

Nur Kamila Ramli
Environmental Technology Division,
School of Industrial Technology,
Universiti Sains Malaysia,
Penang 11800,
Malaysia

Nur Anis Ahmad
Environmental Technology Division,
School of Industrial Technology,
Universiti Sains Malaysia,
Penang 11800,
Malaysia

Masoom Raza Siddiqui
Advanced Materials Research Chair,
Chemistry Department,
College of Science,
King Saud University,
Riyadh,
Saudi Arabia

Moonis Ali Khan
Chemistry Department,
College of Science,
King Saud University,
Riyadh,
Saudi Arabia

Mohd Rafatullah
Environmental Technology Division,
School of Industrial Technology,
Universiti Sains Malaysia,
Penang 11800,
Malaysia

Riti Thapar Kapoor
Amity Institute of Biotechnology,
Amity University Uttar Pradesh,
Noida 201 313,
Uttar Pradesh,
India

N. M. Nurazzi
Bioresource Technology Division,
School of Industrial Technology,
Universiti Sains Malaysia,
Penang 11800,
Malaysia

M. N. F. Norrrahim
Research Centre for Chemical Defence,
Universiti Pertahanan Nasional
Malaysia (UPNM),
Kem Perdana Sungai Besi,
Kuala Lumpur 57000,
Malaysia

M. R. M. Asyraf
Engineering Design Research Group (EDRG),
Faculty of Mechanical Engineering,
Universiti Teknologi Malaysia,
Johor Bahru,
Johor 81310,
Malaysia

M. S. Samsudin
Environmental Technology Division,
School of Industrial Technology,
Universiti Sains Malaysia,
Penang 11800,
Malaysia

Mohammad Qutob
Environmental Technology Division,
School of Industrial Technology,
Universiti Sains Malaysia,
Penang 11800,
Malaysia

Syahidah Akmal Muhammad
Environmental Technology Division,
School of Industrial Technology,
Universiti Sains Malaysia,
Penang 11800,
Malaysia

Norli Ismail
Environmental Technology Division,
School of Industrial Technology,
Universiti Sains Malaysia,
11800 Penang,
Malaysia

Nur Syamimi Selamat
Environmental Technology Division,
School of Industrial Technology,
Universiti Sains Malaysia,
11800 Penang,
Malaysia

Amir Talebi
Schahverdy Entwicklung und Konstruktion
Maschinenbau,
Baden-Württemberg 79274,
Germany

Siti Baidurah
Bioprocess Technology Division,
School of Industrial Technology,
Universiti Sains Malaysia,
Penang 11800,
Malaysia

Debbie Dominic
Bioprocess Technology Division,
School of Industrial Technology,
Universiti Sains Malaysia,
Penang 11800,
Malaysia

Ooi Wen Ching
Bioprocess Technology Division,
School of Industrial Technology,
Universiti Sains Malaysia,
Penang 11800,
Malaysia

Lim Wai See
Bioprocess Technology Division,
School of Industrial Technology,
Universiti Sains Malaysia,
Penang 11800,
Malaysia

Mohd Asyraf Kassim
Bioprocess Technology Division,
School of Industrial Technology,
Universiti Sains Malaysia,
Penang 11800,
Malaysia

Soma Nag
Chemical Engineering Department,
National Institute of Technology Agartala,
Tripura,
India

Joydeep Das
Chemical Engineering Department,
National Institute of Technology Agartala,
Tripura,
India

Snehannita Sarkar
Chemical Engineering Department,
National Institute of Technology Agartala,
Tripura,
India

Kousish Saha
Chemical Engineering Department,
National Institute of Technology Agartala,
Tripura,
India

Sandeep Roy
Chemical Engineering Department,
National Institute of Technology Agartala,
Tripura,
India

Tanvir Arfin
Air Pollution Control Division,
CSIR-National Environmental Engineering
Research Institute (CSIR-NEERI), Nehru Marg,
Nagpur 440020,
India

Nikhila Mathew
Air Pollution Control Division,
CSIR-National Environmental Engineering
Research Institute (CSIR-NEERI), Nehru Marg,
Nagpur 440020,
India

Abha Tirpude
Air Pollution Control Division,
CSIR-National Environmental Engineering
Research Institute (CSIR-NEERI), Nehru Marg,
Nagpur 440020,
India

Anupama M. Pillai
Air Pollution Control Division,
CSIR-National Environmental Engineering
Research Institute (CSIR-NEERI), Nehru Marg,
Nagpur 440020,
India

Pabitra Mondal
Air Pollution Control Division,
CSIR-National Environmental Engineering
Research Institute (CSIR-NEERI), Nehru Marg,
Nagpur 440020,
India

Wajih Jamal
Department of Zoology,
Faculty of Life Sciences,
Aligarh Muslim University,
Aligarh,
Uttar Pradesh,
India

Tijo Cherian
Department of Ocean Studies and Marine
Biology,
Faculty of Life Sciences,
Pondicherry University,
Puducherry,
India

Shibin E.
Department of Ocean Studies and Marine
Biology,
Faculty of Life Sciences,
Pondicherry University,
Puducherry,
India

R. Mohanraju
Department of Ocean Studies and Marine
Biology,
Faculty of Life Sciences,
Pondicherry University,
Puducherry,
India

A. I. Abd-Elhamid*, AbdElAziz A. Nayl, H. F. Aly

Utilization of Nanomaterials for Remediation of Pollutants

Abstract: Nowadays, environmental pollution with various contaminates is a significant global problem all over the world. Wastewater includes organic and inorganic pollutants, which make it significantly dangerous on all living organisms if they released into the water body. Therefore, these effluents must be treated to eliminate or at least minimize the pollutant contents on the wastewater before emission to the raw water. Several strategies were investigated for efficient treatment of the wastewater. Among them photocatalytic and adsorption appear to be the most valuable method due to their availability, low operation cost, easy to proceed, and nontoxic by-product. To achieve great efficiency of these previous strategies, nanomaterials are highly required. Nanomaterials are the materials with dimension less than 100 nm. The materials at this size have superior physicochemical properties compared with bulk materials such as optical, electrical properties, and high specific surface area which make most of atoms located at the material surface. This feature allows for rapid and efficient interaction between the active sites and pollutants. Moreover, the nanomaterials can be modified for specific use. All of these characters make the nanomaterials extensively used in environmental clean-up than other conventional approaches.

Keywords: Nanomaterials, remediation, wastewater, photocatalysis, adsorption

1 Introduction

Several industrial activities require huge amount of water in their process such as textile, paper, paint, mining, electroplating, pharmaceuticals, steel manufacturing, pigment production, and leather. As a result, the large effluent quantities of various contamination species are produced from these industries including dye, active ingredients, heavy metals, and pesticides. The release of these wastewater into water body (streams, rivers, and lakes) will induce aqua-sphere life and significant impacts on the environment and threat the human health due to the harmful toxicity effect of these

***Corresponding author: A. I. Abd-Elhamid,** Composites and Nanostructured Materials Research Department, Advanced Technology and New Materials Research Institute (ATNMRI), City of Scientific Research and Technological Applications (SRTA-City), New Borg Al-Arab, Alexandria 21934, Egypt, e-mail: ahm_ch_ibr@yahoo.com.

AbdElAziz A. Nayl, Department of Chemistry, College of Science, Jouf University, Sakaka 72341, Saudi Arabia

H. F. Aly, Hot Laboratories Center, Egyptian Atomic Energy Authority, Cairo 13759, Egypt

https://doi.org/10.1515/9783111016825-001

materials. Moreover, the growth and quality of the crop can also be influenced by the contamination presented in water. On the other hand, scarcity of pure water considers as one of the vital problems that the world is facing. Thus, it is very important to find suitable treatment of these wastewater before releasing into water body. Various physical, chemical, and biological strategies were investigated to remove or at least minimize the pollutants from the sacred water. But some of these techniques may possess shortages like high energy consuming, long processing period, high cost, and unpleasant odors and require complicated design and produce toxic wastes or subproducts. Among them advanced oxidation processes (AOP) and adsorption technology seem to be the most promising techniques.

Photocatalysis is defined as AOP for a wide range of pollutant organic and some inorganic compounds and disinfection from wastewater. In which $^{\bullet}OH$ and $O_2^{-\bullet}$ are generated by the exposure of a semiconductor to UV or vis light irradiation; thereafter, these radicals can attack a variety of pollutants and mineralized it simple and nontoxic products. In detail, at exposing the catalyst to light with energy higher than its bandgap, this will generate electron (e^-, in the conductive band–hole; h^+, in the valence band pairs). Thereafter, the surface water molecules interact with the hole and produce a reactive ($^{\bullet}OH$) and H^+. At electron the H^+ and O_2 molecule will promote the formation of superoxide radical anions ($^{\bullet}OOH$) species. These generated species ($^{\bullet}OH$ and $^{\bullet}OOH$) are primary oxidizing species that can provide mineralization of the target compounds.

Adsorption is an episode that occurs at the solid surface with common behavior for interaction with various pollutants. When an aqueous environment having absorbable material (adsorbate) is mixed with a solid (adsorbent) with active sites, attraction forces of the liquid–solid result in transferring the adsorbate from the bulk of solution to the active surface sites. Followed by interaction among the adsorbent and the binding site, this surface tendency of adsorbate toward the adsorbent is called adsorption. In interior of the molecules, all the bonding (ionic, covalent, or metallic) of the building units of the material is coupled by another in the same molecule. On the contrary, the atoms located at the adsorbent surface have a vacant site enabled to interact with adsorbates. Depending on the nature of adsorbent–adsorbate bond, the adsorption mechanism can be described as physicsorption, chemisorption, or electrostatic adsorption. The adsorption process is excessively applied for cleaning of wastewater from organic and inorganic contaminates and attracts high interest from the scientists. This is attributed to simple operation and low cost and no side pollution and requires land. Recently, the use for economic adsorbents with excellent contaminated capturing quantities has densified. Local materials such as wastes derived from agricultural and industrial can provide inexpensive source adsorbents.

For great use of these previous techniques, nanomaterials were required. Nanomaterials can be defined as those ultrafine materials with size not exceeded than 100 nm. For the same substance, the features of nanomaterials significantly alter from those of bulk materials. The atom diameter lies between 0.15 and 0.6 nm; thus a large section of the atoms the nanomaterial located its surface. So that the characters of that surface layer of the nanomaterials enhanced markedly over those of the bulk material. This impact will highly improve different physical and chemical properties of the nanomaterials. Moreover, the interaction of these nanomaterials and another one of their interface becomes significant. Thus, the usage of the nanomaterials in treating the pollutants in the wastewater will be an excellent choice. In this chapter, we will explain the role of different nanomaterials in the wastewater treatment using two techniques: photocatalysis and adsorption.

2 Photocatalysis

Photocatalysis is a process in which the light-induced reaction is sped up by means of heterogeneous catalysts for breaking the contaminated particles into less toxic compound. Photocatalyst is a semiconductor material, such as ZnO, TiO_2, zinc sulfide (ZnS), CdSe, Fe_2O_3, MoS_2, and WO_3, that have the ability to conduct the electricity at room temperature at exposure to light source. In the semiconductor, the energy difference among the conduction band (CB) and the valence band (VB) is known as the band gap. When a photocatalyst is irradiated with light source of the suitable wavelength (photon energy > band gap energy), the electrons (e^-) in the VB will absorb the energy of photons and excited to CB. This will generate vacancies, called holes (h^+). The photogenerated electron-hole pairs converted to free charge carriers, transfer to the photocatalyst surface, and share in hole-mediated oxidation and electron-mediated reduction.

In order to extend the range of light absorption and increase their photocatalytic activity of the semiconductor, element doping and heterostructure construction are investigated. This phenomenon is attributed to the generation of new energy levels close to the CB, which result in narrowed band gap energy of the doped photocatalysts. Heterostructure construction can directly improve the absorption irradiation of the photocatalyst matrix by physically or chemically bonding a sensitizer that absorbs visible light. Table 1 lists the various photocatalysts, pollutants, pollutant concentrations, light sources, degradation efficiencies, and contact times.

Table 1: Various photocatalysts, pollutants, pollutant concentrations, light sources, degradation efficiencies, and contact times.

Photocatalysis	Pollutants	Pollutant concentration	Light source	Degradation efficiency (%)	Time (min)	References
ZnO NPs	MB RR120		Sunlight	90	150	[1]
Ag-ZnO	MB 4-NP	15 ppm 40 ppm	UV–visible light	98 100	60 6	[2]

Table 1 (continued)

Photocatalysis	Pollutants	Pollutant concentration	Light source	Degradation efficiency (%)	Time (min)	References
ZnO/CuO	RhB	10 ppm	Sunlight	98%	180	[3]
ZnHAp	Ciprofloxacin ofloxacin	20 ppm	UV A–B–C light (HPK lamp 125 W, 200–600 nm	97.98	180	[4]
Zn-Fe-Sb **Zn-Fe-Ti** **Zn-Fe-Ca**	AO	0.063 mM	Sunlight	98	20 80 90	[5]
ZnS QD	Amaranth dye	40 μM	UV light	85	120	[6]
PVDF/ZnO	AZG MG	10 ppm	Sunlight	85% 90%	120 240	[7]
APTES/TiO_2	MB		Artificial solar light	31	360	[8]
$MgTiO_3$	Lomefloxacin	10 ppm	Visible light (30 W LED)	83	150	[9]
PVDF-TiO_2/ CNT/$BiVO_4$	BSA		Visible light	97	150	[10]
CuAl2O4-3 wt%Ag	MB	7 ppm	450 W Xenon lamps	97	80	[11]
ZC-40	Diazinon Malathion Glyphosate Chlorpyrifos	10 ppm	Visible light (180 mW cm^{-2})	95.2 90.1 85.4 97.4	60	[13]
CNW-gCN/ACF	GLP Cr(VI)	30 200	Light (8,500 lumens)	99.3 77.2	60 150	[14]
CuO–MgO	MB		Visible light	77	150	[15]
BSO_hv	RhB Phenol	5×10^{-6} 5×10^{-6}	LED source	100 23	4 h 8 h	[16]
$Sr_2Nb_2O_7$-Nb_2O_5-$SrCO_3$	MG	50 ppm	Visible light	81	60	[17]
CdSe-MSA	MB		Visible light	80	60	[18]

2.1 Zinc-based materials

Zinc oxide (ZnO) is a superior semiconductor substance that provides a suitable exciton binding energy (60 meV) and an appropriate bandgap (3.37 eV) for the use of photocatalytic process. At the presence of suitable light source, ZnO produces electron-hole pair. Consequently, these photocharge carriers will react with water to yield reactive oxygen species (ROS), and these ROS raid and decompose the organic pollutants. Ramesh et al. [1] used green and low-cost *Phyllanthus niruri* plant extract for the biopreparation ZnO NPs. They found that the as-synthesized ZnO NPs shows superior photocatalytic activity of the methylene blue (MB) dye and Reactive Red 120 (RR 120) under sunlight irradiation. They refer this behavior due to the improved surface area and wide absorption band of the NPs. Moreover, the ZnO NP can inhabit the oxidative damage of cells by remoting the release of ROS. Farooq et al. [2] anchored silver (Ag) in ZnO as an excellent photocatalyst by applying the extract of *Capparis decidua* plant as green capping agent. The catalytic performance of Ag-ZnO was tested by the reduction of 4-nitrophenol, degradation MB, and adsorption Cr(VI) ion from water under UV–visible light in a semibatch reactor. They conclude that Ag assembled on ZnO improve photocatalytic affinity by adjusting the probability of drop back of electrons from conductive to valence band. Moreover, according to Langmuir, Ag-ZnO can adsorb Cr(VI) with maximum adsorption capacity reach to 23.98 mg/g. Truong et al. [3] successfully prepared ZnO/CuO nanomaterials by a facile sol–gel technique and tested for the photodegradation of rhodamine B (RhB). They noted improvement in the absorption of visible light with copper dose. Moreover, the ZnO/CuO nanocomposite can photodecompose 10 ppm RhB (98%, 180 min, 0.1 g/L). The scavenging investigation pointed that the degradation of RhB involves •OH. El Bekkali et al. [4] studied the effect of combination of porous hydroxyapatite, derived from phosphate rock, with ZnO (ZnHAp) to enhance the photocatalytic decomposition of antibiotics in water and to gin the by-products. The results showed that the coexistence of zinc oxide with porous apatite a key factor endowing high photocatalytic and antibacterial affinities.

Chani et al. [5] prepared ternary metal-oxides composites of Zn-Fe-Ca, Zn-Fe-Sb, Zn-Fe-Ti, and Zn-Fe-Co for acridine orange (AO) decomposition under light. The obtained data revealed that 98% of the AO degraded on Zn-Fe-Sb (20 min), Zn-Fe-Ti (80 min), and Zn-Fe-Ca (90 min). Therefore, the photocatalyst Zn-Fe-Sb was further employed for the degradation several organic dyes. Correa-Vargas et al. [6] prepared ZnS QDs as photocatalyst to decompose Amaranth dye in aqueous environments. ZnS QDs can successfully degrade (85%, 120 min) under UV light. Parangusan et al. [7] prepared hydrophobic fiber composite by blending ZnO nanofiller with polyvinylidene fluoride (PVDF/ZnO). The photocatalytic potential of hybrid membrane was assessed by employing azocarmine G (AZG) and MG dye under sunlight. The obtainings explored that the nanofiber composite can degrade (85%, 120 min) AZG and MG dye (90%, 240 min).

2.2 Titanium-based photocatalyst

TiO_2 as a photocatalyst is broadly applied in various fields such as wastewater treatment, building materials, agriculture, and air cleaning. The band gap of TiO_2 3.23 eV requires high photon energy for activation. Thus, several methods were investigated to minimize the TiO_2 band gap energy. Among them grafting the TiO_2 with metals or nonmetals to lowering its band gap energy improved its photoactivity under visible irradiation light. Sienkiewicz et al. [8] prepared TiO_2 nanomaterials grafted with 3-aminopropyltriethoxysilane (APTES) by applying solvothermal modification and calcination at 800–1,000 °C in inert atmosphere for visible-light photocatalytic application. They observed that the addition of Si and C-atoms in the APTES/TiO_2 photocatalysts retard the conversion of anatase phase to the rutile phase and the crystalline growth of both polymorphous compositions of TiO_2 overheating. Therefore, the thermal treatment of APTES/TiO_2 provided more pore volume and specific surface area compared with the control materials. Moreover, variation of positive charge on the TiO_2 surface to negative after the pyrolysis encourages MB adsorption. Hence, photocatalytic properties of TiO_2 are reduced. This is due to the all thermally treated photocatalysts resulted after functionalization with APTES exhibit an improved dye degradation percent than the reference samples. The excellent photoactivity was recorded for nanomaterials treated at 900 °C because of preformed specific surface area than materials annealed at 1,000 °C and a high density of vacant sites located on the TiO_2 surface analogized with pyrolyzed samples at 800 °C. Sneha et al. [9] studied the photocatalysis of lomefloxacin by utilizing magnesium titanate ($MgTiO_3$) under visible light (30 W LED). The band gap energy of $MgTiO_3$ 3.09 eV. Reactive species (electron-hole, hydroxyl, and superoxide radicals) play an important role in the photocatalytic mechanism. Sisay et al. [10] fabricate hybrid nanocomposite membranes composed of PVDF anchor with TiO_2 and/or $BiVO_4$ nanomaterials and/or CNTs in different content for decaying of bovine serum albumin (BSA) under visible light. All treated PVDF membranes showed excellent hydrophilicity in comparing with reference one. Moreover, the grafting of membrane with nanoparticles (NPs) will improve its hydrophilicity, inhabit, and attach BSA to the surface. This causes a slight decrease in BSA rejections, but maintained at 97% in all cases.

2.3 Silver-based nanophotocatalyst

One of the most valuable strategies to enhance the photocatalytic efficiency is immobilizing of a noble metal like Ag nanoparticle semiconductors surface. When the hybrid semiconductor is irradiated, there is a rapid transfer of electrons among the semiconductor and the metal nanoparticle. As a result, an electric field will be built-in in the space charge area, and this will lead to improve the separation e–h pair generated while light irradiation. Additionally, Ag immobilized can induce extra electrons via the surface plasmon resonance phenomenon that can be contributed into photochemical activity. Behnamian et al. [11] synthesized a photocatalyst applying organic acid-aided sol–gel strategy. The acid acts as a complexing, reducing, and protecting agent in fabricating multicomponent nanocomposites. The results derived from XRD pointed that the crystalline phase of Ag and $CuAl_2O_4$ is constructure after thermal treatment of the samples at 800 °C. Moreover, the crystal size and average particle size grow with Ag content. The findings showed that nanocomposite ($CuAl_2O_4$-3 wt%Ag) exhibits excellent photocatalytic percent (97%) and appropriate stability after multiruns. Asgari et al. [12] prepared graphene oxide (GO), mixed with chitosan, and anchored with (Ag NPs) (AGC) NPs. Mixtures of polyacrylonitrile and AGC NPs were synthesized in different ratio of 0.5–10.0 wt% and applied to prepare nanofibers membrane through electrospun. The NFs (10.0 wt%) exhibit the excellent cytotoxicity and antibacterial performance in both with/without light irradiation conditions. But the activity in the light is significant. The NF (5.0 wt%) recorded the best cell viability of 80% within 24 h and 60% within 48 h using L929 cells and 12 and 13 mm resistance zone diameter (IZD) for *E. coli* and for *S. aureus*, respectively, after light irradiation for 24 h.

2.4 C_3N_4-based nanophotocatalyst

C_3N_4 nanomaterials have attracted more attention attributed to its excellent electronic and optical features. Moreover, it exhibits suitable bandgap make it studied as a semiconductor photocatalyst. Furthermore, it is widely applied as a symbolic material attributed to its properties like cost effect, simple prepared, availability, stability, sustainability, and high oxidizing power. Rittika et al. [13] synthesize a novel $Zn_3V_2O_8$/g-C_3N_4 Z-scheme nano-heterojunction for diazinon photodecomposition under the visible light (180 mW/cm^2) as a source of irradiation. The ratio of photocatalyst component was optimized as $Zn_3V_2O_8$/40 wt% g-C_3N_4 (ZC-40) composites that can decompose 95.2% diazinon (DZN) in 60 min and shows excellent stability after several reuse. Rittika et al. [13] explained the superior photocatalytic efficiency of the photocatalyst as a mixed induce of Z-scheme mechanism, excellent absorption of visible light, charge separation, oxygen vacancies, and metallic redox mediator. Moreover, they noted that the ZC-40 showed high decomposition performance toward other pesticides. Alam et al. [14] prepared efficient photocatalyst composed of activated carbon fiber (ACF) anchored with $CoNiWO_4$ (CNW) and g-C_3N_4 (gCN) for glyphosate (GLP) and Cr(VI) removal from aqueous medium. The results obtained from individual systems provide that GLP (30 ppm) completely oxidized within and 200 ppm of Cr(VI) reduced after 150 min at 1 g/L dose of CNW-gCN/ACF. Under optimal experimental conditions the prepared composite presented high photocatalytic activity for binary pollutants than that in individual system. Studies under flow conditions, CNW-gCN/ACF provides oxidation and reduction affinity similar to prevalent under batch conditions.

2.5 Other nanophotocatalyst

Sackey et al. [15] synthesized CuO–MgO nanocomposite employing date pits agro-waste. XRD analysis data explored the crystalline structure of CuO nanoparticles (monoclinic, 26.7 nm) and MgO nanoparticles (cubic, 21.4 nm). The molecular dynamics assess exhibited that MB dye possesses a high hydrophobic behavior with a great tendency to react with the CuO–MgO and then attach to the surface. Consequently, the experiments affirmed that bio-nanocomposite CuO–MgO showed significant degradation for MB. Shabalina et al. [16] prepared bismuth silicate (BSO) nanomaterials by applying laser technique in liquid phase via dual step: (1) separated nanosuspensions of Bi and Si were generated through ablating their metallic precursor in aqueous solution and then blended BSO; (2) posttreatment of the synergic suspension was fabricated under similar parameters as in first stage (1) (BSO-hν). Thereafter, their photocatalytic affinity toward (RhB) and phenol was investigated. The findings showed that the posttreated sample BSO-hν exhibited high stability and catalytic activity. They refer this behavior to the additional laser treatment led to improve interaction among Bi- and Si-containing species and form Bi–O–Si bonds. Nazarlou et al. [17] fabricate $Sr_2Nb_2O_7$-Nb_2O_5-$SrCO_3$ nanocomposites using an economic ultrasonic-coexisted solvothermal approach employing $Sr(NO_3)_2$ and Nb_2O_5 precursor materials at stoichiometric Sr:Nb molar ratio. The photocatalytic potential of the prepared nanomaterials was examined for the decomposition of MG in aqueous environment under visible light source. The composite can decompose 81% of MG under condition of H_2O_2, 40 mg catalyst, 60 min, visible light, and 80 mL of 50 ppm MG. Mohamed et al. [18] are successful in using mercaptosuccinic acid (MSA) in modification of CdSe nanocrystals for photocatalytic decay of MB under visible light and sunlight irradiations aqueous solution. The prepared material showed adsorption properties toward MB with maximum capacity (27.07 mg/g in 10 min), which may be referred to electrostatic attraction and H-bonding. Moreover, they conclude that adsorption performance of CdSe-MSA nanomaterials enhanced the dye photocatalytic degradation percent under visible light (80%, 60 min). They refer this high efficiency due to small crystal size (3.7 nm), the surface area, and structural disorders (selenium vacancies, Se) of CdSe nanophotocatalyst, which improve the composite response toward the visible light and their photocatalytic affinity.

3 Adsorption

Adsorption is the ability of certain materials (adsorbent) to attract the dissolved pollutants (adsorbate) onto their surface. The adsorption system is composed of an adsorbent, adsorption media, and an adsorbate; this system is controlled by the tendency between the adsorbent and adsorbate. Where available binding sites, permeability, types of interactions, and the surface of the adsorbent are affected on the process.

Adsorption is considered more effective technique for water recycling based on low cost, flexibility, simple operation, clarity of conception, no toxic by-product, affinity toward contaminates, and not cause substances damaging.

Adsorbent is a solid material employed to eliminate the undesired species that can hazard the environment from liquid or gas. The adsorbent should be characterized by large surface area, strong interaction among binding sites and contaminated species, high selectivity for the target molecule, and easy to recover. At present, scientists are seeking for cost-effective adsorbents with less processing and are easily available. The most generally utilized adsorbents in the adsorption process are activated carbon, polymer resins, clays, agricultural waste, and industrial waste. Table 2 summarized various adsorbent, pollutants, adsorption capacities, removal percent, and equilibrium time.

Table 2: Various adsorbents, pollutants, adsorption capacities (Q_e), removal percent (%*R*), and equilibrium time (*t*).

Adsorbate	Pollutant	Q_e (mg/g)	%*R*	*t* (min)	References
PTh-Fe_3O_4	Pb (II), Cd (II) Cr(VI)	1.638 1.531 1.427		60	[19]
Fe_3O_4@SBA-15-Gd	Cu(II) Pb(II)	344.82 303.03		50	[20]
$MnFe_2O_4$-Cys	Pb(II)	137.45		6 h	[21]
CD-Fe_3S_4	Cr(VI)	220.26			[22]
RIFP-FeONPs	Cr(VI)	16.393			[23]
cIONPs	Ar(V)	121.4		4 h	[24]
$MF_{0.1}$ **$MF_{0.2}$** **$MF_{0.33}$**	As(III) As(V)	51.48, 26.06 100.53, 43.44 103.94, 45.52		30	[25]
MnO_2@NiFe@DE	MO MB TC		100 100 83	60 90 270	[26]
MFC@Mn-NH_2-UiO-66@IL	MB	260.42	98	90	[27]
nZVI@mSiO_2	Pb(II) Cd (II) As(V)	372.2 105.2 115.2		40	[29]
EBNs **PB**	Tl(I)	382.38 42.3		60	[30]
BC@Fe/Ni	Cr(VI)	160.76	99.62	20	[31]
AKB-600	MB	150.60	94.61	60	[32]

Table 2 (continued)

Adsorbate	Pollutant	Q_e (mg/g)	%*R*	*t* (min)	References
nPPAB	CIP	142.86	89.94	420	[33]
bPP		5.92	84.74	360	
M-BC	Iodine	43.65 mmol/g		30	[34]
h-NC-800	Cd(II)	356.4		90	[36]
rGO-Fe/Ni	Sb(III)	1.976		5 h	[38]
	Sb(V)	3.324			
HSMSMs	U(VI)	822		120	[40]
HSMSMs-AO		877			
SiO_2	Triclosan caffeine	2.74	76	60	[41]
		0.75	48	30	
BIS	Anthraquinone	334	94		[42]
PVDF/ATP-CDs	Hg^{2+}	0.99			[44]
	Cu^{2+}	0.73			
	Fe^{3+}	0.60			
Fe-PVDF	EB	434.78			[45]
TA/BiOBr/PVDF	MB		99.6		[46]
	MO		97.7		
	RhB		99.8		
	TC		90.7		
	Bis A		85.5		
M-PSF	CR		99.86		[47]
	Humic acid		99.81		
$g-C_3N_4/ZnO/PSF$	RG 19		99.9		[48]
	RY 160		85.5		
$W-MoS_2$@SCA	Pb(II)	169.02		75	[49]
rGO/phenolic nanomesh	CR		98		[50]
SBR	Cd (II)	1.10, 0.03			[51]
BFR	Cu (II)	0.03, 0.01			
Crude NC	Acid red 131		63	30	[52]
$Mn_5P_4O_{20}-H_8$	Pb(II)	1,510		480	[53]
	Cr(III)	201			
	Fe(III)	300			
PPy-BT	Mo(VI)	100.17		120	[54]
H-TNTs	MV	106	88	180	[55]

Table 2 (continued)

Adsorbate	Pollutant	Q_e (mg/g)	%*R*	*t* (min)	References
$Mg(OH)_2/MgO$	CIP	1,789	97	30	[56]
Au/CB6-BOFs	CR	2,178.36		120	[57]
PEI-TNTs	V(V) Cr(VI)		1,280.36 980.39	60 80	[58]

3.1 Magnetic-based materials

Based on the need to excellently recover the adsorbent materials at the end of the adsorption process, the requirement to synthesize magnetic nanomaterials has been manifestly increased. Kumar et al. [19] modified Fe_3O_4 nanoparticles with polythiophene (PTh-Fe_3O_4) for efficient adsorption of heavy metal (Pb(II), Cd(II), and Cr(VI)) ions from water environment. Hassanzadeh-Afruzi et al. [20] prepare magnetic nanomaterials (Fe_3O_4@SBA-15-Gd) by employing guanidine (Gd) to modified mesoporous Santa Barbara amorphous-15 (SBA-15)/Fe_3O_4 for the elimination of Pb(II) and Cu(II) from aqueous solutions. The prepared adsorbent owes to the form of bidentate interacting agent, Gd groups that will enhance the adsorption tendency toward heavy metals. Hence, at optimized conditions the maximum removal capacity of Cu(II) and Pb(II) was 344.82 mg/g and 303.03 mg/g, respectively. Gao et al. [21] modified $MnFe_2O_4$ nanoparticles with L-cysteine ($MnFe_2O_4$-Cys) using high-gravity technology for the capturing of Pb(II) from water. They found that the crystallinity and magnetization of $MnFe_2O_4$-Cys engineered via impinging stream-rotating packed bed reactor ($MnFe_2O_4$-Cys-R) were improved and analogized using a stirred reactor. Moreover, maximum removal capacity of the $MnFe_2O_4$-Cys-R was 137.45 mg/g equaled to 1.98 times of free cystine $MnFe_2O_4$-R (69.50 mg/g). Kong et al. [22] modified Fe_3S_4 nanomaterials with β-cyclodextrin (CD) via one-step simple method and employed for Cr(VI) removal. The prepared modified adsorbent showed excellent chelating capacity with Cr(VI) (220.26 mg/g) than Fe_3S_4. Moreover, the CD-Fe_3S_4 composite possessed good selective removal of Cr(VI) from the mixture of other ions. More importantly, CD-Fe_3S_4 captured effectively Cr(VI) from real wastewater resulted from electroplating. The mechanism of CD-Fe_3S_4 removal for Cr(VI) could be explained as surface adsorption/reduction. Prema et al. [23] biofabricated iron oxide nanomaterials employing *Rosa indica* flower petal extracts (RIFP-FeONPs) for the capture of Cr(VI). According to Langmuir, RIFP-FeONPs showed maximum adsorption capacity toward Cr(VI) 16.393 mg/g. Diephuis et al. [24] prepared iron oxide nanoparticles (IONPs) with various synthetic strategies (thermal decomposition or coprecipitation) and analogized with commercial IONPs to monitor the degree of aggregation. The aggregation status of the original materials after assembly onto the sand was assessed by SEM. Based on the dynamic studies, IONPs immobilized on the sand to get rid of while treatment of water. To overcome that, they concluded using cIONPs clusters, prepared by a solvothermal technology. According to Langmiur, cIONPs presented capture saturation (121.4 mg/g) higher than the other IONPs (11.1 mg/g, thermal decomposition; 6.6 mg/g, coprecipitation; and 0.6 mg/g, commercially available IONPs). Uddin et al. [25] prepared $MgFe_2O_4$ with various Fe and Mg ratios by easy one-step solvothermal technique and used for arsenic elimination from aqueous solution. After that, the ratios of Mg:Fe = 10:90; $MF_{0.1}$, 20:80; $MF_{0.2}$ and 33:67, $MF_{0.33}$ were fabricated. The obtained results showed that the increasing Mg amount constrict the pore size, volume, and magnetization and improve surface area and pH_{PZC}. Arsenic adsorption results were fitted with Freundlich isotherm model and maximum removal saturations As(III) (51.48, 100.53, and 103.94 mg/g) and As(V) (26.06, 43.44, and 45.52 mg/g) by $MF_{0.1}$, $MF_{0.2}$, and $MF_{0.33}$, respectively. Dai et al. [26] prepared magnetic binary-core@shell nanomaterials (MnO_2@NiFe@DE) by applying hydrothermal strategy over two separated steps for the elimination of organic pollutants via oxidation and adsorption mechanism. At low pH condition, the as-prepared nanomaterial completes the removal of MO and MB and 83% tetracycline (TC). The removal mechanism is based on synergetic physical adsorption (electrostatic, anion exchange, and H-bonding) and the oxidation by $^{\bullet}OH$. Another class of magnetic nanoparticles, Zr-MOF composites based on Mn-anchored magnetic hierarchical porosity and ionic liquid (IL) postmodification (MFC@Mn-NH_2-UiO -66@IL), was investigated by Lu et al. [27] for dual adsorption/Fenton catalysis of MB in aqueous solution. The adsorption process highly improved due to electrostatic attraction, H-bonds, and π–π stacking interactions among [BMIM][PF_6] and MB. Thereafter, Zr-MOF and Fe_3O_4 quickly decompose H_2O_2 to generate $^{\bullet}OH$ for the decaying of MB. The dual function of the prepared composite achieves high kinetic removal efficiency of 98% and 90 min. Moreover, Li et al. [28] employed hydrothermal and coprecipitation techniques to grow ZIF8 and/or Fe_3O_4 on various functional engineered nanomaterials (ENMs) to prepare a group of ZIF8@ENMs nanomaterial. Under neutral condition, the Fe_3O_4-ZIF8@ENMs amongst the prepared composites achieve the maximum removal capacity of 441.7 mg/g and this is due to dense OH groups. While Teng and coworkers [29] coated zero-valent iron (nZVI) with mesoporous silica to prepare core-shell nanoscale (nZVI@$mSiO_2$) for the adsorption of heavy metals. At pH 5.0, the nZVI@$mSiO_2$ showed maximum capture saturation (mg/g) for Pb(II) (372.2), Cd(II) (105.2), and As(V) (115.2). The dissolved iron during the adsorption process describes the important role of the mesoporous silica ($mSiO_2$) in maintaining the stability of nZVI. The interfering studied with competing ions showed synergetic enhancing the removal efficiency Cd (II) and As(V).

3.2 Biochar and carbon-based materials

Biochar is derived from pyrolysis of agriculture waste and considers a sustainable solution to clean water from various contaminants. The cost of biochar preparation can be highly reduced by applying biomass source derived from agricultural wastes.

Consequently, the adsorbent materials required for the preparation of biochar can further develop the efficiency of the treatment process. Haris et al. [30] prepared biochar with thin-layered nanosheets structure derived from (wheat straw) via multisteps (treating the biomass, pyrolysis in nitrogen atmosphere, and finally, exfoliation at flash heat) for thallium(I) adsorption from water. The results showed that BET surface area and pore size of as-prepared exfoliation biochar nanosheets (EBNs) are much higher than the original wheat-straw biochar (PB). At pH 7.0, EBNs had a maximum saturation capacity of T1(I) (382.38 mg/g, ≈9 folds) higher than the PB. Moreover, EBNs maintained excellent selectivity for T1(I) in the presence of competing cations and organic materials. Moreover, after five recycling runs high removal ability (>93%) of EBNs was observed. The EBNs can remove ~90% of T1(I) from river water. Xing et al. [31] utilized biochar (BC) as a support for Fe/Ni bimetal nanocomposite (BC@Fe/Ni) for the effective removal of Cr(VI) from water. The results showed that porous structure of BC exhibit a suitable support for Fe^0 that will minimize the agglomeration of Fe^0. Moreover, the adsorption performance enhanced in order (BC@Fe/Ni, BC, and Fe/Ni) and the good removal was achieved at Ni ratio 5%. As a result, 99.62% Cr(VI) was removed at pH 2.0. Electrostatic attraction and reduction mechanisms mainly involved on the removal of Cr(VI) using BC@Fe/Ni. Zhou et al. [32] prepared innovated N-doped biochar (AKB) nanocomposite base on raw kelp root pyrolysis at 600–1,000 °C and were employed for cationic dye removal from water. The porosity assessment showed that AKB pyrolysis at 600 °C (AKB-600) provided a honeycomb-like microstructure with higher specific surface area. FTIR and XPS analyses exhibited that AKB with numerous surface functional groups, such as AOH, C@O, and ACONHA, leading to high kinetic removal percent of dye (94.61%, 60 min). Electrostatic attraction and physical adsorption explained adsorption process. Hamadeen and Elkhatib [33] prepared nanostructured activated biochar (nPPAB) from pomegranate peels (PP) for antibiotic ciprofloxacin (CIP) removal. The results described that Langmuir model explained well equilibrium with maximum adsorption capacity for CIP 142.86 mg/g which is 26.85 times more than PP. H-bonding, π–π interaction, hydrophobic, and electrostatic interactions explored the removal mechanisms of CIP by nPPAB. The ability of nPPAB for CIP captures from real sample applying batch (89.94%) and packed-bed reactor (84.74%). Obey et al. [34] prepared biochar derived from Matamba shell (M-BC) after 2 h pyrolysis at 600 °C. The abundant surface oxygen composition shows sufficient chelating ability for iodine (43.65 mmol/g) via van der Waals, π–π, and π-stacking interaction.

Moreover, another carbonous material was investigated by Vieira et al. [35], where they synthesized silica@polymer core@shell spheres structure followed by thermally pyrolysis at 800 °C to yield spheres of SiO_2@CSs. Thereafter, the silica content was removed using alkali solution to produce hollow spheres carbon (CSs). They observed that the ratio of ethanol/water (E/W) used for SiO_2 determine the size of SiO_2@CSs and CSs particles. Diclofenac and venlafaxine were used to study the adsorption characters of CSs. Furthermore, Xu et al. [36] synthesized series of hierarchical nitrogen-doped carbon (h-NC-750, h-NC-800, and h-NC-850) through pyrolysis of the ZIF-8 particles in NaCl crystals at 750, 800, and 850 °C in Ar atmosphere. The experimental results showed that 1 g of h-NC-800 (N: 15.5%) can remove 356.4 mg of Cd(II), which was efficient by 12.5% in case of h-NC-750 and 48.3% at using h-NC-850. This may be explained by the harmonized edge- to graphitic-nitrogen ratio that tight the binding energies among Cd(II) and h-NC. Moreover, the h-NC-800 can remove (>92%) of Cd(II) from real water sample and five regeneration times.

3.3 Graphene-based materials

GO is obtained by strong oxidation of graphite. It is 2D materials composed from single carbon layers with thickness of one C-atom. Oxygen function groups (–OH and C–O–C) located at either sides of the basal plane and the COOH groups present at the edges. Reduced graphene (rGO) is prepared by the reduction of GO using suitable reducing agent. The layered structure of both GO and rGO makes them an excellent support for another nanomaterials. El Shahawy et al. [37] extracted chitosan from shrimp shellfish mixed with AgNPs/GO in the ratio of 3: 1 for the preparation of AgNPs/GO/chitosan nanocomposite for Fe(III) and Cr(VI) ions' elimination. The optimal adsorption conditions were adjusted at contact time (80 and 30 min), pH (4 and 6), and adsorbent dose (0.1 and 0.22 g 100 mL^{-1}), initial concentration of 50 and 40 ppm, and stirring at 150 and 250 rpm of Cr(VI) and Fe(III), respectively, at room temperature (30 °C). Chen et al. [38] prepared (rGO-Fe/Ni NPs) nanocomposite and applied for the capture of Sb(III) and Sb(V). When referred to individual rGO the composite presented high removal efficiency.

Moreover, rGO-Fe/Ni showed a wide range of adsorbent where it shows suitable adsorption capacity (mg/g) toward Sb (1.59), As (2.61), Pb (2.41), and Cd (1.25) from real mining sample. Finally, after reusing four times the composite exhibited adsorption efficiency 72.7% for Sb(III). Liu et al. [39] simulated the aging processes of microplastics (MPs) and GO in a real environment (sunlight exposure) by ultraviolet (UV) irradiation and thermal treatments, respectively. Thereafter, the efficiency aging step was evaluated by studying the removal efficiency of CIP on series of MPs [polypropylene (PP), polyamide (PA), and polystyrene (PS)]. The obtained results showed that the aged materials present removal saturation reach to two-fold unaged samples. Moreover, at grafting of MP with GO or rGO, the adsorption capacity was highly enhanced. Where the removal capacity of TC was improved by ~336% at using aged PP-GO and ~100% in case of aged PP-rGO. This considerable adsorption performance is due to the sharing of oxygenated functional groups.

3.4 Silica-based nanoadsorbent

Silica nanomaterials own a significant specific surface area, high hydrophilicity, porous structure, and simply functionalized with other active groups, making them widely used in several applications. Chen et al. [40] prepared silica nanomeshes (HSMSMs) using high shear strategy to form nanobubble for uranium adsorption. HSMSMs exhibited an excellent uranium capture saturation of 822 mg/g from seawater. Compared to HSMSMs, the modification with amidoxime (HSMSMs-AO) induced good capture for U(VI) (877 mg/g) from simulated seawater. Tejedor et al. [41] synthesized SiO_2 nanoparticles from rice husk as a precursor using acid leaching (acetic acid) followed by 2 h burning at 700 °C for removing caffeine and triclosan aqueous media. The results showed that dose, adsorption efficiencies, and maximum adsorption capacities for triclosan were 8 g/L, 76%, and 2.74 mg/g and caffeine were 24 g/L, 48%, and 0.75 mg/g. Patel et al. [42] prepared bioinspired silicas (BIS) and chemically modified for dye scavenging. While 0% adsorption was observed for the commercial sorbent, BIS exhibited up to 94%. According to Langmuir BIS presented excellent capture saturation (334 mg/g), with quick removal in acidic medium, suitable thermal stability, and good reuse.

3.5 Polymer membrane nanocomposite

The addition of nanoparticles (NPs) and polymer membrane will lead to enhancement of the membrane hydrophobicity or hydrophilicity, improve their stretching, as well as alter the membranes surface charge. Accordingly, different properties of the composite membrane like the rejection ability, low permeability, selectivity, biological fouling, scaling, and stability will be improved. As a result, it will achieve reduction in the required energy and enhance-scale water treatment system. Chabane et al. [43] modified poly(vinylidene fluoride) (PVDF) membrane with silica gel and montmorillonite clay (maghnite) using precipitation-phase inversion approach for copper and nickel removal of from water. Zhao et al. [44] successfully fabricate a PVDF/ATP-CDs composite membrane by grafting carbon dots (CDs) with attapulgite (ATP) present in the PVDF/ATP membrane. The membrane characterized with dual functions (adsorption and detection) of heavy metal ions. Due to the super-hydrophilicity of CDs, the wettability and permeability of composite membrane will be enhanced, which make it suitable for water treatment application. The modification with ATP highly enhances the saturation capacity of the membrane toward Hg^{2+} (0.99 mg/g), Cu^{2+} (0.73 mg/g), and Fe^{3+} (0.60 mg/g). On the other hand, the interaction among the heavy metal ions with the CDs presented at the membrane surface will quench the CDs fluorescence for the assessment of the target ion. Shin et al. [45] removed anionic dye using chitosan-grafted iron oxide-modified PVDF (Fe-PVDF) membranes. At optimal conditions, Fe-PVDF provided a maximum saturation capacity of 74.6 mg/g for Evans blue (EB), which will be highly improved (434.78 mg/g) in acid solution. The mechanism of the removal process can be described on the basis of the electrostatic attraction among the adsorbent positively charged (chitosan and Fe) the negatively charged sulfate groups in the dye species. Moreover, simple washing with alkaline solution can efficiently regenerated Fe-PVDF.

Hence, the membrane can be reused many cycles without a remarkable decrease in adsorption efficiency. Zhang et al. [46] designed an efficient tannic acid (TA)/BiOBr/PVDF) membrane by simple in situ precipitation of BiOBr with (PVDF) followed by self-immobilizing of TA and halloysite nanotubes to improve the efficiency of membrane hydrophilicity and anti-fouling. The TA/BiOBr/PVDF membrane achieved removal percent for MB (99.6%), MO (97.7%), RhB (99.8%), TC (90.7%), and bisphenol A (85.5%). Moreover, modification step highly improved the blended contaminates flux and water contact angle of the resulted membrane. In addition, the prepared membrane exhibited appropriated stability for reusing and antipollution activities.

Chandra et al. [47] prepared nanofiller of $MnCo_2O_4$ for preparation polysulfone (PSF) blended (M-PSF) matrix membranes. PSF nanocomposite membranes were fabricated at different $MnCo_2O_4$ ratio (0–1.5 wt%) via a nonsolvent-induced phase-separation process. The anchoring with nanoparticle will dense oxygen species in the membrane matrix; as a result, affinity of the anchored membrane toward water will enhance its various properties. Where the anchored membranes can provide greater pure water flux (PWF) (220 L/m^2 h) and flux recovery ratio (FRR) (88%) when treated with BSA protein which is higher compared to pure PSF membrane (30 L/m^2 h PWF and 35%FRR). Moreover, the grafted membrane can reject 99.86% of Congo red (CR) and recover 99.81% humic acid. Vatanpour et al. [48] anchored the PSF membrane with prepared g-C_3N_4 and hydrothermal prepared g-C_3N_4/ZnO nanomaterials for efficient separation of the dye and lowering the organic fouling. Due to high content of –OH and –NH reactive groups present in g-C_3N_4/ZnO, it will improve various (g-C_3N_4/ZnO/PSF) membrane functions such as porosity, hydrophilicity, negativity, permeability, fouling resistance, and minimize the contact angle. The rejection efficiency was also improved for dyes (99.9% for Reactive green 19 and 85.5% for Reactive yellow 160).

Qiu et al. [49] synthesized an innovated 3D nanocomposite, W-MoS_2@SCA, by embedded widened interlayer spacing MoS_2 nanoparticles (W-MoS_2) in the porous networking of sulfonated cellulose aerogel (SCA). The adsorption results present that the nanocomposite possesses fast adsorption kinetics (75 min) and high selectivity removal for Pb(II). This may be attributed to combine the Donnan membrane impact rise by the positive charge of sulfonic group of SCA and the high tendency of MoS_2 toward PB (II) ions. Dynamic experiment showed that excellent applicability is obtained with a treatment of 9,000 kg wastewater/kg sorbent applying the China wastewater. Interestingly, Pb(II) was not detected in the first 7,000 kg effluent. Moreover, the exhausted sorbent easy recovered by the NaCl-Na_2EDTA binary solution and keep on its activity even after 10 adsorption–desorption cycles. Lan et al. [50] mixed porous (rGO) nanosheets with phenolic/polyether nanosheets, vacuum filtrated to compose laminated structures, finally, treated with acid to leach polyether to produce porous rGO/phenolic nanomesh membranes. Therefore, ternary channels were formed [pores (30–60 nm), rGO nanosheets; mesopores (9 nm) of phenolic nanomeshes; interlayer nanochannels (sub-nanometer)]. Moreover, the phenol with OH can polymerize, causing enhancement in the hydrophilicity and strength the structure of membranes. Therefore, the membranes showed ultrafast water flow while maintaining preformed rejection toward CR dye (98%).

3.6 Recycled material-based adsorbent

The safe recycling of industrial solid wastes for preparing low-cost nanomaterials and their efficient apply for water cleaning are of high concern from environmental and economic points of view. Lashen et al. [51] utilized sugar beet processing (SBR) and brick factory waste (BFR) to prepare two novel nanomaterials for the decontamination of water and soil from Cd and Cu. The SBR captured up to 99% and 91% of Cu and Cd detected in water and provided quick and higher adsorption capacity (g/kg) = 1111.1 for Cu and 33.3 for Cd than achieved by BFR. Moreover, SBR can reduce up to 57% for Cu and 86% for Cd and BFR up to 36% for Cu and 68% for Cd of the metals in the soil. The capture tendency of SBR over BFR could be related to its alkalinity (pH = 12.5), CO_3 ratio (82%), reactivity of hydroxyl –OH and Si–O groups, and high surface area. Maslennikov et al. [52] study the preparation of nanocellulose (NC) from recycled paper sludge (RPS) via two steps: (1) ozonation, which promote an enough delignification, and (2) hydrolysis with 64% w/w sulfuric acid. Commercial NC can remove 63% within 30 min of 0.1 ppm acid red 131 dye.

3.7 Other efficient adsorbents

Deng et al. [53] prepared an innovated compound $Mn_5P_4O_{20}$-H_8 via a simple one-pot solvothermal process. The composite owns a 3D network cross-linked by [PO_4] tetrahedra and [MnO_6] octahedra. The as-prepared $Mn_5P_4O_{20}$-H_8 composite shows maximum removal capacity toward Pb^{2+} (1,510 mg/g), Cr^{3+} (201 mg/g), and Fe^{3+} (300 mg/g). This is referred to the dual effect of surface (OH) group trapping and lattice ion exchange. Moreover, $Mn_5P_4O_{20}$-H_8 possessed suitable regeneration-reuse without significant degradation. Wang et al. [54] prepared an innovated adsorbent based on polypyrrole capping of bentonite (PPy-BT) polymerization technique for high capturing of Mo(VI) from aqueous environment. The results showed that PPy and BT could retard the agglomeration of each other and therefore led to an excellent exposure of active sites to pollutant species. Thus, at pH 4.0, Mo(VI) will be adsorbed with maximum capture saturation of 100.17 mg/g at 25 °C. Moreover, the high recovery capability (>89.3%) of PPy-BT was observed and resued five times. Santos et al. [55] synthesized hydrogen titanate nanotubes (H-TNTs) for the adsorption of methyl violet (MV) dye from water. The density functional theory (DFT) explained that the adsorption proceeded through weak dispersion interactions $C - H \cdots O$ and $- H \cdots Ti$. More interesting, interactions were also known in the H-TNTs, such as intramolecular hydrogen bondings $O–H \cdots O$. Therefore, the H-TNTs can remove 88% of

MV within 180 min. Moreover, according to Langmuir model H-TNTs exhibit maximum adsorption capacity 106 mg/g. Falyouna et al. [56] prepared composite formed from magnesium hydroxide/magnesium oxide nanorods [$Mg(OH)_2/MgO$] for effective and quick capture of CIP from water. The composite presented high tendency to remove CIP from aqueous medium (97% of 200 mg/L of CIP was removed within 30 min by using 0.1 g/L of $Mg(OH)_2/MgO$ at neutral pH and room temperature). According to Langmuir 1 g of as-prepared nanomaterials is capable to adsorb 1,789 mg of CIP. FTIR spectra of the spent $Mg(OH)_2/MgO$ explored that the adsorption mechanism proceeded via a complexation and electrostatic interaction with (COOH) and ($-NH_3^+$) group, respectively.

Yang et al. [57] synthesized Au/CB_6-BOFs (BOFs namely *closo*-dodecaborate-based supramolecular organic frameworks) and employed as adsorbent and catalyst for CR and MO. The removal efficiency of Au/CB_6-BOFs on CR is significant with the removal capacity 2,178.36 mg/g. By the addition of sodium borohydride, the Au/CB_6-BOFs act as a catalyst for reduction the dye molecules. The adsorption/degradation efficiency of Au/CB_6-BOFs maintained without significant attenuating at least five times.

Li et al. [58] a novel composite (PEI-TNTs) was investigated by locating polyethyleneimine (PEI) onto TNTs. The initial solution pH highly induces the adsorption performance of the nanocomposite. Where the saturated adsorption capacity for V(V) reached 1,280.36 mg/g at pH 3.0 and for Cr(VI) reached 980.39 mg/g at pH 2.0 referring to the Langmuir model. In addition, PEI–TNTs exhibit high selectivity at presence of coexisting anions and cations. Further, the PEI–TNTs could be recovered and reused for five cycles without significantly decreasing its capturing efficiency.

4 Conclusion

With the increase in urbanization, the consumption of water increased, and the sources of pollution increased as well as the amount of wastewater. On the other hand, climatic changes caused a decline in water sources. Thus, it was necessary to use effective methods to remove these pollutants from wastewater to be reused for different purposes. Photolysis and adsorption are considered to be the most effective ways to remove pollutants by safe methods. Where, in photocatalysis, light is used in the presence of nanomaterials to break down pollutants into nontoxic materials. While, in adsorption, the pollutants move from the aqueous solution to the surface of the nanomaterials and then they can be disposed of safely.

References

[1] Ramesh P, Rajendran A, Ashokkumar M. Biosynthesis of zinc oxide nanoparticles from *Phyllanthus Niruri* plant extract for photocatalytic and antioxidant activities. Inter J Environ Analyt Chem 2022. 10.1080/03067319.2022.2041004.

[2] Farooq M, Shujah S, Tahir K, Nazir S, Khan AU, Almarhoon ZM, Jevtovic V, Al-Shehri HS, Hussain ST, Ullah A. Ultra efficient 4-Nitrophenol reduction, dye degradation and Cr(VI) adsorption in the presence of phytochemical synthesized Ag/ZnO nanocomposite: A view towards sustainable chemistry. Inorg Chem Commun 2022, 136, 109189.

[3] Truong TT, Pham TT, Truong TTT, Pham TD. Synthesis, characterization of novel ZnO/CuO nanoparticles, and the applications in photocatalytic performance for rhodamine B dye degradation. Environ Sci Poll Res 2022, 29, 22576–22588.

[4] El Bekkali C, Labrag J, Oulguidoum A, Chamkhi I, Laghzizil A, Nunzi J-M, Robert D, Aurag J. Porous ZnO/hydroxyapatite nanomaterials with effective photocatalytic and antibacterial activities for the degradation of antibiotics. Nanotechn for Environ Eng 2022, 7, 333–341.

[5] Chani MTS, Khan SB, Rahman MM, Kamal T, Asiri AM. Sunlight assisted photocatalytic dye degradation using zinc and iron based mixed metal-oxides nanopowders. J King Saud Univer – Sci 2022, 34, 101841.

[6] Correa-Vargas KD, Luciano-Velazquez J, Nash-Montes VI, Bailon-Ruiz SJ. Photocatalytic activity of zinc sulfide quantum dots on amaranth dye. MRS Adv 2022, 7, 295–299. https://doi.org/10.1557/s43580-021-00154-6.

[7] Parangusan H, Bhadra J, Ahmad Z, Al-Maadeed ASMA, Al-Mohannadi Abdulaziz MAA, Al-Thani N. Electrospun PVDF/ZnO based composite fibers for oil absorption and photocatalytic degradation of organic dyes from waste water. Fib Poly 2022, 23, 5, 1217–1224.

[8] Sienkiewicz A, Rokicka-Konieczna P, Kusiak-Nejman AWE, Morawski AW. Artificial solar light-driven APTES/TiO_2 photocatalysts for methylene blue removal from water. Molecules 2022, 27, 947. https://doi.org/10.3390/molecules27030947.

[9] Sneha Y, Shivamurthy Yashas R, Thinley Y, Jijoe SP, Shivaraju HP. Photocatalytic degradation of lomefloxacin antibiotics using hydrothermally synthesized magnesium titanate under visible light-driven energy sources. Environ Sci Poll Res. https://doi.org/10.1007/s11356-022-20540-3.

[10] Sisay EJ, Ver´ G, Pap Z, Gyulavari´ T, Agoston´ A, Kopniczky J, Hodúr C, Arthanareeswaran G, Arumugam GKS, Laszlo Z. Visible-light-driven photocatalytic PVDF-TiO2/CNT/BiVO4 hybrid nanocomposite ultrafiltration membrane for dairy wastewater treatment. Chemosphere 2022, 307, 135589.

[11] Behnamian AE, Serate D, Tolentino Z, Niazi H, Mostafaei A. Synthesis, characterization, and photocatalytic activity of $CuAl_2O_4$–Ag nanocomposite for water treatment yashar. Ceram Intern xxx, xxxx, xxx. https://doi.org/10.1016/j.ceramint.2022.06.103.

[12] Asgari S, Ziarani GM, Badiei A, Setayeshmehr M, Kiani M, Pourjavadi A. Electrospun Ag-decorated reduced GO-graft-chitosan composite nanofibers with visible light photocatalytic activity for antibacterial performance. Chemosphere 2022, 299, 134436.

[13] Thakur PR, Sharma S, Kumar A, Sharma G, Ghfar AA, Naushad M, Stadler FJ. Fabrication of a Z-scheme $Zn_3V_2O_8$/g-C_3N_4 nano-heterojunction with high interfacial charge transfer for superior photocatalytic removal of diazinon pesticide under visible light. Appl Nanosci. https://doi.org/10.1007/s13204-021-02338-w.

[14] Alam U, Pandey K, Verma N. Photocatalytic oxidation of glyphosate and reduction of Cr(VI) in water over ACF-supported $CoNiWO_4$-gCN composite under batch and flow conditions. Chemosphere 2022, 297, 134119.

[15] Sackey J, Akbari M, Tandjigora N, Cloete KJ, Bashir AKH, Morad R, Maaza M. Industrial textile removal using date pit assisted CuO-MgO nanocomposite: Molecular dynamics and biosynthesis analysis. J King Saud Univer – Sci 2022, 34, 101840.
[16] Shabalina AV, Fakhrutdinova ED, Golubovskaya AG, Kuzmin SM, Koscheev SV, Kulinich SA, Svetlichnyi VA, Vodyankina OV. Laser-assisted preparation of highly-efficient photocatalytic nanomaterial based on bismuth silicate. Appl Surf Sci 2022, 575, 151732.
[17] Nazarlou ZD, Kafi-Ahmadi L, Khademinia S. Ultrasonic assisted solvothermal synthesis of a novel $Sr_2Nb_2O_7$-Nb_2O_5-$SrCO_3$ nanocomposite as a reusable and superior photocatalyst for malachite green degradation under visible light irradiation. J Austr Ceram Soci 2022, 58, 333–346.
[18] Mohamed NBH, Ouni S, Bouzid M, Bouzidi M, Bonilla-Petriciolet A, Haouari M. Synthesis and preparation of acid capped CdSe nanocrystals as successful adsorbent and photocatalyst for the removal of dyes from water and its statistical physics analysis. Environm Sci Poll Res. https://doi.org/10.1007/s11356-022-20990-9.
[19] Kumar N, Joshi NC. Potential of PTH-Fe3O4 based nanomaterial for the removal of Pb (II), Cd (II), and Cr (VI) ions. J Inorg Organomet Poly Mater 2022, 32, 1234–1245.
[20] Hassanzadeh-Afruzi E, Esmailzadeh F, Asgharnasl S, Ganjali F, Taheri-Ledari R, Maleki A. Efficient removal of Pb(II)/Cu(II) from aqueous samples by a guanidine-functionalized SBA-15/Fe_3O_4. Sep Purif Techn 2022, 291, 120956.
[21] Gao Y, Qi G, Yan W, Guo D, Jiao W, Liu Y. Preparation of l-cysteine modified $MnFe_2O_4$ nanoparticles based on high-gravity technology and application for the removal of lead. J Environ Chem Eng 2022, 10, 107193.
[22] Kong L, Yan R, Liu M, Xu J, Hagio T, Ichino R, Li L, Cao X. Simultaneous reduction and sequestration of hexavalent chromium by magnetic β-Cyclodextrin stabilized Fe_3S_4. J Hazard Mater 2022, 431, 128592.
[23] Prema P, Nguyen V-H, Venkatachalam K, Murugan JM, Ali HM, Salem MZM, Ravindran B, Balaji P. Hexavalent chromium removal from aqueous solutions using biogenic iron nanoparticles: Kinetics and equilibrium study. Environ Res 2022, 205, 112477.
[24] Diephuis WR, Molloy AL, Boltz LL, Porter TB, Aragon OA, Duron R, Crespo D, George LJ, Reiffer AD, Escalera G, Bohloul A, Avendano C, Colvin VL, Gonzalez-Pech NI. The effect of agglomeration on arsenic adsorption using iron oxide nanoparticles. Nanomaterials 2022, 12, 1598. https://doi.org/10.3390/nano12091598.
[25] Uddin M, Jeong Y-K. Application of magnesium ferrite nanomaterials for adsorptive removal of arsenic from water: Effects of Mg and Fe ratio. Chemosphere 2022, 307, 135817.
[26] Dai X, Rao J, Bao Z, Li K, Feng L, Song D, Zhao L, Li W, Liu X, Yi S, Dong P, Zhan Y. Magnetic double-core@shell MnO_2@NiFe@DE as a multifunctional scavenger for efficient removal of tetracycline, anionic and cationic dyes. J Coll Interf Sci 2022, 628, 769–783.
[27] Lu R, Wang C, Chen Y, Tan L, Wang P, Feng S. Il-functionalized Mn(II)-doped core–shell Fe_3O_4@Zr-MOF nanomaterials for the removal of MB from wastewater based on dual adsorption/Fenton catalysis. New J Chem 2022, 46, 8534.
[28] Li B, Xu C, Yu D, Qi Z, Wang Y, Peng Y. Enhanced phosphate remediation of contaminated natural water by magnetic zeolitic imidazolate framework-8@engineering nanomaterials (ZIF8@ENMs). J Coll Interf Sci 2022, 613, 71–83.
[29] Ma Q, Teng W, Sun Y, Chen Y, Xue Y, Chen Y, Zhang C, Zhang H, Fan J, Qiu Y, Fu R. Multi-component removal of Pb(II), Cd(II), and As(V) over core-shell structured nanoscale zero-valent iron@mesoporous hydrated silica. Sci Tot Environ 2022, 827, 154329.
[30] Haris M, Usman M, Su F, Lei W, Saleem A, Hamid Y, Guo JK, Li Y. Programmable synthesis of exfoliated biochar nanosheets for selective and highly efficient adsorption of thallium. Chem Eng J 2022, 434, 134842.

[31] Xing X, Ren X, Alharbi NS, Chen C. Biochar-supported Fe/Ni bimetallic nanoparticles for the efficient removal of Cr(VI) from aqueous solution. J Mol Liq 2022, 359, 119257.

[32] Zhou Y, Li Z, Ji L, Wang Z, Cai L, Guo J, Song W, Wang Y, Piotrowski AM. Facile preparation of alveolate biochar derived from seaweed biomass with potential removal performance for cationic dye. J Mol Liq 2022, 353, 118623.

[33] Hamadeen HM, Elkhatib EA. New nanostructured activated biochar for effective removal of antibiotic ciprofloxacin from wastewater: Adsorption dynamics and mechanisms. Environ Res 2022, 210, 112929.

[34] Obey G, Adelaide M, Ramaraj R. Biochar derived from non-customized matamba fruit shell as an adsorbent for wastewater treatment. J Biores Bioprod 2022, 7, 109–115.

[35] Vieira ALS, Ribeiro RS, Ribeiro ARL, Ribeiro AM, Silva AMT. Hollow carbon spheres for diclofenac and venlafaxine adsorption. J Environ Chem Eng 2022, 10, 107348.

[36] Xu C, Zhao W, Ma H, Wang H, Wang Y. Tailored design of well-defined hierarchical nitrogen-doped carbon *via* salt-confined strategy for selective Cd(II) adsorption. Chem Eng J 2022, 446, 137222.

[37] El Shahawy A, Mubarak MF, El Shafied M, Abdulla HM. Fe(III) and Cr(VI) ions' removal using AgNPs/GO/chitosan nanocomposite as an adsorbent for wastewater treatment. RSC Adv 2022, 12, 17065.

[38] Chen W, Lin Z, Chen Z, Weng X, Owens G, Chen Z. Simultaneous removal of Sb(III) and Sb(V) from mining wastewater by reduced graphene oxide/bimetallic nanoparticles. Sci Tot Environ 2022, 836, 155704.

[39] Liu N, Yu F, Wang Y, Ma J. Effects of environmental aging on the adsorption behavior of antibiotics from aqueous solutions in microplastic-graphene coexisting systems. Sci Tot Environ 2022, 806, 150956.

[40] Chen O, Xue X, Liu Y, Guo A, Chen K, Yin J, Yu F, Zhu H, Guo X. Shear-induced fabrication of SiO2 nano-meshes for efficient uranium capture. J Hazard Mater 2022, 438, 129524.

[41] Tejedor J, Guerrero VH, Vizuete K, Debut A. Environmentally friendly synthesis of silicon dioxide nanoparticles and their application for the removal of emerging contaminants in aqueous media. J Phys: Conf Ser 2022, 2238, 012005.

[42] Patel H, Routoula E, Patwardhan SV. Decontamination of anthraquinone dyes polluted water using bioinspired silica as a sustainable sorbent. Silicon 2022, 14, 1235–1245.

[43] Chabane M, Tavolaro A, Russo F, Chiappetta G, Dahmani B, Figoli A. Preparation and characterization of PVDF hybrid membranes by embedding nanomaterials as silica gel and clay: Evaluation of the removal efficiency of nickel and copper ions from water. Desalin Wat Treat, 2022, t255, 13–26. 10.5004/dwt.2022.28317.

[44] Zhao H, Zhang D, Sun H, Zhao Y, Xie M. Adsorption and detection of heavy metals from aqueous water by PVDF/ATP-CDs composite membrane. Coll Surf A: Physicochem Eng Asp 2022, 641, 128573.

[45] Shin J-H, Yang JE, Park JE, Jeong S-W, Choi S-J, Choi YJ, Jeon J. Rapid and efficient removal of anionic dye in water using a chitosan-coated iron oxide-immobilized polyvinylidene fluoride membrane. ACS Omega 2022, 7, 8759–8766.

[46] Zhang R, Yang Z, Hu Z, Zhao C, Zeng K, Yu J, Cai L, Chen Z, Jiang J. Modification of PVDF membranes using BiOBr precursor in-situ deposition and tannic acid self-assembly for effectively removing organic pollutants. Appl Surf Sci 2022, 599, 153888.

[47] Chandra L, Jalalah M, Alsaiari M, Balakrishna RG, Harraz FA. Comprehensive analysis of spinel-type mixed metal oxide-functionalized polysulfone membranes toward fouling resistance and dye and natural organic matter removal. ACS Omega 2022, 7, 4859–4867.

[48] Vatanpour V, Khadem SSM, Dehqan A, Paziresh S, Ganjali MR, Mehrpooya M, Pourbasheer E, Badiei A, Esmaeili A, Koyuncu I, Naderi G, Rabiee N, Abida O, Habibzadeh S, Saeb MR. Application of g-C_3N_4/ZnO nanocomposites for fabrication of anti-fouling polymer membranes with dye and protein rejection superiority. J Membr Sci 2022, 660, 120893.

[49] Qiu S, Zhang H, Nie D, Wang W, Nie G. Designing a 3D-MoS2 nanocomposite based on the Donnan membrane effect for superselective Pb(II) removal from water. Chem Eng J 2023, 452, 139101.
[50] Lan Q, Liu L, Wu Y, Feng C, Ou K, Wang Z, Huang Y, Lv Y, Miao Y-E, Liu T. Porous reduced graphene oxide/phenolic nanomesh membranes with ternary channels for ultrafast water purification. Comp Commun 2022, 33, 101216.
[51] Lashen ZM, Shams MS, El-Sheshtawy HS, Slaný M, Antoniadis V, Yang X, Sharma G, Rinklebe J, Shaheen SM, Elmahdy SM. Remediation of Cd and Cu contaminated water and soil using novel nanomaterials derived from sugar beet processing- and clay brick factory-solid wastes. J Hazard Mater 2022, 428, 128205.
[52] Maslennikov A, Peretz R, Vadivel VK, Mamane H. Recycled Paper Sludge (RPS)-derived nanocellulose: Production, detect wat treat. Appl Sci 2022, 12, 3077. https://doi.org/10.3390/app12063077.
[53] Deng M, Liu W, Liu Q, Sun D, Huang F. A new compound $Mn_5P_4O_{20}H_8$ achieving efficient heavy metal removal to the ppb level through a dual chemisorption–ion exchange pathway. Environ Sci Nano 2022, 9, 1146.
[54] Wang L, Wang M, Muhammad H, Sun Y, Guo J, Laipan M. Polypyrrole-bentonite composite as a highly efficient and low-cost anionic adsorbent for removing hexavalent molybdenum from wastewater. J Coll Interf Sci 2022, 615, 797–806.
[55] Santos D, Treméa R, Lorençon E, Batista PR, Coral L, Bassetti F. Removal of methyl violet dye by adsorption process on hydrogen titanate nanotubes: Experimental-theoretical study. Water Air Soil Pollut 2022, 233, 166.
[56] Falyouna O, Bensaida K, Maamoun I, Ashik UPM, Tahara A, Tanaka K, Aoyagi N, Sugihara Y, Eljamal O. Synthesis of hybrid magnesium hydroxide/magnesium oxide nanorods [Mg $(OH)_2$/MgO] for prompt and efficient adsorption of ciprofloxacin from aqueous solutions. J Clean Product 2022, 342, 130949.
[57] Yang Z, Liu X, Xia S, Ding O, Liu H, Wang Z, Zhang L, Zhang H. Au/Boron organic frameworks for efficient removal and degradation of azo dye pollutants. Colloids Surf A Physicochem Eng Asp 2022, 646, 128884.
[58] Li H, Huang Y, Luo Q, Liu J. The simultaneous reduction and adsorption for V(V) and Cr(VI) anionic species in aqueous solution by polyethyleneimine cross-linked titanate nanotubes. Sep Purif Techn 2022, 299, 121682.

Arniza Ghazali*

Reconceptualizing Industrial By-Products: Biomass Circularity for People and Nature

Abstract: Commodities from food crops and nonfood crops are the all-time demand for as long as the Earth is sheltered. Being the leading contributor of global edible oil, palm oil is a significant source of serendipity, leaving behind a string of rich residues routable as the wealth factor for countries like Indonesia, Thailand, and Malaysia. The rubber-related sector is another affluent contingent, marking rubber glove makers Forbes-listed wealthiest 50 in Malaysia. The unified use of rubber plantation residue has popularized global rubber producers as aesthetic furniture manufacturers, beating Greece and Romania. This study reviews the composition and applications of the residual biomass for wealth seekers to understand and expedite the domiciling commodities in a responsible, resource-efficient manner. In doing so, circularity in human capital management and natural resource management is set as a target. The need to exploit artificial intelligence for informed process selection and sustainability-aligned decision-making is also demonstrated. The strategy is to eradicate false assumptions about sustainability and insight into measurable sustainable criteria enabled by the emerging Internet of things. Awareness and deep apprehension of sustainability are key drivers to catalyze the change in practices. Coverage on biomass management for striking the balannce between circularity and advanced technology wraps the chapters for insights into circular bioeconomy actualization.

Keywords: Biomass, circularity, resource-efficient, biominerals

1 Introduction to biomass and circularity

Recognized as the physique of living things with measurable ash, organic mass, and dry mass, biomass is the symbol of the planet's wealth. Derivable from the sea, ground, and air, biomass is the gift of nature, exploitable during its functional period and postharvest. Among them are the flora, fauna, geological structures, and their derivatives. This chapter focuses on the botanical biomass of oil palm and rubberwood origin due to their prominent export revenue contribution [1–3], commodity-loaded residues, and the immeasurable research interest [1–13] placed in them.

***Corresponding author: Arniza Ghazali,** Renewable Biomass Transformation Cluster, Division of Bioresource Technology, School of Industrial Technology, Universiti Sains Malaysia, 11800 USM Penang, Malaysia, e-mail: arniza@usm.my

https://doi.org/10.1515/9783111016825-002

Hevea brasiliensis' latex is an example of a commodity exudate marking the region as a prominent rubber producer. Thailand and Indonesia led the top five global rubber producers, with Vietnam, India, and China outranking Malaysia [14]. The rubber-producing countries enable global medical, sports, defense, construction, engineering, automotive, and infinite other goods supplies. Besides the source of global goods and products, isoprene in rubber latex has been synthesized in nonrubber-producing countries to supplement the natural rubber supplies. Rubberwood makes the attractive biomass of the rubber plantation at its ultramaturity age devoid of latex. Post-latex-producing years, rubberwood becomes the valuable biomass feedstock for furniture making, adding another multibillion-dollar export revenue (Figure 1a) to the rubber-producing countries.

Isoprene (2-methyl-1,3-butadiene) is polymerized to polyisoprene with a growth possibility of 1 million molecular weight [15] in comparison to natural rubber's 1–2.5 × 10^6, made up of regular isoprene repeat units [16]. The synthetic rubber forming via solution polymerization, however, resulted in four cis- and trans-isomeric variations, which was the hindrance factor to commercial production until the 1950s. High-purity (>90%) *cis*-1,4-isoprene with a desired set of properties comparable to natural rubber gives compositional uniformity for reproducible curing effects. The feature allows blending with ethylene propylene diene monomer and solution styrene butadiene rubber (Figure 1), offering better thermo-durability on top of tensile and tear improvements [16]. Further elastomeric augmentation achieved via various additives is critical to tailor synthetic rubber to specific applications.

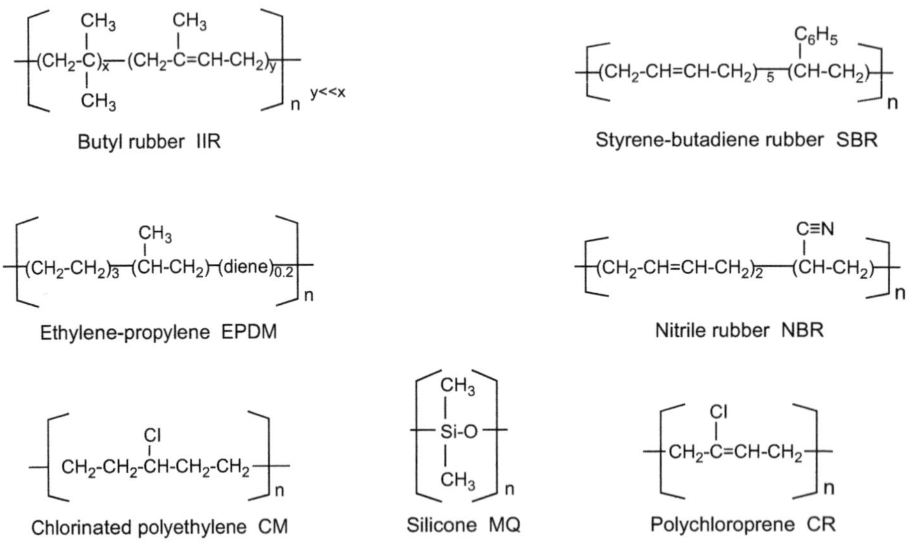

Figure 1: Examples of elastomers forming synthetic rubber (source: [16] 3-s2.0-B9,780,815,514,3,4350,0038-f02-01a-9,7808,1551,4343.gif (471 × 610) (els-cdn.com)).

Natural latex's high structural regularity makes natural rubber prone to spontaneous crystallization upon stretching and at low temperatures. Post-harvest, the latex is subject to several processing stages. Natural latex is modifiable in several ways. While blending with polypropylene yields thermoplastic natural rubber, adding 10–40% process oil produces epoxidized natural rubber. Oil-extended natural rubber is a form of modified latex apart from the DNR obtained by the deproteinization of latex [16].

Latex in rubberwood is collected in fresh liquid form or hardened to a "cup lump." By careful 1.5 mm scission into the rubberwood bark, the exudate ejaculating from the complex latex vessels [17] is valued by its purity, curing rate, viscosity, viscostability, and oxidation resistance.

Post-latex-producing years, the felled *Hevea brasiliensis* becomes the feedstock materials in the furniture-making sector. Malaysia's rubberwood furniture alone secures US $2 billion in export value, making the rubber sector a perfectly circular industry, where residues are properly routed (Figure 2a) as commodities of a different downstream segment.

In the bigger picture, the circularity of the rubber-related sector is evident from the multibillion-ringgit export revenue from both sides of the biomass circularity science (BCS) model in Figure 2a. Further deliberation revealed that H, I, and J as commodities will potentially move the biomass to soil biomineralization post-functional years as plant-growing substrates.

Beating the export revenue from rubber and rubber products is the edible oil palm products. Breeding from the oil palm milling sector, *Elaeis guineensis* biomass structures make up various biomass structures like the empty fruit bunches (EFBs), the oil palm fronds (OPFs), and the oil palm trunk. While EFB is generated throughout the year at the palm oil milling factories, OPF is the biomass arising from the scheduled pruning of the palm tree. Post-fruiting years, the trunk of *Elaeis guineensis* is felled (see I in Figure 2b).

Figure 2 presents the BCS model for rubber and palm oil economic sectors emphasizing the left loop, where by-products of the primary harvest are routed as commodities for direct market demand and those potentially further processed for market.

Despite stagnantly revolving in the research areas, the use of EFB for a similar growth substrate demonstrated by Mamimin and colleagues had effectively reduced the inorganic contents [10], and thus reduced the residue related to the combustion process. Inorganics are also irrelevant to bioethanol production, and discarding the component before a fermentation process could increase yield and accelerate the processing steps by the indirect demineralization of EFB following the uptake of minerals by the mushroom. The specific duration of growth prepares the biomass for sugar extraction and conversion to bioethanol more readily. Not only had the process extended the functional life of EFB but also generated income from mushroom production as envisaged for rubber smallholders [9], which is the timely need for the sector's sustainability and countermeasure for the food scarcity issue facing the world.

In both rubberwood and EFB cases, beyond the functional period as a growth substrate, the biomass will be more ready for carbon fixation. The process readied the biomass for conversion (of briquettes) to biofuel, where high-carbon backbone contents lead to a higher calorific or energy value, ranging from 10 to 13 MJ/kg or 5 kcal/g.

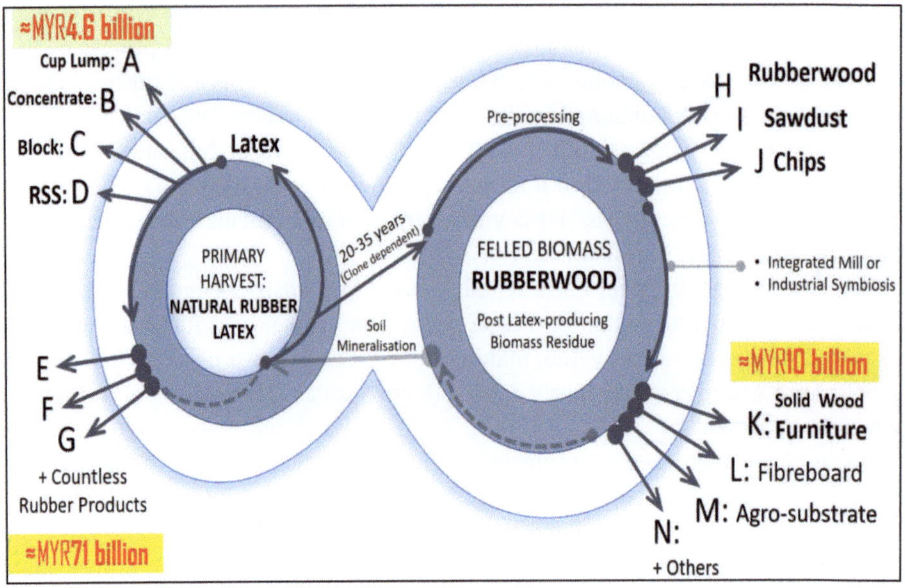

(a)

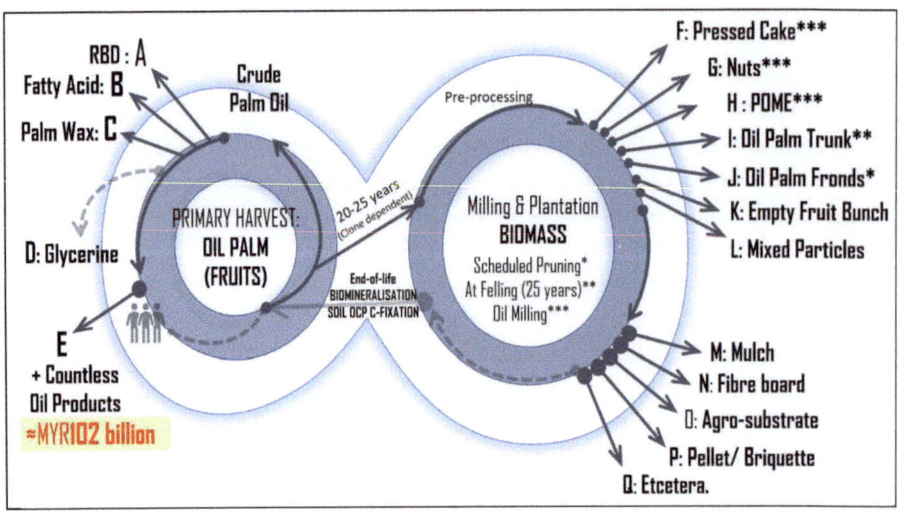

(b)

Figure 2: Biomass circularity science (BCS) model versus 2021/2022 segmented earnings for (a) rubberwood and (b) oil palm biomass for reconceptualizing by-products as commodities via utilization routing (adapted from Ghazali and Zbiec [9]).

To understand the relationship between circularity, sustainability, and prosperity, Section 2 discusses the constituents of the commodity-loaded biomass from the wealth creation industries in Table 1. As palm oil and rubber industries serve the global food and nonfood resource chains, Table 1 presents Malaysia's export revenue for the commodities.

Table 1: Revenue from agri-commodity.

Established agrosector		Hectarage (×1,000 ha)	Revenue (MYR, billion)	Revenue from residue (MYR, billion)
1.	Palm oil-based products	5,730 [18]	102 [1, 19]	–
2.	Rubber	400 [20] – 455 Active plantation	71 [2, 20] Rubber products	8–9 [3]
		1,1372 [21] Total	4.568 [22, 23] Natural rubber	–
3.	Rice	670 [24]	2.44 [18]	–

2 Biomass from the highest earning sectors

2.1 Organocomponents – priced by purity

Analogous to isoprene in rubberwood, the vast organic elements such as tocopherol acetates and carotene in palm oil create value for the edible oil commodity. The rubber and palm oil-based products create prosperity [24], proven by the 10% icons of the Forbes-listed richest people in Malaysia [25].

The 500–700 ppm carotenoids serving as a fingerprint of palm oil [26], vanilla, chlorophyll, lycopene, and isoprene are among the extracts in extreme demand, and that study was extensively done to determine the chemical origin, structure, and synthesis. Today, synthetic rubber production increases at a 10% rate, while natural vanilla extractable from the oil palm biomass is supplemented by its synthetic, semisynthetic petroleum-based [27] counterparts. With the rising demand, the price of pure biovanillin has risen to US $11,000 per kg, which has now become the wealth factor to Madagascar Bourbon vanilla smallholders [28]. Today, vanilla is known for its nine flavor wheels, toasty, spicy, foresty, and off-flavor that are sulfury and medical, besides the basic tastiness notes imparting the sweet, creamy, cooked, or denatured flavor. Vegetative notes with waxy and floral hues are among notes derived from the host of chemical components besides the key, 4-hydroxy-3-methoxy benzaldehyde [27].

In plants, vanillin is part of thc lignin macromolecule. Vanilla extracts of ligno-origin are, however, considered artificial vanilla. The microbial transformation of such natural precursors as ferulic acid to generate vanillin, on the contrary, is considered a biosynthetic pathway, and thus, natural origin [27]. Vanilla of EFB origin [29] is an example of the possible industrial symbiosis between the pulp-producing mill and the flavor and fragrance industry. Although supplementary synthetic vanilla is a need of the sector, the conversion of pulping black liquor (BL) lignin to vanilla (Figure 3) is controversial, given the resource intensiveness of the multistep process. The call for a giant company akin to Nestle to go for an "All-Natural" campaign has also shrunk the market for synthetic vanilla. However, these are robust proofs of industrial symbiosis, marking the heart of circular bioeconomy and resource efficiency.

Soda EFB Pulp Black liquor

↓

+ Nitrobenzene (Oxidation)

↓

Autoclave 165°

↓

Extraction in Chloroform ($CHCl_3$)

↓

Dissolution in dichloromethane (CH_2Cl_2)

↓

Vanilla

Figure 3: Process flow for vanilla extraction from soda pulping black liquor post-fiber extraction from EFB.

In normal practice, sodium in sulfite BL is a catalyst for synthetic gas (syngas) production. The commodity-based gaseous products originate from the organic components of the botanical biomass feedstock.

The principal building blocks are the carbon backbone, known as the organic structures, almost always fused uniquely to the inorganic or micromineral. Table 2 and Table 3 list the approximate quantities of the constituents that coexist in the botanical structures.

Besides the deliberated latex, such tree exudate as gum also offers a broad spectrum of uses. Acacia gum, for instance, finds food-based and nonfood-based applications. Being aromatic, frankincense is a highly valued exudate from various *Boswellia* species, marking the prominence of Somalia, Oman, and Hadhramaut, Yemen.

2.2 Inorganic: tiny jewels with big impact

Plants, as a producer of complex carbon compounds [42], are dysfunctional without the inorganic elements. The inorganic constituents are absorbed from the soil in the form of ions, enter the plant structure through the root systems, and grow in specialized cells

Table 2: Examples of botanical organic constituents.

Constituents	Function	Rich sources	Benefits and applications
1. Chlorophylls	Solar capture and starch synthesis [30]	Harvest of botanical parts	Antioxidant
2. Tannin	Defense against pathogen [31] and structural luster	Technical biomass Chestnut (*Castanea sativa*) and sumac [32]	Antioxidant, therapeutics, leather coloring, and promote the health of digestive system [32]
3. Isoprene	Defense mechanisms	Latex exudates of *Hevea brasiliensis* and dandelion [9]	Cushioning agent
4. Vanilla (lignin)	Antimicrobials [27].	*Vanilla planifolia*	Flavanoid, climate and soil curing [28]
5. Anthocyanins	Colors attract pollinators, protect against stressors [33]	Petals of *Clitoria ternatea*	[34]
6. Wax (E/B extractives)	Prevent dehydration	*Elaeis guineensis* vessels, *Cymbopogon citratus* leaves	Glazing agent [8, 35]
7. Gums: botanical adhesives as a mixture of salts, minerals, and polysaccharides	Protection against mechanical and microbial attack [36]	Seeds of *Cassia obtusifolia*	Gum powder E499, E427 gelling agent, cosmetics [36], potential biosemiconductors [37]
8. Frankincense (aromatic gum resin) [38]		*Boswellia*	Aromatherapy, antioxidant [39]
9. Eucalyptol	Defense mechanisms		Therapeutics, insect repellent
10. Ascorbic acid	Mineral regulation	Fruits	Antioxidants, immune system booster
11. Cellulose	Structural elements		
12. Protein	Biosynthesis/growth control [27]		Cell rejuvenation
13. Carotene	Fingerprint to oil characterization [26] Recipient of excess energy [30]	Carrots, pumpkins	Antioxidants, vision rejuvenation

Table 2 (continued)

Constituents	Function	Rich sources	Benefits and applications
14. Serotonin	Colors attract pollinators, protection against stressors [40]	Banana	Antioxidant, antidepressant
15. Cernitin		Pollen of rye (*Secale cereale*)	Antioxidants: elixir for nocturia – a urogenital disorder [41]

[43] to form various shaped crystals. The ions travel to the plant through a vessel known as the xylem and land in biofactories where they grow by the command from the plant's DNA.

Silica bodies are examples of inorganic constituents in plants. Studies on botanical silica in the oil palm fruit bunch vessel structure show evidence of silica growth [8, 44, 45] to the size ranging from nanometer scale [46] to 20 mm granules (Figure 4). While Figure 4a presents the skeletal silica devoid of the hydrocarbon biostructure, Figure 4b presents the mature silica bodies on the petiole of the OPFs apart from the unique siliceous thread shown in figure 5. The strong affinity of silica toward other metals allows silica to serve the plant's detoxification mechanism by attracting excess metals to it [8].

(a) (b)

Figure 4: The two forms of siliceous structures in oil palm biomass: (a) skeletal silica granules [45] and (b) silica bodies engulfed in the adsorbed minerals.

Inorganics also form different shapes and functions depending on the cell where they originate. The lucid position by Currie and Perry [47] rationalizes the encountered skeletal silica [44] in scientific studies.

The function of siliceous structures in plants has yet to be fully understood. Currie and Perry's [47] in vivo and in vitro studies found the link between orthosilicic acid and siliceous structures in plants. Figure 6 presents the biophysical and biochemistry governing siliceous structural buildup in living cells giving rise to silicification.

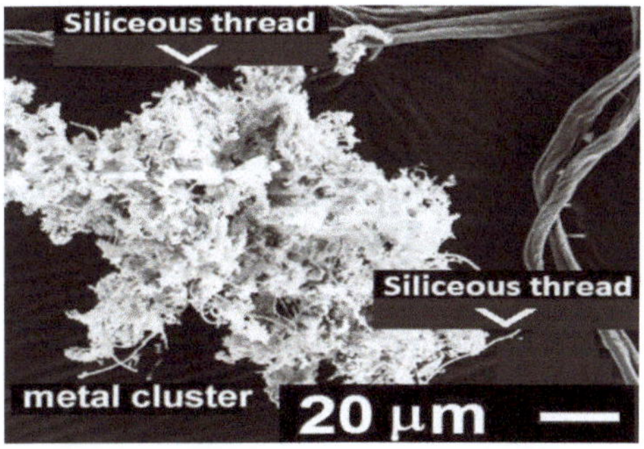

Figure 5: Siliceous thread amid alpha-cellulose.

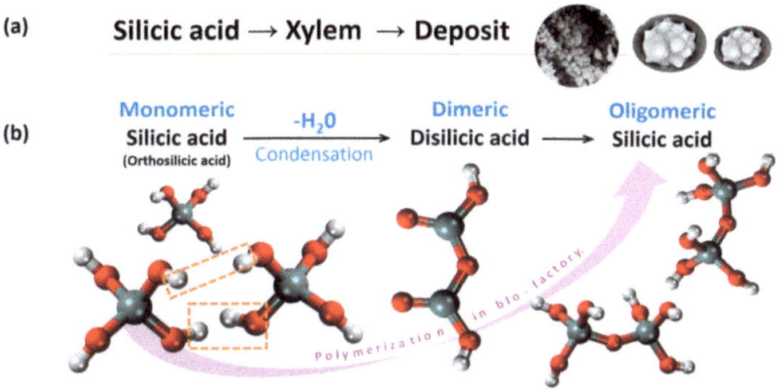

Figure 6: Biosystemic reactions leading to (a) physical deposition of silicic acid intake from the roots and (b) biopolymerization of the siliceous skeleton.

Siliceous structures vary in composition and shape depending on the localized cell pH and the customized function. Plants are made up of 0.1–10% (o.d.) siliceous structures, and their growth is enabled by soils of pH above 9 where silicification occurs [47]. In plants, the pH varies with the varying cell compartments, justifying the occurrence of the siliceous skeleton in some cells, granules in others, and mineral-deposited SiO_2 originating from the physical deposits of the root-absorbed vascular surface as in Figure 6a.

Numerous analyses consider silica as the essential element in plants, offering structural rigidity and allowing plants to grow by height, to harness light, which in turn enhances photosynthesis. By inhibiting dehydration, siliceous biostructures provide plants with drought tolerance. Efficiency in mineral and food transport is achieved through the siliceous structures sculptured with nanopores [35] serving for gaseous and water passage.

Postharvest of the stalk for the food market, lemongrass leaves containing siliceous bodies and calcium-rich crystals were the glazing agent for wood and nail. Figure 7 shows the zero-waste potential with the recommended customized use of *Cymbopogon citratus*.

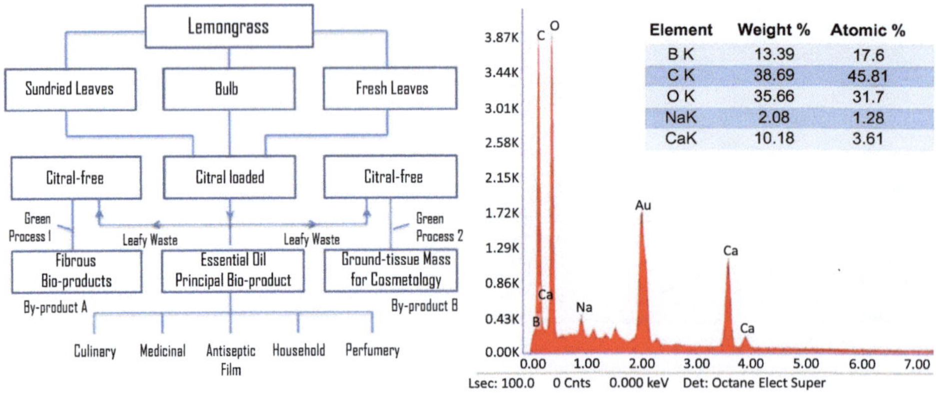

Figure 7: Lemongrass commodity maximization taking its high calcium content into consideration.

Silica harvested from the rice husk is an equivalently functional renewable source of biomaterials. Besides its longstanding use as a pottery glazing agent, the recent Solvay's "breakthrough silicate process" reveals the possibility of a 50% CO_2 emission cut by substituting biosilica for the fossil-based fillers used in tires [48].

Other botanical inorganic components (and their roles in living plants) are shown in Table 3. These functionalities are not limited to the living rubberwood, lemongrass, and oil palm but to plants, in general.

The high micronutrient and macronutrient indicate that the non-hydrocarbon residues from biogas or biofuel production can be directly returned to the soil for biomineralization or further processed and purified as human and animal supplements. Via the integrator business model, the biogas producer is also called as the "biocoal" or "biochar" producer. Their use in soil-less growth substrate (Figure 8) has also proven practicality for a vertical garden. Besides plant nutrition, a specific hydrocarbon-to-mineral ratio will also cater for the needs of the construction formulation.

The many harvestable components in residual biomass demonstrate the serendipity to be expedited for humankind and nature empowerment. The rich composition and various applications are the gateways for zero-waste use of the plants, with one botanical part being harvested as food and the others as a medicinal agent, therapeutics, manicure, cosmetology, furniture glazing, and perfumery industry.

When it comes to industrial processes targeting the isolation of such organic components as cellulose, these minerals impose competing reactions that consume the chemicals and reduce process efficiency. Tabulated are the challenges related to the presence of metals in EFB that reduce their suitability for high-end product making. Table 4 lists the industrial issues imposed by the inorganic components in botanical structures. The obstacles in pulping and related processing are shown in Table 4.

The usefulness and undesirability of the components forming the building blocks of biomass signal the need for compositional analysis as the first step to decide on the

Figure 8: Biochar growth enhancement manifested from lustrous leave blades and unexpected flowering and budding frequencies despite the tiny growing pod.

Table 3: Examples of botanical inorganic and organometal constituents.

Elements	Function
Al	Cell integrity, photosynthesis
Au	Found in certain cyanogenic plants, interact with cell protein [49] In field applications, Au nanoparticles exhibit antiseptic effects [50]
Ca (calcium oxalate) [43]	Protection against herbivory attack, Ca regulation [43], 2 μg/g in EFB [8]
Cu	0.8 μg/g in EFB [8]
Fe	Cofactor for many botanical enzymes [51] Chelated for controlled reactivity [51].
K	Macronutrient; photosynthesis: starch, ATP, protein synthesis. Regulates opening and closing of stomata [52], photoprotection [42]
Mg	Bound in chlorophylls a and b in chloroplast and serves the photosynthetic function [42], 4 μg/g in EFB [8]
Mn	Micronutrient for plant metabolism, cofactor for oxygen-evolving complex, catalysts in water photosplitting reaction [53]
Ni	Enzyme cofactor (urease activation), chlorophyll enhancer, germination at optimal Ni level [54, 55]
Si	Structural element [8, 44] Detoxification agent, 1 μg/g of skeletal and granular bodies in EFB [8]
Se	Curb heavy metal toxicity, photodurability, and chlorophyll protector [56]
N	Plant's macronutrient, plant growth [52]
Zn	Bound to organic acids and Zn^{2+} [57], organometallic form plant metabolism and enzymatic (protein) coworker [58]

Table 4: Industrial challenges associated with metals coexisting with the organic commodity.

Metal elements	Industrial issues
1. Fe (iron)	Production of colored complex upon reaction with lignin [59]
2. Mn (manganese)	Agent for brightness reversion [59] Catalyst for decomposition of hydrogen peroxide [60] in APMP pulping liquor
3. Ca (calcium)	Restriction of swelling [59, 61] Color reversion [59] Source of scaling problem [62]
4. Cu (copper)	Agent for brightness reversion [59]
5. Al (aluminum)	Color reversion [63]
6. SiO_2 (silica) and skeletal silica	Difficulty in chemical recovery from liquor [64, 66] Abrasion of metals frequently applied to paper [67] Abrasion damage and lowered cutting blade life during converting and printing operations [65]
7. Extractives	Consumer of pulping chemicals [68] Form organic deposits on equipment [69, 70]

choice of a feedstock. While yield is a critical factor for process economy, multistage processing needs a reasonable limit to avoid unnecessary complexity and risk. Along with treatments and processes are the by-products from competing or side reactions which affect the purity of the desired product. Solid and liquid forms of by-products require treatment before safe discharge.

In an ideal resource-efficient practice, the by-product of a process can be routed as a feedstock of another process. Such an ideal situation is known as industrial symbiosis, which actualizes circularity.

Section 3 demonstrates the option for oil aplm waste utilization with focus on the abundant oi palm friut brunches (EFB) yet aachieving circuolariy, unlike the rubber sector. Highlights on the practicality of the low-risk nd low-cost process for striking a balance between green (or circular) pratice and techology are highlighted. Gaps predefined by the models in figure 2, thus set foot to the deliberation on the strategic tool for simulation of sustainabilityaligned decision-making.

3 Circularizing the utilization scheme

Emission reduction, delayed end-of-life, waste reduction, and low-emission targets [71–75] are the criteria for resource efficiency synonymously recognized as circularity or circular economy. Rooting to the twentieth century [71, 76], circularity is today the "nature-powered" economy [77], setting compliance for waste minimization as wastes to symbolize inefficiency and emissions cut through redesign of processes [73], more efficient use of natural resources, and lower the dependence on the depleting fossils.

Tracking and cutting greenhouse gas (GHG) emissions are mandatory for the world to pursue industrialization sustainably. The booming global demand for goods [78] and services arising from population growth elevates global warming when the heat-trapping effects of GHG [78, 79] (CO_2, mainly) [80] derail from the control radar. In addition to destructive acidification effects [81–84], methane (CH_4) emission following biomass degradation corrodes the environment [85, 86], perturbing the ecosystem further. Stratospheric GHG and the ensuing heat effects cause uncontrolled evaporation [87, 88] manifesting as droughts. Ice-melting, which raises sea levels and causes flash floods, polarizes the form of environmental disasters, impacting also agricultural productivity [89–91].

Circularity is strategized for humankind to further industrialization while co-living harmoniously with nature. The product's extended functional life is ensured by jiving into the criteria of recyclability, durability, and reparability [73, 74] for reusability and fit for repurposing and critical account of burdens imposed on nature [92].

Rerouting of industrial and plantation residues is equivalently important and this is a priority for the highest earning sectors (Table 1) [1–3, 21–22, 93–95], particularly in resuming economic growth [93] responsibly. Whether by an integrated system or industrial symbiosis, the fruit of reconceptualizing wastes as a commodity is portrayed in Figure 2. The all-year primary harvest, natural rubber latex, offered multibillion dollar export values to rubber-producing countries worldwide. The BCS model demonstrating Malaysia's billion-dollar rubber and rubber product export [1–3, 20–22, 26, 95] translating to a multibillion downstream felled rubberwood furniture provides the impression of a circular economy.

The biomass circularity model (BCS) marking labor intensiveness also alerts the possible inefficiency and profit intractability [96, 97] ill-affecting the upstream segment. The model, in this light, points to the essentiality of tracking human resource-efficient management. The fate of the 95% contributor of the billion dollar calls for strategic coordination of rubber production [98]. BCS can therefore identify the preyed smallholders in the constant challenge of dwindling rubber prices. Initiatives for income diversification [9] were proposed to help smallholders navigate the current economic turbulence productively.

In the context of palm biomass and rubber-related biomass, the subsequent section describes the low-risk and high-risk utilization schemes in terms of capital requirement. While humble processes refer to the low-risk, low-capital utilization routes, advanced processes denote the high-risk, large-scale, capital-demanding processing.

To embrace the sustainability assessment resembling the life cycle assessment, LCA ISO14040:2006 [73, 92], a set of measurable sustainability criteria (MSC) was proposed. Carbon footprints (CF), ecological footprints (EP), and water footprints (WF) are marked on the framework. While the exact footprint values are for future scrutiny, this chapter is devoted to showing the simplified simulation for triangulation possibly achieved by big data analytics (BDA) processing.

3.1 Humble processes

The biomass in focus is the oil palm fruit bunch due to the tremendous attention it has enticed. The BCS model in Figure 2 reveals the forms of biomass commodities generated as oil milling by-products. EFB remains the most attractive biomass due to the all-year availability but is reputed as the factor for valley-of-death (VOD) projects.

Based on the BCS model in Figure 2, the sector has a great potential for mixed use of commodities to slim down the size of the waste but mixed use requires a simple, reproducible process, less sensitive to the effects of small parameter fluctuations. Figure 9a shows the folded feedstock-process-product (FPP) framework as a bird's eye view of the EFB utilization strings.

The low-risk process shown on the hanging left region of the framework is mostly a chemical-free process. Production of burning residue for use as fertilizer is a common practice, despite the emission and an unknown level of impact on human health. Burning can execute more humane manner by performing the process in a semi-closed system for controlled fume dispersion and possibly absorbing the particulate matter and other associated volatiles by a trapping mechanism. Recommended in the framework is the use of OPB-derived activated carbon as an adsorbent.

Also regarded as simple low-cost, low-risk processes are shown on the right. Ash of EFB, subject to chemical-free processes, also finds less delicate applications. Included in the low-risk production scheme is briquette production with detailing in Figure 10a.

Simulation of analytics application shown as CF, EF, and WF markings representing the MSC is hypothetically marked for insight into the type of assessment BDA could generate. If the footprint data could be autogenerated in the exhibit alongside the capital and process strings, sustainability-aligned assessment could be performed, particularly for an informed decision-making process. Together with the footprint data, the total social cost involved in the project can also be estimated.

Comparison between the frameworks in Figure 10a and b reveals the relative capital demand with biomass processing complexity. While the footprints in the production of briquettes are inevitable, the huge footprints involved in bioethanol production is rest assured.

Attempts to venture into the capital demanding pulp production through chemical processes (fiber extraction, Figure 10b), which seem halfway the bioethanol string, had proven a setback. Figure 11 presents the location of the commercial-scale operation that failed due to challenges with EFB continuous feedstock. Today, sustainable sourcing may pose a different form of challenge to EFB utilization due to the hurdles in complying with the roundtable for sustainable palm oil stringent requirements.

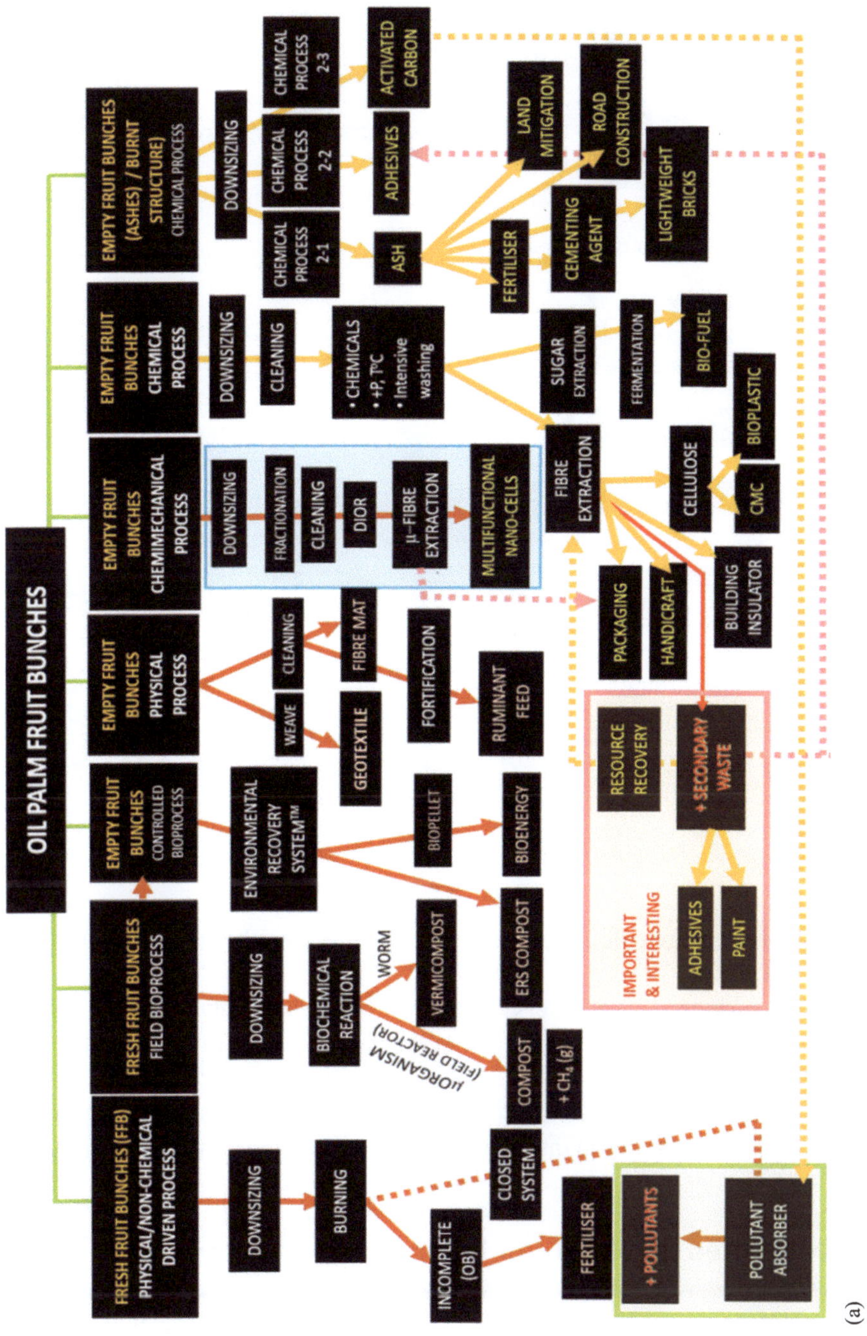

Figure 9: Process strings for oil palm empty fruit bunch (EFB) utilization availed in research.

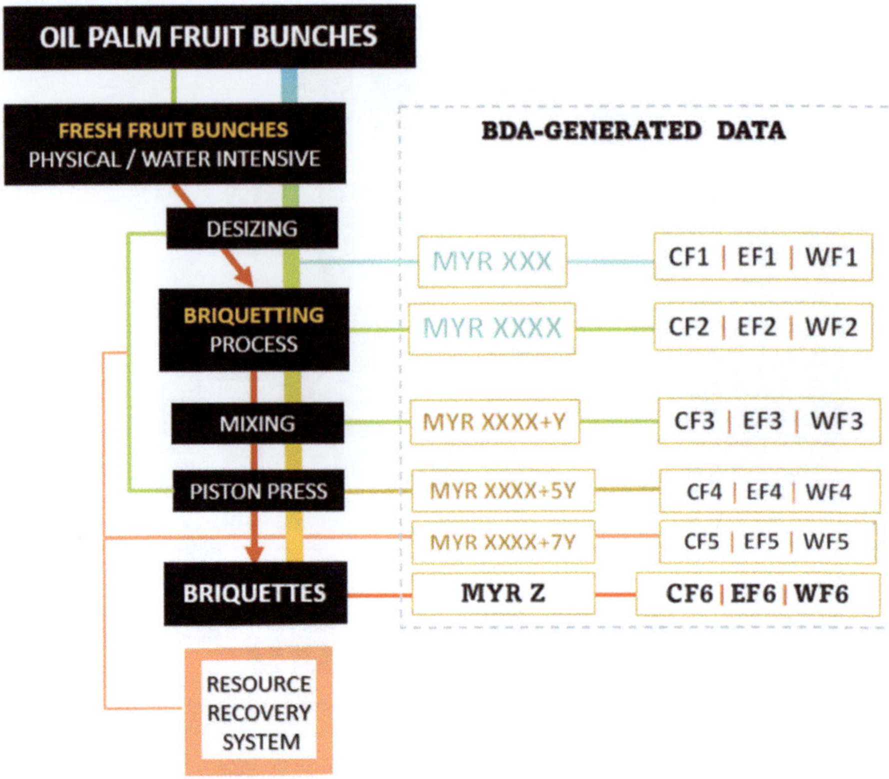

(b)

Figure 9 (continued)

The facilities today resume different economic activities like the fate of the assessed biorefinery VODs [100].

Other uses such as absorption of pollutants revealed as involving minimal processing [100] seem logically more viable due to the reduced risk, at least from the zero need for clean feedstock, which in turn increases the water intensiveness.

The subsequent section dedicates the LCA alternative for sustainability assessment to allow capital ventures to intelligently select a project for investment.

3.2 Sophisticated FPP: cold call for commercialization

Sophisticated processes are the term denoting processes requiring a huge million to billion-dollar investment capital for a large-scale production hitting a reasonable rate of investment return. By the principles of scale, market power, and quality management [99], the high production scale proportionality to profit applies on conditions of high

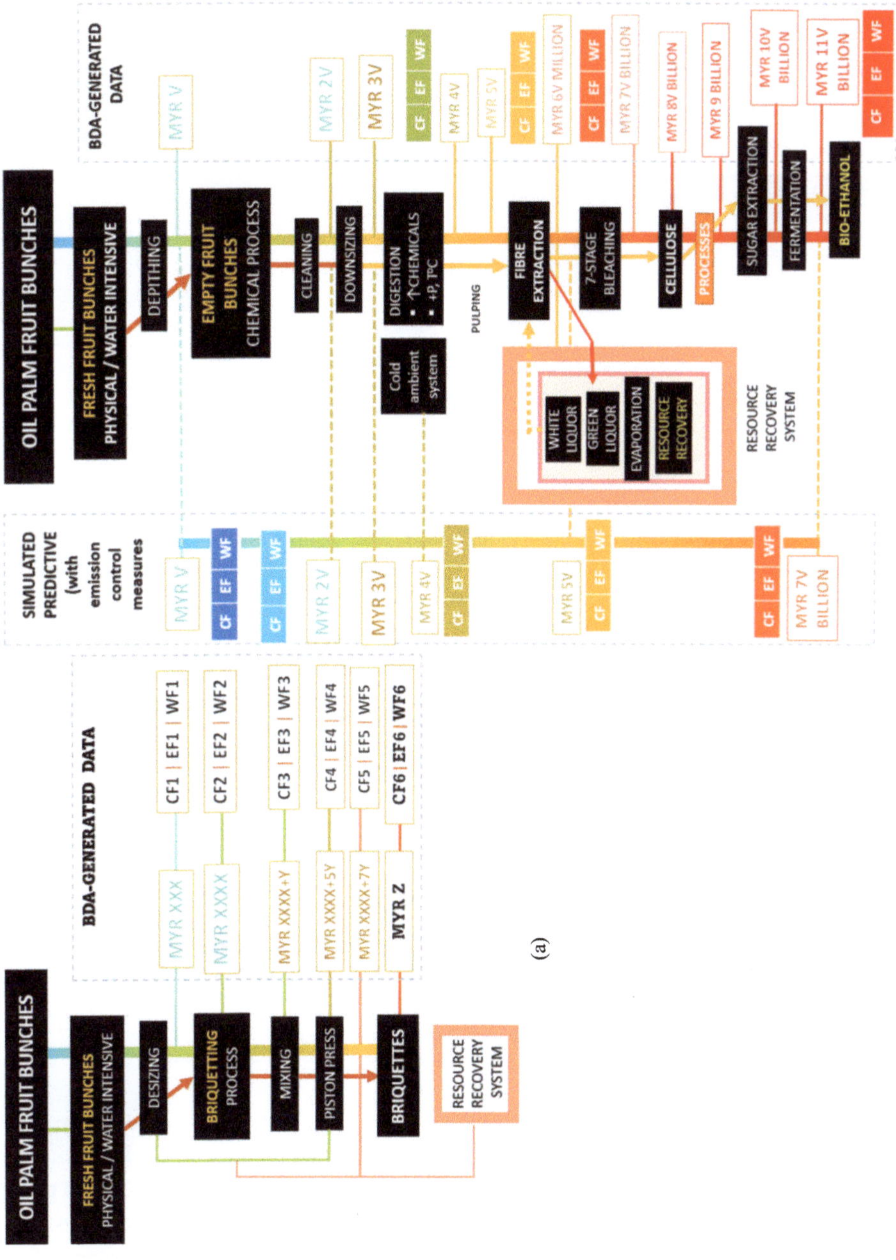

Figure 10: Process strings for oil palm empty fruit bunches (EFB): (a) conversion to briquettes and (b) processes leading to bioethanol production for visualization of complexity.

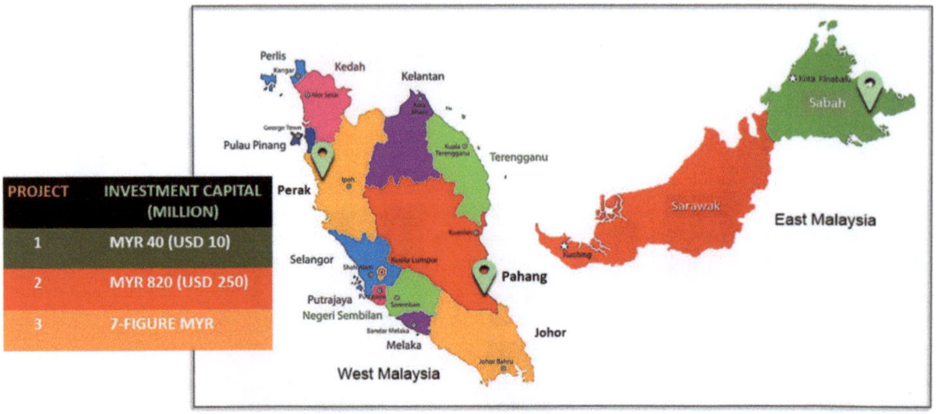

Figure 11: Three pulp mill closures attributable to feedstock issues.

purchases-to-sales ratio. Bioethanol production based on the sugar breakdown scheme with pulp as feedstock is an example of a huge-scale project. Targeting isolation of one set of components among aldo-sugars extends the fiber extraction process to several purification and breakdown processes, resulting in a complex process series portrayed by the long string in the FPP framework (Figure 10b). As history recorded, capital intensiveness poses an extremely high risk, and one miscalculation can lead to VOD [100] analogous to the 200 VOD biorefineries in the United States (Table 5).

Table 5: VOD factors disclosed by project insiders.

Project	Investment capital (MYR)	Issue
1.	40 million [101]	1A: Feedstock 1B: Lab-commercial outcome gap
2.	820 million (staggered sum) [102]	2A: Feedstock
3.	Unknown capital	3-x: Multiple factors

Figure 12 demonstrates the hypothetical effects of incorporating desilication in the pulp production process. Based on the theoretical and industrial claims, silica dissolving in the alkaline white liquor causes an increase in liquor viscosity, inefficiency in the pulping reaction, and challenges in the resource recovery systems. Incorporating a desilication process taxing MYR X, therefore, allows an efficient fiber extraction process, less black liquor (BL) viscosity, and hence, energy-saving pulping process. The payback is marked as –X in the MYR 5 V pulping overhead at the origin.

Further predictable savings may be reaped at the resource recovery steps. Due to less viscosity and reduced (or elimination of) scale buildup, maintenance costs can be trimmed by reducing process shutdown for the removal of silica deposits on the recovery

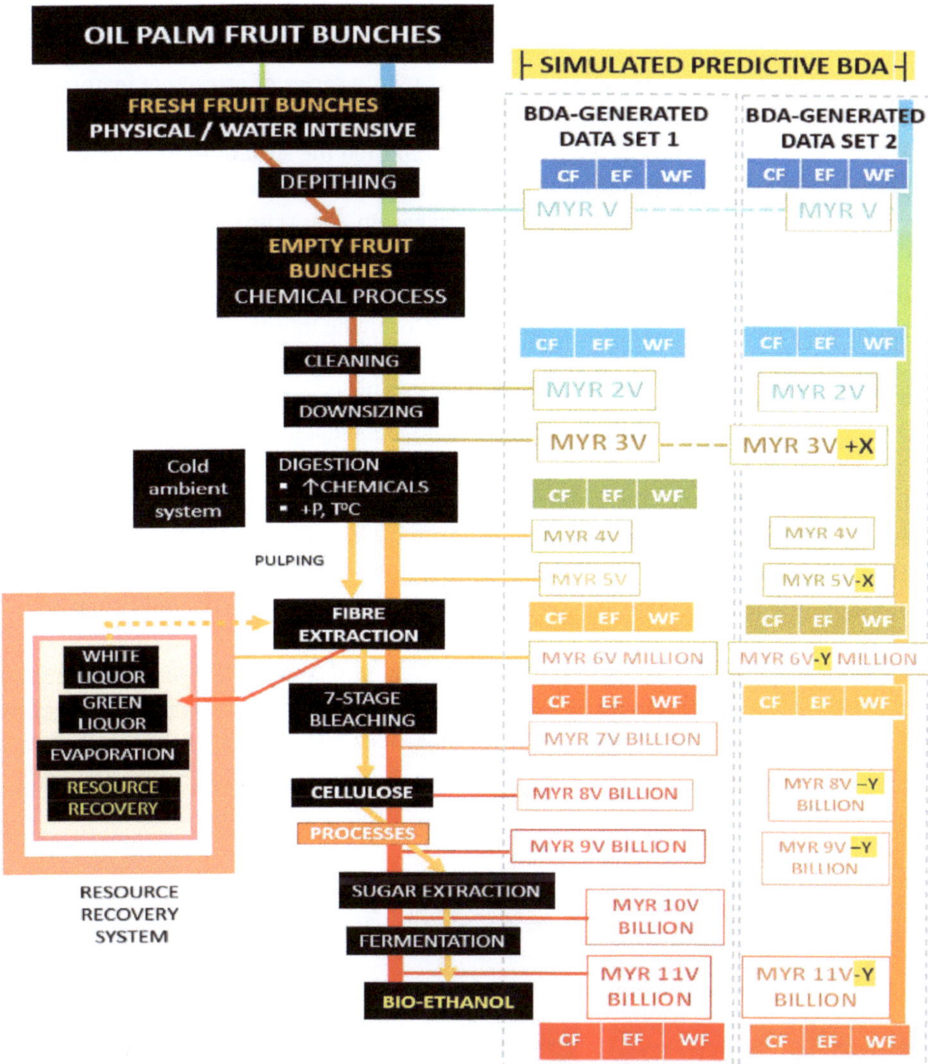

Figure 12: Predictive analytics may simulate cost shift and footprint reduction accurately via triangulation data processing. On a framework, hypothetical cost reduction is indicated as –Y.

boiler. The ensuing MYR 6V-Y is projected to impose a carry-over profit to the overall bioethanol production process of MYR 11V-Y. With a further reduction in emissions and WF, EF, and CFs, a reduction in social cost is also projected. The triangulation capacity of BDA would enable more accurate footprint and cost shift analysis, hence a close-to-perfect guide for informed decision-making for resource efficiency and process sustainability.

Without BDA, other alternative fiber extraction processes with significant differences are easily compared. Section 3.3 deliberates on the assessment.

3.3 Sandwiching the scale polarity

A less capital-intensive fiber extraction process with a series of attractive features is at present given by an alkaline peroxide mechanical pulping (APMP) process. Operating full scale in North America in 1989, APMP has reported the sustainable solution for the polluting chlorine bleaching in the earlier century. The adoption of alkaline peroxide at an optimized condition allowed pulping and bleaching to occur in one process. The alkali premixing with hydrogen peroxide fascinatingly produces dioxydanyl radicals (DIORs) that dearomatize the lignin, creating photostable radial-modified molecular fragments that are not only photostable but also promote interfiber bonding.

On EFB, APMP produces high paper quality comparable to a chemical pulp. The recyclability was found limited, however, prompting further refinement of the fibrous mass for nanofunctional cell production. Not aimed at bioethanol production, APMP functional mass production is compared to the processes of producing cellulose nanofibers.

The most common of the groups are nanofibrillated cellulose, nanowhiskers, and nanocrystals. These nanobiomaterials are principally cellulose obtained via a multistep process indicated A to I in Figure 10b, where the EFB feedstock is concerned.

Employing a simpler fiber extraction technique such as APMP could shorten the process. As extensive refining is performed, nanofunctional cells offering nanoenabling effects in grain coating and paper surface modification [34, 46] are generated.

Figure 13 presents the comparison of process-choicees leading to nanoscale cell production analogous to an extended EFB-to-celluloseprocess string in Figure 10b in contrast EFB-to-Nanocells string in Figure 9. The apparent alternative routes to product development need innovation to lower the risk of commercialization failure.

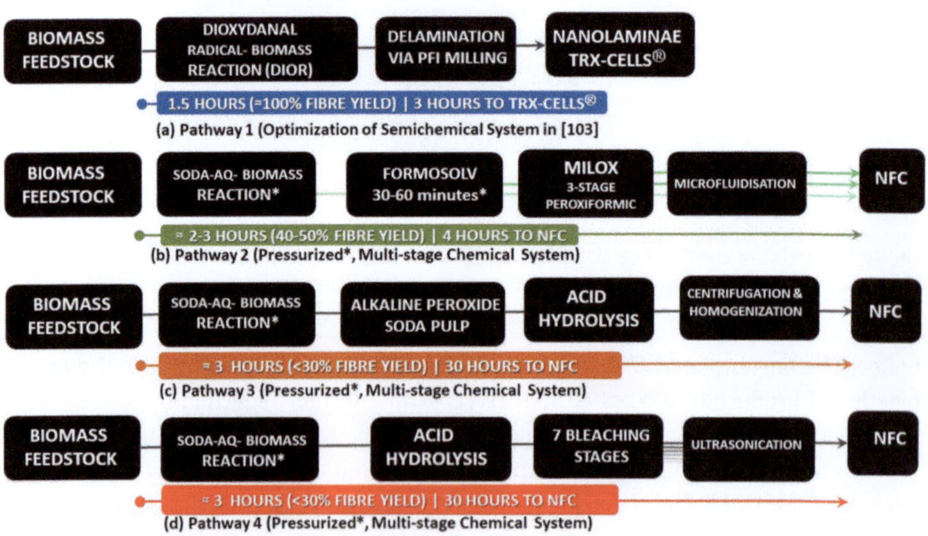

Figure 13: Process differences between DIOR and other systems.

The rapid brightening process and weakening of the chromophoric sites allowed the primary fiber cell walls to be collapsed at the milling stage. Without fiber purification from lignin residue that would otherwise demand the seven bleaching stages (Figure 12), DIOR extension with mechanical refining offers an economic option for nanoscale functional cell production. The subsequent sections detail the unique properties of the nanofunctional cells carrying the trade name TRX-cells®.

3.3.1 TRX-cells®: nanotech made simple

The nanofunctional cells obtained from pathway 1 (Figure 13) originate from the cell walls collapsing at the DIOR weakened sites based on the steps marked in the blue box in Figure 9. Peeled by subsequent refining, the cells are not entirely cellulose but a mixed mass of nanolaminae containing microfibril still held by the noncellulosic, hereby denoted as radical-altered chromophoric sites, RADIALS. Figure 14 presents the concept of cell wall deconstruction.

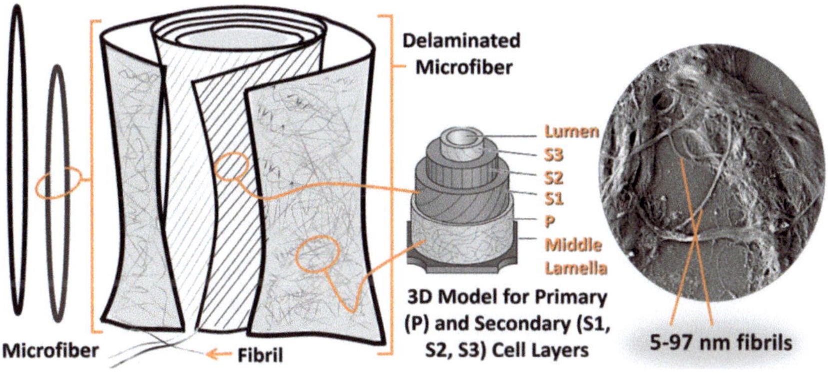

Figure 14: Deconstruction of the dechromophorized EFB mass.

Dip-coating papers with the nanofunctional cells (TRX-cells®) enhanced the paper surface hydrophilicity and improved inkjet (Canon PG47) printability by up to 80%.

While stray of laserjet prints are significantly higher than those exhibited on inkjet prints, the TRX-cells® coat helped render a 44–53% reduction of stray on A4 and imperfect paper webs. The enhanced print coverage comparable to those of commercial A4 implies an increased toner affinity to the paper surface following TRX-cells® coating. On certain 100% fiber paper, toner splashing implies the hot melt particle repulsion forces toward the paper surface plausibly due to charge incompatibility. However, on an extremely nonuniform surface of papers with high fiber bundles, laserjet prints improved with the snug-fitting of TRX-cells® into the micropores. Figure 15a and b illustrates the occurrence.

The minimal stray alongside high coverage values of 112% and 93% for their inkjet print coverage (cf. 86%:13 and 100%:17 percentage coverage-to-stray values) suggests TRX-cells® compatibility for inkjet prints, giving rise to the better continuum of PG47 Canon inkjet (Figure 15c and d).

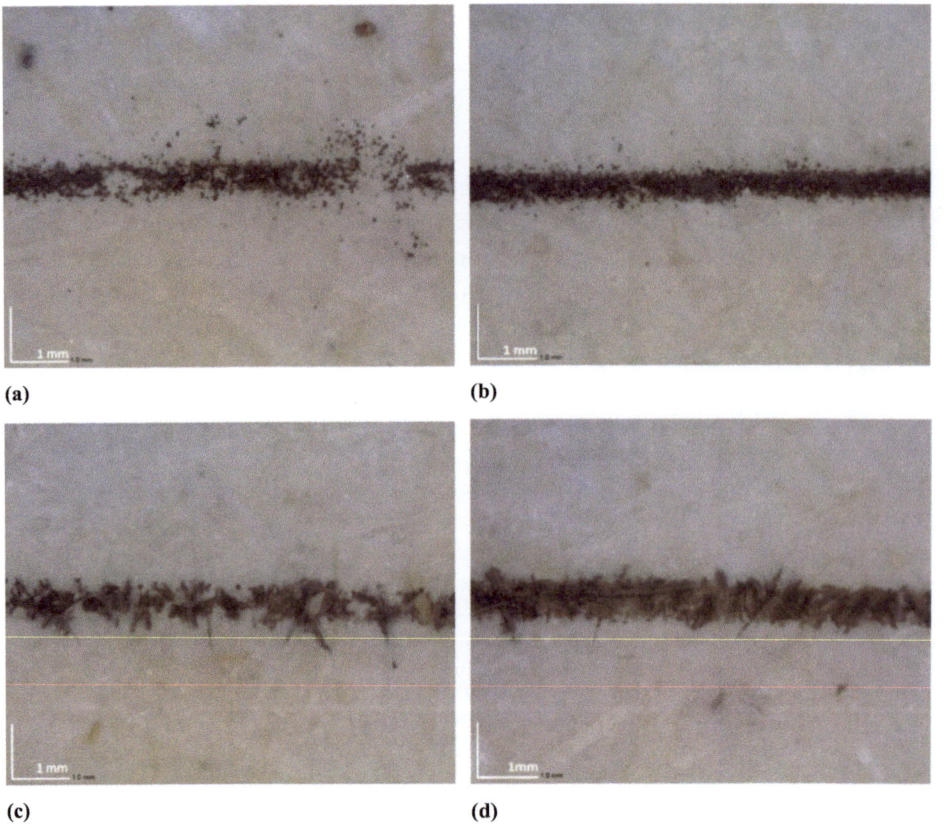

Figure 15: Paper substrates printed with laserjet prints (a) without and (b) with TRX-cells® and paper substrates printed with inkjet prints (c) without and (d) with TRX-cells®.

The influence of TRX-cells® coating in rendering an improved printability is associated with the 18–40% increase in surface roughness, implying the desirable R_q, value ranging from 0.098 to 0.119 μm [46]. With the TRX-engineered surface, the 0.1 μm surface roughness optimality for great print quality demonstrates the industrial guideline [105] for personalized coating formulations for the specific paper surface. Good print may not necessarily demand surface smoothing.

Analysis of TRX-cells®-coated paper surface reveals evidence of an altered capillary action arising from an increase in the depth of ink trails by the insertion of TRX nanofibrils into paper micropores. The narrowing micro-to-nanoscale capillary trails curbed inkjet from bleeding in the horizontal and vertical directions, thus reducing the possibility of ink penetration to paper overleaf. The surface modification effects signal capital savings by stray and bleed reductions, thus, the precision printing applications.

TRX-cells® captured on the surface of commercial A4 papers (Figure 16a) portrays the principal mechanism of printability enhancement. TRX-cells® formed a canopy (S) on the protruding $CaCO_3$ filler, enhancing the coated surface infinity for inkjet printers.

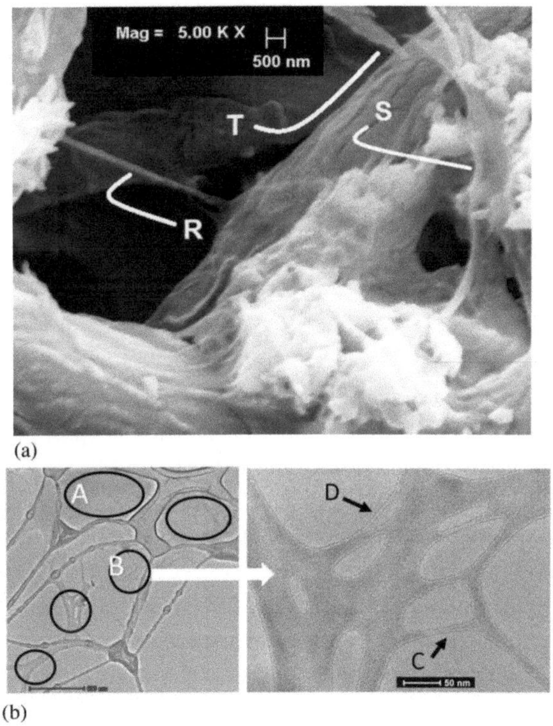

Figure 16: TRX-cells® forming (a) canopy on inorganic structures used as fillers in commercial A4 paper. The capillary bridges (R) matched to the isolated TRX-cells® in (b) demonstrate the nanoscale surface coverage.

Under the electron voltage of the energy-dispersive x-ray, charging of TRX-cells® exhibits a "soft" cell layer with 100 nm approximate thickness. Compacting the cells into the 60 gsm hand sheets (made of 100% TRX-cells®) yielded a translucent sheet of TAPPI opacity of 72 ± 4%, typical of the densified building block of the nanoscale elements.

Also evident from Figure 16 is the 170 nm width hanging tube forming as a result of TRX-cells® curling. Denoted as R (Figure 16a), the structure could also act as an additional capillary bridge, catching the picoliter ink droplets and precisely flowing around its tubing. In contact with microvoids, the coating cells tended to slip into the voids, leading to the occurrence of dendrites of fibrillary ends. The dendritic masking effects defined the trails for inks to flow and, thus, the edges of the print.

The stray reduction was also largely the effect of nanoscale biominerals contained in TRX-cells®. Detected from the high-resolution transmission electron microscope hyphenated with energy-dispersive analysis of X-ray, the nanoscale particles were the spots enticing toner particles [106] upon surface charging during laserjet printing. The minuscule size of the nanominerals promoted a confined print line mimicking the programmed images with high accuracy. At the microscopic level (Figure 15b), uniformly distributed toner particles formed well-defined edges and significantly reduced stray not possibly achieved with the conventional inorganic filler.

The presented pieces of evidence lead to contemporary understanding of translucent cell being a mix of cellulose and noncellulosic materials noticed to have also promoted interfiber bonding, evident from the sticking of translucent cells to the virgin pulp surface, which made it unlikely delineated from the finely masked microfibers, unless “folded” or tangled. The primary advantage of translucent cells notably lies in their nanometer-scale thickness and, thus, flexibility to occupy surface void, ruling out shape [107] as the previously reported prerequisite for successful surface filling.

All combined, the canopy, capillary bridges, masking, dendritic filling of microvoids by the fibrillary ends, and nanoscale alterations of the paper surface, define the limits to ink penetration and bleed.

TRX-cells® are the potential practical complement of commercial inorganic fillers in the papermaking line. The reduction could simultaneously curb the occupational safety risks associated with nanoinorganic fillers, raising the concerns addressed in earlier investigations [108–111]. Besides safe packages, the cells’ potential use in precision printing could improve not only print but also the print economy.

TRX-cells® affinity for pigments was further analyzed in the study of parboiled rice coating with and without *Clitoria ternatea* pigment. The reason for choosing parboiled rice was its reputable health claim attributed to the 90% predominance of resistant starch type III (RSIII) with 8% protein and a negligible portion of lipids [112]. The constituents favorably support the human digestive system [113, 114] by effectively absorbing calcium and iron [112], and improving gut flora and colon health.

Deliberation on the nanoenabling effects of TRX-cells® on the overall nanofunctionalized *C. terneata*-coated parboiled rice production process signals a resource-efficient and zero-waste management possibility principally engendered by the “affinity” factor. Figure 17 presents the relative affinity between delphinidin anthocyanin (DA), RSIII, radials in TRX-cells® and the coating reservoir containing the pigment suspension.

The stronger attraction between DA and the RSIII prevented the adhesion of the TRX-cells- immobilized DA. As a result, the pigment immobilized on the nanocells (Pinacell) could peel easily from the polystyrene surface holding the Pinacell reservoir. The effects allowed ease in the storage the reuse of the Pinacell flakes. In a similar vein, fresh Pinacell could also be dry peeled from the polystyrene surface for the market of instant Pinacell flakes. Cost-wise, the fresh Pinacell would be marketed multiple times higher than the flakes from the spent Pinacell liquor.

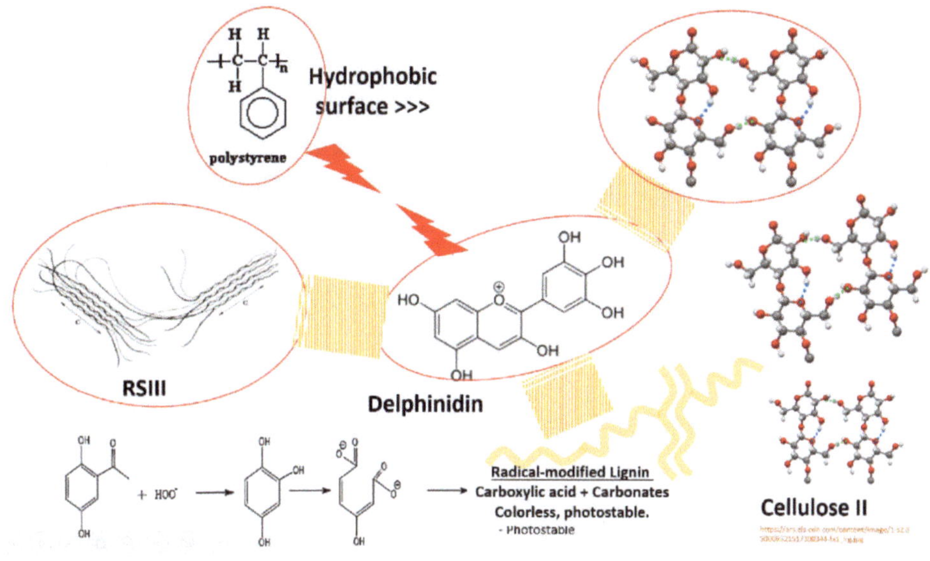

Figure 17: Relative affinity of modified lignin in TRX-cells® indicated as a yellow bar, resistant starch type III (RSIII) and delphinidin.

Another attractive feature of the study was the extended shelf life of the Pinacell-coated parboiled grains. Also due to the relative affinity, DA "rushed" into the RSIII structure (Figure 18), leaving the TRX-cells on the grain surface (Figure 18). The antiseptic effects that had penetrated deep into the grain core plus the thermoblanket (Figure 19) effects of TRX-cells® during drying were the factors [34] responsible for the extended grain shelf life.

Color uniformity as an added aesthetic value, which is crucial in consumers' first impression, the overall zero-waste possibility, and the extended shelf life of the Pinacell-coated grains are the resounding factors for the commercialization of Pinacell-coated "Instant Kerabu Rice" production. Longevity, a vital feature to comply with transportation and storage logistics, is critical in product marketability [34].

Efficient pigment penetration offering dual effects (antimicrobial and ultimate quality preservation) is also critical for preventing wastage of the antioxidant pigment during rice cooking, particularly steaming. The ease of delaminating the dried Pinacell spent liquor from a polystyrene surface signals an excellent promise for zero-waste *Instant Kerabu Rice* production. The maximized Pinacell–polystyrene surface tension avails the possibility of commercializing two grades of *C. ternatea* delphinidin flakes as a natural coloring agent – a timely substitute for the blue artificial food colorant.

Butterfly Pea anthocyanin chauffeured by TRX-cells™ in a travel package called "pinacell."
Charting the journey was made possible with the sharp eyes of XHR-FESEM
that tracked for atomic-level markings.

Figure 18: Pigment penetration into parboiled rice grain (reproduced from [34]).

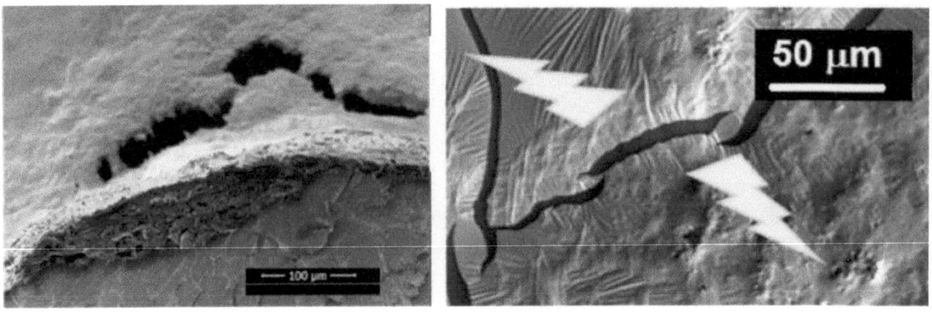

(a) (b)

Figure 19: Pigment-rice interface: (a) close-up look at the cross section of pigment-coated rice and (b) three-quarter view of the rice (left) in the vicinity of crack and surface tension of pigment smear at the "point of defect" (right) (reproduced from [34]).

3.4 Circularity from humble research perspectives

Not purely cellulose, the produced TRX-cells® may be a humble material from a humble 3-hour process with excellent nanoenabling properties. A glance at the simplicity of the process grants the nanoenabling agent an economically more viable nanotechnology in comparison to the more convincing others. Assessment of resource efficiency for sustainability-aligned assessment is a critical step before commercialization.

TRX-cells® commercialization allows small and medium-scale operation due to the possible rerouting of every by-product to practical utilization. The processes allow circularity and zero-waste possibility. Selecting nanofunctional food as an industrial symbiont, the processing routes in Figure 20 exhibit eight marketable commodities. Production 1, production 2, and production 3 can operate with open business model, welcoming collaboration and sharing of external ideas and goods supplies. Alternatively, an integrated business model is adoptable by a sustaining industry to limit the dependence on external supply chains.

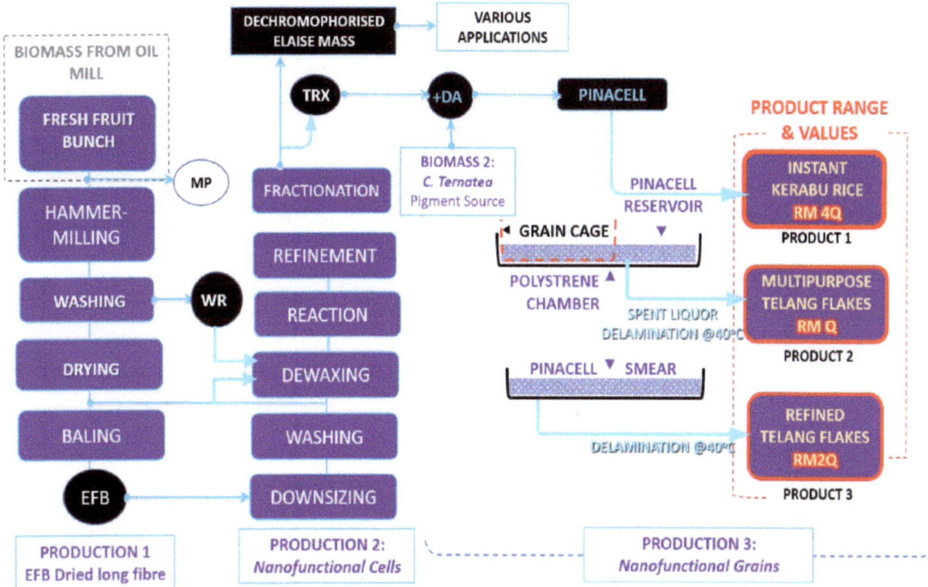

Figure 20: Zero-waste management of resources in *Instant Kerabu Rice* production with by-products from the individual step showing the possibility of utilization. Until the scaling-up takes place, the real challenges remain hidden.

4 Conclusion

Biomass is the source of serendipity for smart use and responsible exploitation for sustainability. Owing to the loose definition of sustainability acceptable in many academic discourses, scaling up production for the commercialization of research products requires a robust assessment. In Europe, life cycle assessment (LCA) has been a part of ISO14040:2006 for sustainable product labels. The setback, however, is the time- and resource-intensive protocol. LCA also requires a significant level of expertise, making access to laymen a possible hurdle. The future potential of Internet of things exploitation through analytics is capable of triangulation for an informed decision-making process. The approach is especially important in navigating the time of economic turbulence, where miscalculations leading to VOD instances need avoidance. As revealed by the FPP framework, processes involving one component isolation create risks akin to capital loss, making such an investment a close resemblance to gambling. Data and process detailing (conditions, e.g., temperature, pressure, water use, and power) from one process may be applied to other processes, suggesting that the more frameworks are created, the more accurate the analytics triangulation will be. The application projected for broad use in process and experiment assessment at an unlimited education level will educate more effectively and lift the nation to a dignified level by opting for unbiased decisions.

Assessed from the FPP framework, TRX-cells® production seems less demanding than other nanotechnology processes for nanobiomaterials production, offering a balance between green parctices and technology. Given the rich sources of the inorganics and organics in biomass (Sections 2.1 and 2.2), in general, process selection needs to weigh the competing reactions, and hence, the by-products from the processes. While the isolation of individual components created complexity and risks, mixed use of biomass via simpler processes shed light on the benign pathway to circularity for people and nature.

Acknowledgement: The study is sponsored by The Ministry of Higher Education (MOHE) of Malaysia through Fundamental Research Grant Scheme FRGS/1/2019/STG07/USM/02/8, commensurate with the United Nations' Sustainable Development Goal 12, Responsible Production and Consumption.

References

[1] Ministry of Plantation and Industrial Commodity. Rubber News. Getah: SMR20 price expected to exceed RM6.50 per kg in a few months, says MRB DG October 3, 2022. https://www.mpic.gov.my/mpi/en/rubber-news

[2] Jaafar F. Rubber Exports Hit RM71 Billion in 2021. The Malaysian Reserve, Economy. Malaysian Reserve. https://themalaysianreserve.com/2022/03/14/rubber-exports-hit-rm71bil-in-2021/

[3] World Wildlife Fund. Key Messages: Natural Rubber Report. September 2021. 2021. https://www.worldwildlife.org/publications/2021-wildlife-progress-report

[4] Law KN, Daud WRW, Ghazali A. Morphological and chemical nature of fiber strands of oil palm empty-fruit-bunch (OPEFB). BioResources 2007, 2, 3, 351–362.

[5] Rafatullah ML, Ahmad MR, Ghazali A, Sulaiman O, Danish M, HashiM R. Oil palm biomass as a precursor of activated carbons: A review. Crit Rev Environ Sci Technol 2013, 43, 11, 1117–1161.

[6] Dermawan YM, Ghazali A, Zukeri MRHM, Kamaluddin NH, Wanrosli WD. Alkaline peroxide synergy with mechanical refining as factor in the development of EFB paper properties. Adv Mater Res 2014, 832, 487–493.

[7] Danish M, Ahmad T, Hashim R, Zukeri MRHM, Ghazali A, Sulaiman O. Characterization and adsorption kinetic study of surfactant treated oil palm empty fruit bunches. Desalin Water Treat 2015, 57, 20, 1–14.

[8] Ghazali A, Wan Daud WR, Law K-N. Alkaline peroxide mechanical pulping of oil palm lignocellulosics Part 2: EFB responses to pretreatments. Appita J 2006, 59, 1, 65–70.

[9] Ghazali A, Zbiec M. Rich dad and poor dad: Biomass circularity sciences empathizing rubber smallholders. J Adv Res App Sci Eng Technol 2022, 29, 1, 207–222.

[10] Mamimin C, Chanthong S, Leamdum C, O-Thong S, Prasertsan P. Improvement of empty palm fruit bunches biodegradability and biogas production by integrating the straw mushroom cultivation as a pretreatment in the solid-state anaerobic digestion. Bioresour Technol 2021, 319, 124227.
[11] Fattah IYA, Shawkataly AK, Hossain MS, Aziz AA, Davoudpour Y, Dngani R, Bhat A. Exploration of a chemo-mechanical technique for the isolation of nanofibrillated cellulosic fiber from oil palm empty fruit bunch as a reinforcing agent in composites materials. Polymers 2014, 6, 2611–2624. doi: 10.3390/polym6102611.
[12] Padzil FRM, Lee HS, Asaari ZMA, Ching HL, Abdullah LC. Potential of oil palm empty fruit bunch resources in nanocellulose hydrogel production for versatile applications: A review. Materials 2020, 13, 5, 1245. doi: 10.3390/ma13051245.
[13] Ayu RS, Abdan K, Harmaen AS, Zaman K, Isma T, Liu Q, Ilyas RA, Lee CH. Characterization study of Empty Fruit Bunch (EFB) fibers reinforcement in Poly(Butylene) Succinate (PBS)/Starch/Glycerol composite sheet. Polymers 2020, 12, 1571. doi: 10.3390/polym12071571.
[14] NationMaster. Top Countries for Natural Rubber Net Production. 2023. https://nationmaster.com/nmx/ranking/natural-rubber-net-production
[15] Lockwood GB. Gas chromatography | Terpenoids reference module in chemistry, molecular sciences and chemical engineering. 2013, available online 9 September 2013.
[16] Ciullo PA, Hewitt N. Compounding materials in the rubber formulary. William Andrew Applied Science Publishers, 1999.
[17] Riches JPR, Gooding EGB. Studies in the physiology of latex. I. Latex flow on tapping-theoretical considerations. New Phytol 1952, 51, 1–10.
[18] Statista. Size of areas planted for palm oil Malaysia 2012–2021. 2022. https://www.statista.com/statistics/1198337/malaysia-size-of-areas-planted-forpalm-oil/
[19] Statista. GDP from rice industry 2020. 2022. https://www.statista.com/statistics/952735/malaysia-gdp-from-rice-industry/
[20] Bernama. About 400 000 hectares of rubber plantations in Malaysia inactive, with inconsistent production, says minister. Malay Mail. March 13, 2022. https://malaymail.com/news/money/2022/03/13/about-400000-hectares-of-rubber-plantations-in-Malaysia-inactive-with-incon/2047219
[21] Malaysian Rubber Council. Industry Overview. 2022. https://www.myrubbercouncil.com/industry/malaysia_production.php
[22] Tiseo I. Value of natural rubber exports from Malaysia 2016–2021. Statista. August 12, 2022. https://www.Statista.com
[23] Tunku-Yahya TM. Crop Diversification in Malaysia. https://www.fao.org/3/x6906e/x6906e08.htm#:~:text=Agricultural%20or%20crop%20diversification%20is%20practiced%20in%20Malaysia.,crops%20grown%20by%20the%20private%20and%20public%20sectors.
[24] Ng J, Hamdan MA. Cover Story: The plight of Malaysia's rubber smallholders. The Edge Malaysia. September 10, 2020. https://www.theedgemarkets.com/article/cover-story-plight-malaysia-rubber-smallholders#.Yx_xb1pSFPO.whatsapp
[25] Kamali and Ho. Malaysia's 50 richest Forbes. Malaysia's 50 Richest List (forbes.com).
[26] Ng MH, Chu YM. Improved method for the qualitative analyses of palm oil carotenes using UPLC. J Chromatogr Sci 2006, 54, 4, 633–638. https://doi.org/10.1093/chromsci/bmv241.
[27] Mallage NJ, Moller BL. Vanillin -bioconversion and bioengineering of the most popular plant flavour and its De Novo biosynthesis in the vanilla orchid. Mol Plant 2005, 1, 40–57.
[28] Bomgardner MM. The problem with vanilla. CandEn ACS 2016, 4, 36, cen.acs.org/.
[29] Ibrahim MNM. Vanilla from black liquor. Bulletin Kimia. 2013. http://eprints.usm.my/id/eprint/2035
[30] Son M, Pinnola A, Schlau-Cohen GS. Observation of dissipative chlorophyll-to-carotenoid energy transfer in light-harvesting complex II in membrane nanodiscs. Nat Commun 2020, 11, 1295.
[31] Wurger G, McGaw LJ, Eloff JN. Tannin content of leave extracts of 53 trees used traditionally to treat diarrhoea is an important criterion in selecting species for further work. South J Bot 2014, 90, 114–117.

[32] Camara Argentna de Productores de Extractos de Quebracho, CAPEQ Tannin – A gift of nature tannins.org.
[33] Liu Y, Tikunov Y, Shouten RE, Marcelis RFM, Visser RGF, Bovy A. Anthocyanin biosynthesis and degradation mechanisms in solanaceous vegetables: A review. Front Chem 2018, https://doi.org/10.3389/fchem.2018.00052.
[34] Ghazali A, Mohd Salleh R, Ahmad MQ, Azhar NH, Malik MFA. Capturing anthocyanin immobilization on rice through the ultra-high resolution electron lenses. Malays J Microsc 2021, 17, 2, 20–31.
[35] Ghazali A, Azhar NA, Wahab NF, Zaini MAA, Mohd Nor SB, Muhammad N. Recalcitrant structures in lemongrass leaf blades – needs for a systemic process analytics. Int J Adv Res Technol Innov 2020, 2, 3, 1–12.
[36] Encyclopedia Britannica Gum, 2023. https://www.britannica.com/technology/gum-adhesive
[37] Elzain EMI. Abdalbasit AM Analytical Techniques for New Trends in Gun Arabic (GA) Research In Gum Arabic Plant Gum – an overview 2018. https://www.sciencedirect.com/topics/agricultural-and-biological-sciences/plant-
[38] Encyclopedia Britannica. Frankincense. 2018. https://www.britannica.com/topic/frankincense
[39] Butie A. Why do we love Frankincense essential oil? Aromahead Institute. School of School of Essential Oils Studies, 2021.
[40] Lauren AE, Erland LAE, Susan J, Murch SJ, Russel J, Reiter RJ, KS Saxena PK. A new balancing act: The many roles of melatonin and serotonin in plant growth and development. Plant Signal Behav 2015, 10, 11, e1096469.
[41] Pizzorno JE, Joiner-Bay H. Chapter 12 – Benign Prostatic Hyperplasia. The Clinician's handbook of natural medicine. 3rd edition. 2.16. doi: org/10.1016/B978-0-7020-5514-0.11121-X.
[42] Trankner M, Tavakol E, Jakil B. Functioning of potassium and magnesium in photosynthesis, photosynthate translocation and photoprotection. Physol Plant 2018, 163, 3, 414–431.
[43] Francheschi NPA. Calcium oxolate in plants – formation and function. Annu Res plant Biol 2005, 56, 41–71.
[44] Ghazali A, Wan Rosli WD, Law KN. Pre-treatment of oil palm biomass for alkaline peroxide pulping. Cell Chem Technol 2009, 43, 7–8, 329–336.
[45] Ghazali A, Ghazali S, Zukeri MRHM, Wanrosli WD. Silica in EFB – an infinite amazement. Malays J Microsc 2012, 8, 70–74.
[46] Ghazali A, N A, Mahmud SH, Khairudin MF, Ahmad I, Rafatullah M, Zaini MA, Yusof Y. Delaminated cells for nano-enabled inkjet printability. Cellul Chem Technol 2021, 55, 9–10, 1071–1081.
[47] Currie HA, Perry CC. Silica in plants: Biological, biochemical and chemical studies botanical briefing. Biomol Ann Bot 2007, 100, 1383–1389. doi: 10.1093/aob/mcm247.
[48] Eur Rubber J. Solvay Expands Italian Site to produce rice-based-silica 2023, 18 Jan., www.europran-rubber-journal/article/2092720/solvay-expanding-italian-plant-plant-to-produce-rice-based-silica.
[49] Girling CA, Peterson PJ. Gold in plants. Gold Bull 1980, 13, 151–157. https://doi.org/10.1007//BF03215461.
[50] Can M. Green gold nanoparticles from plant-derived materials: An overview of the reaction synthesis types, conditions, and applications. Rev Chem Eng 2019, https://doi.org/10.1515/revce-2018-0051.
[51] Schmidt W, Sebastien Thomine S, Buckhout TJ. Editorial: Iron nutrition and interactions in plants. Front Plant Sci 10, 1670. doi: 10.3389/fpls.2019.01670.
[52] Xu X, Du X, Wang F, Sha J, Chen Q, Tian G, Zhu Z, Ge S, Jiang Y. Effects of potassium levels on plant growth, accumulation and distribution of carbon, and nitrate metabolism in Apple Dwarf rootstock seedlings. Front Plant Sci 2020, 11, 904. doi: 10.3389/fpls.2020.00904.
[53] Alejandro S, Holler S, Meier B, Peiter E. Manganese in plants: From acquisition to subcellular allocation. Front Plant Sci Plant Nutr 2020, 11, 300. https://doi.org/10.3389/fpls.2020.00300\.

[54] Ghazanfar S, Komal A, Waseem A, Hassan W, Iqbal RJ, Toor S, Asif M, Saleem IA, Khan SU, Tarar ZH, Nazar S, Rehman H-U, Ahmed MI, Rebi A. Nat Volatiles Essent Oils 2021, 8, 5, 13457–13469.
[55] Patra A, Pradhan SN, Dutta A, Mohapatra KK. Nickel the ultra-micronutrient: Significant for plant growth and metabolism. Food Sci Rep 2020, 1, 35–37.
[56] Lanza MGDB, Reis ARD. Roles of selenium in mineral plant nutrition: ROS scavenging responses against abiotic stresses. Plant Physiol Biochem 2021, 164, 27–4.
[57] Ovca A, Elteren JT, Falnoga I, Selih VS. Comparison of Zinc species in two specimens of edible plants and their fate in the human gastrointestinal tract. Pure Appl Chem 2012, 84, 2, 301–311.
[58] Hassan MU, Aamer M, Chattha MU, Haiying T, Shahzad B, Barbanti L, Nawaz M, Rasheed A, Afzal A, Liu Y, Guoqin H. The critical role of zinc in plants facing the drought stress. Agriculture 2020, 10, 0396. doi: 10.3390/agriculture10090396.
[59] Prasakis J, Sain M, Daneault C. Metal management improves peroxide bleaching of TMP. Tappi J 1996, 76, 10, 161.
[60] Vuorinen T, Heikkila M. Mechanism of transition metal ion catalyzed decomposition of hydrogen peroxide in pulp bleaching. Proc. 28th EUCEPA Conference, Lisbon 2003.
[61] Zhao X, Odberg L, Risinge G. Beating of wheat-straw pulp: Dissolved carbohydrates and lignins, fibre swelling and fines generation. Tappi J 1992, 153.
[62] He Z, Ni Y, Zhang E. Comparison of oxalate formation from Spruce TMP: Conventional peroxide bleaching process versus the PM process. J Pulp Pap Sci 2003, 29, 12, 391.
[63] TAPPI – TAPPI test methods 1996–1997. Tappi Press, Atlanta USA, 1996.
[64] Pan X-J, Santo Y, Ito T. Atmospheric acetic acid pulping of rice straw II: Behavior of ash and silica in rice straw during atmospheric acetic acid pulping and bleaching. Holzforschung 1999, 53, 49.
[65] Tutus A, Eroglu H. A practical solution to the silica problem in straw pulping. Appita 2001, 56, 2, 111.
[66] Siegle S. Current development in formic acid non-wood pulping. PAPTAC 90th Annual Meeting 2004.
[67] Wise LE, Murphy M, D'Addieco AA. Chlorite holocellulose – its fractionation and bearing on summative wood analysis and on studies on the hemicelluloses. Pap Trade J 1946, 122, 2, 35.
[68] Hutholm T, Robertson L, Lonnberg B, Kettunen A, Huericson K. Impregnation in alkaline pulping. Proc. Tappi Pulping Conference 1997, 897.
[69] Higgins HG. Australian development in paper science and technology. In: Arthur J. C. Ed. Cellulose and fibre science developments: A world view. ACS Symposium Series 50 American Chemical Society. Washington D. C, 1977, 135.
[70] Hillis WE, Sumimoto M. Effects of extractives on pulping. In: J.w R Ed. Natural products of woody plants. Springer Verlag, Berlin, 1989, 880.
[71] Chaturvedi AK, Gaurav KJ, Gupta P. Circular economy the many circuits of a circular economy. 2017.
[72] Wautelet T. The concept of circular economy: Its origin and its evolution. Working Paper. 2018. https://doi.org/10.13140./TG.2.2.17021.87523
[73] Pena C, Civit B, Gallego-Schmid A, Druckman A, Caldeira-Pires A, Weidema B, Mieras E, Wang F, Fava J, Canals LM, Cordella M, Arbuckle P, Valdivia S, Fallaha S, Motta W. Using Lifecycle assessment to achieve a circular economy. Int J Life Cycle Ass 2020, 1–2. https://doi.org/10.1007/s11367-020-01856-z.
[74] Duch GJ. Circular economy: Definition, importance and benefits. European Parliament Article 14, 2022, 41, 201510201ST005603. https://www.europarl.europa.eu/pdfs/news/expert/2015/12/story/20151201STO05603/20151201STO05603_en.pdf.
[75] Charlene V, Sweeney J, Murphy F. Space, time, and sustainability: The status and future of life cycle assessment frameworks for novel biorefinery systems. Renew Sust Energ Rev 2022, n. pag.
[76] Lange L. Business models, including higher value products for the new circular, resource-efficient biobased industry. Front Sust Perspect March 15 2022, https://doi.org/10.3389/frsus.2022.789435.
[77] Centre for International Forestry Research. Bioeconomy solutions. Circular Bioeconomy 2022. https://www.cifor.org/cbe

[78] Fai SZ, Yoong CC. Physical and chemical characterization of oil extracted from *Citrofortunella microcarpa, Hibiscus sabdariffa* and *Artocarpus heterophyllus* seeds. Prog Energ Environ 2022, 22, 1–12.

[79] Shameer S, Viswanath B, Bin L. Oxalate-carbonate pathway – conversion and fixation of soil carbon – a potential scenario for sustainability. Front Plant Sci 2020, 11, 591297. https://doi.org/10.3389/fpls.2020.591297.

[80] United States Environmental Protection Agency. Overview of Greenhouse Gases. Updated May 16, 2022. https://www.epa.gov/ghgemissions/overview-greenhouse-gases

[81] Stager C. What happens after global warming? Nat Educ Knowledge 2012, 3, 10, 7.

[82] United States Geological Survey. Acidification of Earth: An assessment across mechanisms and scales. December 1, 2016. https://www.usgs.gov/publications/acidification-earth-assessment-across-mechanisms-and-scales

[83] Ouyang Z, Chen L, Wu Y, Lei R, Chen B, Feely R, Anderson L, Zhong W, Lin H, Polukhin A, Zhang Y, Zhang YY, Bi H, Lin X, Luo Y, Yanpei Z, Chen J, Cai W-J. Climate change drives rapid decadal acidification in the Arctic Ocean from 1994 to 2020. Science 2022, 377, 1544–1550. https://doi.org/10.1126/science.abo0383.

[84] Lang KC, See TL, Tan J, Shariff AM, Halim HNA. Life cycle assessment of potassium lysinate for biogas upgrading. Prog Energy Environ 2022, 22, 29–39.

[85] Latõšov E, Loorits M, Maaten B, Volkova A, Soosaar A. Corrosive effects of H_2S and NH_3 on natural gas piping systems manufactured of carbon steel. Energy Proc 2017, 128, 316–323.

[86] United States Geological Survey. (n.d.). How Can Climate Change Affect Natural Disasters www.usgs.gov/faqs/how-can-climate-change-affect-natural-disasters

[87] Konapala G, Ashok MK, Wada Y, Mann ME. Climate change will affect global water availability through compounding changes in seasonal precipitation and evaporation. Nat Commun 2020, 11, 3044. https://doi.org/10.1038/s41467-020-16757-w.

[88] Ali MF, Aziz AA, Williams A. Assessing yield and yields stability of *Hevea* clones in the southern and central regions of Malaysia. Agronomy 2020, 127, 102449.

[89] Thu KM, Myint T. Myanmar's natural rubber policy: Key stakeholders and farmers' participatory approach. FFTC Agricult Pol Platform August 30 2022, https://ap.fftc.org.tw/article/2866.

[90] Silva D. Huge potential for sustainability with natural rubber, experts say. Forest News. June 30, 2022. https://forestmews.cifor.org.78076/huge-potential-for-sustainability-with-natural-rubber-experts-say?fnl=en

[91] Ibrahim A. New strategy needed to help rubber smallholders. New Straits Times. Columnist. September 8, 2022. https://www.nst.com.my/opinion/cplumnist/2022/09/829431/new-strategy-needed-to-help-rubber-smallholders

[92] Muralikrishna V,I, Manickam V. Life cycle assessment. Environ Manage 2017, www.sciencedirect.com/topics/earth-and-planetary-sciences/life-cycle-assessment.

[93] Khin AA, Ling RLB, Sia BK, Fong YC LTK. Challenges of the export for natural rubber latex in the ASEAN market. IOP Conf Ser Mater Sci Eng 2019, 548, 29, 012024. https://doi.org/10.1088/1757-899X/548/1/012024.

[94] Bernama. Palm product export for January-June reaches RM71.02 billion. Sun Daily. September 20, 2022. https://thesundaily.y/business/palm-product-export-for-January-June-reaches-rm7102b-LJ9847782

[95] Workman D. Natural rubber exports by country. 2022. World's Top Export. https://www.worldstopexports.com/natural-rubber-exports-country/

[96] Kodoh J, Mojiol JAR, Lintangah W, Maid M, Kang CL. The contributions of rubber plantation to the socio-economic development: A case study on Kanibongan project-rubber smallholders community in Pitas. Sabah, 2016.

[97] DW Documentary. Rubber tires – a dirty business. [Video]. YouTube. September 18, 2019. https://youtu.be/-fusUxEPwsw

[98] Ali MF, Akber MA, Aziz AA. The dynamics of rubber production in Malaysia: Potential impacts, challenges, and proposed interventions. For Policy Econ 2021, 127, 102449.
[99] Buzzle RD, Gale BT, Sultan RGM. Market share – a key to profitability. Harvard Bus Rev, 1975.
[100] Aui A, Wang Y, Mba-Wright M. Evaluating the Economic Feasibility of Cellulosic Ethanol: A Meta-Analysis of TechnoEconomic Analysis Studies URL: https://www.sciencedirect.com/science/article/pii/S1364032121003865
[101] Cheah R. World's first oil palm -based pulp paper. The Star Online – Nation, November 24, 2008. https://www.etawau.com/OilPalm/PulpPaperMill.htm
[102] MPOC, Malaysia Palm Oil Council. Palm-based Paper Venture, n.d. https://mpoc.gov.my/palm-based-paper-venture/
[103] Ghazali A, Zukeri MRHM, Dermawan YM, Ghazali S, Wanrosli WD. Nanofiber network rooted from the alkaline peroxide treatment of oil palm empty fruit bunches. Adv Mater Res 2014, 832, 500–505.
[104] Iggesund Paperboard Reference Manual Surface Structure and Smoothness pp. 91–94 https://www.iggesund.com/paperboard-reference-manual.aspx (accessed September 12, 2019).
[105] Bangera J. Laser toner cartridge: What's inside? Inkjet Wholesale. https://blog.inkjetwholesale.com.au/printer-education/laser-toner-cartridge-whats-inside/
[106] Asztemborska M, Jakubiak M, Steborowski R, Chajduk E, Bystrzejewska-Piotrowska G. Titanium dioxide Nanoparticles Circulation in an Aquatic Ecosystem. Water Air Soil Pollut 2018, 229, 6, 208.
[107] Buzea C, Pacheco II, Robbie K. Nanomaterials and nanoparticles: Sources and toxicity. Biointerphases 2007, 2, 4, MR17–MR172.
[108] Chaudry Q. Presented at ILSI Europe workshop on nanoparticles migration and packaging lifecycle, the food and environment research agency. United Kingdom, 2012.
[109] Song Y, Li X, Du X. Exposure to nanoparticles is related to pleural effusion, pulmonary fibrosis and granuloma. Eur Respir J 2009, 34, 3, 559–56.
[110] Song Y, Li X, Wang L, Rojanasakul Y, Castranova V, Li H, Ma J. Nanomaterials in humans: Identification, characterization and potential damage. Toxicol Pathol 201, 1.
[111] Derycke V. Parboiling of Rice: Changes in Starch and Protein and Their Relation to Cooking Properties. 2007. https://www.biw.kuleuven.be/m2s/clmt/lmcb/publications/docs/vderycke
[112] Birt DF, Boylston T, Hendrich S, Jane J-L, Hollis J, Li L, McClelland J, Moore S, Phillips GJ, Rowling M, Schalinske K, Scott MP, Whitley EM. Resistant Starch: Promise for Improving Human Health. Adv Nutr 2013, 4, 6, 587–601.
[113] Donohoe DR, Garge N, Zhang X, Sun W, O'Connell TM, Bunger MK, Bultman SJ. The microbiome and butyrate regulate energy metabolism and autophagy in the mammalian colon. Cell Metab 2011, 13, 5, 517–526.
[114] Raigond P, Dutt S, Singh B. Bioactive molecules in food. Springerlink, 2019, 815–846.

Anandkumar J., Jyoti Kant Choudhari, Mukesh Kumar Verma, Jyotsna Choubey, Biju Prava Sahariah*

Advances in Microbial Treatment of Emerging Contaminants

Abstract: The modern and fast lifestyle of the human society has introduced a lot of new substances into the environment. Among these elements, many have long life duration as parent/transformed products and cause negative health impact in single form or in compound forms with other elements. However, due to poor care, identification, or due to the presence in very low quantities, many of these contaminants are often ignored, though their presence can result in serious health issues to humans and to the environment. The scientific community regard these emerging contaminants (EC) as mostly coming from seven categories of origin, namely, pharmaceuticals, personal care products, industrial chemicals, disinfection byproducts, algal toxins, biocides and their metabolites, and bioterrorism and disruption devices. The complexity of ECs, such as their availability at very low concentrations and high hazardous potential to health, demand accurate technology for their treatment. Bioremediation is a highly recognized tool for the protection of the environment where tiny microbes play a significant role. The current article discusses the adverse effects of ECs, the treatment options, and the mechanisms of EC bioremediation for securing and increasing the good health of the environment. A system biology approach for the recognition of numerous potential microbes for the treatment of ECs, with the help of literature and omics tools, is summarized.

Keywords: Emerging contaminants, environment pollution, systems biology, microbial pollution treatment

1 Introduction

1.1 Emerging contaminants

The health of the environment and its population is directly related to the living and nonliving factors present in it. With the revolutionary days of the industrial, commercial

***Corresponding author: Biju Prava Sahariah,** Chhattisgarh Swami Vivekanand Technical University, Bhilai, Chhattisgarh 491107, India, e-mail: biju.sahariah@gmail.com
Anandkumar J., National Institute of Technology Raipur, Raipur, Chhattisgarh 492010, India
Jyoti Kant Choudhari, Maulana Azad National Institute of Technology Bhopal, Bhopal 462006, India
Mukesh Kumar Verma, National Institute of Technology Raipur, Raipur, Chhattisgarh 492010, India
Jyotsna Choubey, Chhattisgarh Swami Vivekanand Technical University, Bhilai, Chhattisgarh 491107, India

https://doi.org/10.1515/9783111016825-003

and agricultural activities, and economy, the human lifestyle has changed drastically. A large number of synthetic products are used in daily life as a matter of livelihood standard, comfort or feasibility, and for protection and economic benefit. Therefore, residuals of these substances, in very low concentrations such as in picogram (pg) to microgram (µg) or milligram (mg) range per unit of soil/water, get discharged through point/diffuse sources. The new group of these synthetic micropollutants, along with few natural ones, recently termed as emerging contaminants (EC), is present in the environment in low quantities, but can consist of large numbers of individuals in its groups [1–5]. Bestowing to the prevalent conditions, the parent compound may undergo biotransformation or photolysis, and the resultant transformed byproduct can possess more persistency and toxicity as well as distribution. The physicochemical properties and the resultant polarity govern the mobility of the ECs in the soil and water, and even encompass groundwater through leaching; hence, they can be present in a site far away from its origin. Presently, ECs are broadly categorized into seven groups as shown in Figure 1, namely,

i) **Pharmaceuticals**, encompassing all the prescribed as well as illicit drugs and their residue
ii) **Personal care products**, including cosmetic, surfactants, domestic biocides, etc.
iii) **Industrial chemicals** such as food additives, pesticides, polychlorinated biphenyl, flame retardants, antimicrobial substances, fire retardants, nanomaterials, plasticizers, etc.
iv) **Disinfection byproducts** from water treatment plants, like nitrosamine, chloramines, halonitromethanes, trihalomethanes, bromonitromethanes, bromotrihalonitromethanes, iodo-trihalomethanes, iodo-trihalomethanes, dihaloaldehydes, MX (3-chloro-4-(dichloromethyl)-5-hydroxy-2(5H)-furanone), iodo-acids, halo acetic acids, etc.
v) **Algal toxins** such as toxics released from cyanotoxins and microcystin
vi) **Biocides and their metabolites** including pesticides
vii) **Bioterrorism and disruption devices**, for example, biological and chemical weapons, antibiotic resistant bacteria (ARB) and genes (ARG), etc.

The ECs from pharmaceutical and personal care product (PCP) are part of a never-ending list. For example, in the case of pharmaceuticals, the medicines/drugs are typically made of active chemicals or biological reagents with targeted points of action. Thanks to continuous research and development of numerous modern sophisticated equipment, thousands and millions of new pharmaceutical products, with significant bioactivity, are released for the betterment of human health and for the treatment of human illnesses and of animals. The broad spectrum of pharmaceutical products is entirely manufactured from synthetic substances/chemicals, with the application of genetic engineering, along with few that are derived from natural origin such as those derived from microbial or plant/animal origin. There is an immense number of identified diseases and a gigantic world of medicines for its recovery in the present scenario,

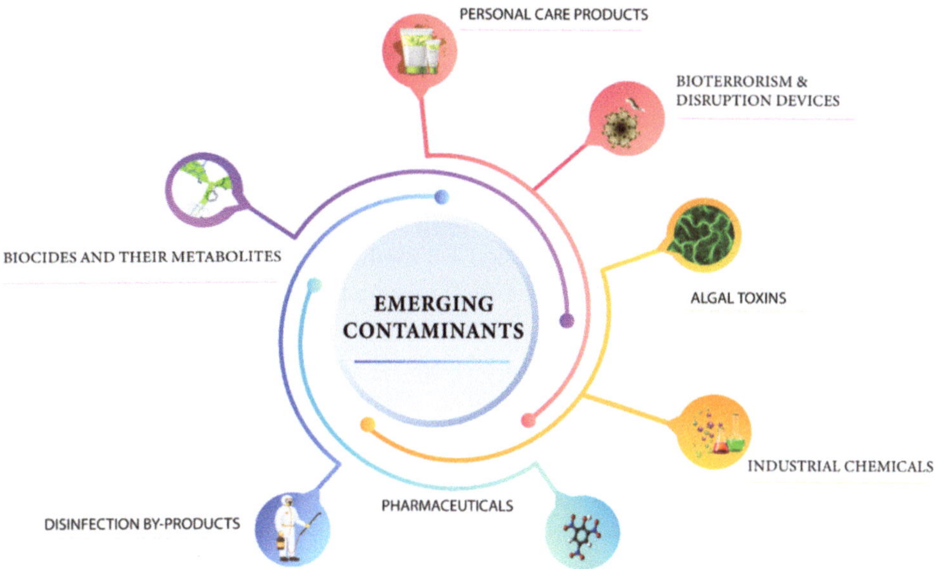

Figure 1: Emerging contaminants in the environment.

which is steadily increasing. The ingredients of the available medicine in today's condition are beyond the capacity of enlisting. Antibiotics, nonsteroidal anti-inflammatory drugs (NSAIDs), analgesics, (painkillers), antipyretics (fever reducers), antiseptics, hormone replacements, contraceptives, statins, mood stabilizers, antidepressants, cytostatic, antiretroviral, anticancer, antidiabetic, and veterinary antibiotics are major pharmaceuticals of utility [6–8]. A fair fraction of these compounds gets released from the patients through their excreta as metabolites and intermediates. With reference to the interaction with various functional groups (e.g., hydroxyl, carboxyl, amine, or sulfhydryl) and polar molecules (e.g., glucuronic acid, acetate esters, carboxamides, sulfates, etc.) of the receptor/patient body, a single administered drug can follow different biotransformations and be released in different forms. Another major concern is the disposal of tons of medicines yearly as waste due to their expiry of potency. Though the medicines lose human treatment standards, they still retain several of their properties and remain active, capable enough to influence environmental factors when they come in contact with the environment. Endocrine disruptors, antibiotics NSAIDs, analgesics, antiretrovirals, anticancer drugs, beta-blockers, and psychoactive drugs are mostly detected in effluent and wastewater treatment plant [1, 6]. A few chemicals that are common components of several pharmaceutical drugs are sulfamethazine, diclofenac, ibuprofen, salicylic acid, carbamazepine, propanolol, fluoxetine, bezafibrate, sulfamethoxazole, iopamidol, trimethoprim, sulfamethoxazole, and diazepam, and some natural hormones like estrones, 17β-estradiol, and the synthetic hormone 17α-ethinyl estradiol. Similar to the huge numbers of pharmaceuticals, PCP items too are widely distributed in huge numbers and quality, which include the continuously evolving beauty products such as, numerous skin/hair care products (variety of sunscreen lotions, moisturizers, oils, shampoos, etc.), fragrance items like deodorants/perfumes, and health care/hygienic products, including, products for hand wash, body wash, sanitizers, etc. One may need several lifetimes to count all their ingredients in the current situation. Sunscreens or UV-filter formulations are either organic (chemicals) or inorganic (minerals), for example 3-(4-methylbenzylidene) camphor (4-MBC), benzophenone-3 (BP-3), 2-ethylhexyl 4-methoxycinnamate (OMC), 2-ethylhexyl 4-dimethylaminobenzoate (OD-PABA), 3-benzylidene camphor (3-BC), homosalate (HMS), 4-aminobenzoic acid (PABA), etc. A few commonly detected plasticizers, for example, phthalates [phthalic acid esters, di-(2-ethylhexyl) phthalate (DEHP), diethyl phthalate (DEP), and dibutyl phthalate (DBP), phosphates, adipates, benzoates, trimellitates, esters of sulfonic acids, sulfonamides, elastomers, etc. are originated from medicines, perfumes, nail polish, paints, adhesives for printers, shampoos, and lotions, and released in wastewater. Thanks to the industrial and agricultural revolution, thousands of modern synthetic chemicals are entering the environment in the form of medicines, PCPs, pesticides, antimicrobial substances, and many adornment items. Different contaminants, such as antipyrine, caffeine, carbamazepine, clofibric acid, diclofenac sodium salt, galaxolide, ibuprofen, tonalide, tylosin, trimethoprim, and sulfamethoxazole, are detected in the drinking water distribution system. The other groups of ECs are not as vast as those of the first three categories; nonetheless, they are also of serious concern and cannot be ignored as they consist of several dynamics. For example, ECs in the group of disinfection byproducts, many times remain inconsiderate, as they are hard to recognize or identify – they may form unexpected compounds with the applied or receiving component for treatment. Disinfection byproducts that evolved from disinfection chemicals are majorly of chlorinated compounds, together with other halogenated chemicals, namely, iodinated trihalomethanes, aldehydes, ketones, halomethanes, hydroxy acids, carboxylic acids, alcohols, keto acids, esters, and even nitrosamines, perflourinated carbons, etc. Similarly, the massive and extensive cellular components in the case of algal toxins and biocides are severely complex EC groups, along with continuously synthesized bioterrorism and disruption devices.

These synthetic products can emanate to the environment in the form of various intermediates and structural unit chemicals, and are mostly polar-to-semipolar organic substances. The unused and even used chemicals find clear way into the environment through the municipal as well as industrial waste treatment units. Due to the poor attention toward the ECs with respect to their identification, scarce information about their threats, low abundance, and differentiated components of origin, ECs are often ignored or recognized but remain unregulated. Hence, the ECs often succeed to pass through conventional treatment units that are not designed/configured for their removal. They can easily thrive in the environment of treatment plants, for example, in water streams through the addition of the treated effluent from manufacturing/municipal waste treatment plant, byproducts of treatment units, for example, as sludge exposed to soil, application/reclamation of treated water, hospital waste, waste/runoff from agriculture/poultry/animal husbandry, direct exposure to air via evaporation and leachates from landfills, etc. These directly lead to contact of aquatic and soil organisms with ECs, resulting in increased surge of the same, with potential threat to the environment, triggering deleterious effects in aquatic and soil organisms as well as human life. With the advancement of sophisticated analytical equipment, and progressive knowledge, detection of these ECs has evolved recently, enabling their identification, research, understanding the limitation of their adverse effect in the environment, and their treatment.

1.2 Impacts of ECs on environment

ECs are mostly of xenobiotic nature, as they are originated from synthetic compounds, along with a few biological factors, for example pathogens with wide range of bioactivity. It implies therefore that ECs remain as complex mixture of pollutants, instead of isolated elements. Thus, there occurs additive or multiplicative responses, according to the EC present, which may be either synergic or antagonic. The effluents discharged from the manufacturing industries and municipal treatment units that are rich in ECs can utterly alter the quality of the receiving water or soil, owing to their reactivity and components. There is also a tendency of being accumulated in soil, as well as their transfer to vegetations grown in the soil and to higher organisms that feed on the crops. As result of these phenomena, the ecological functioning can be severely affected, cause alteration of biotic components, resulting in explosion of diseases, and even mortality.

1.2.1 Impact of ECs on microbes

Variation in the composition of microbial communities and their function is highly attributed to their exposure to ECs. Alteration in the configuration of biofilms and their metabolic functions, spreading of antibiotic/or another components resistance, and including photosynthetic efficiency, are verified when the microbes are located in higher concentration of ECs. The ECs – sulfadiazine and ciprofloxacin – can enhance the abundance of resistance of bad gene, Fusobacteria and sulfamethoxazole genes, while inducing mexA expression, a gene encoding for an efflux pump of these antibiotics, resulting in a drastic effect on the bacterial ability to adhere and invade host cells, as seen for *Burkoldheria cepacian*, etc. [2, 9] Antibiotics (such as penicillin, sulfonamides, and tetracylines) can cause significant resistance among bacterial pathogens. Increase in resistance to ciprofloxacin, gentamycin, polymyxin B, rifampin, tetracycline, ciprofloxacin + gentamicin, and ciprofloxacin + tetracycline occurred in *Pseudomonas aeruginosa*, after its exposure to bromoacetamide, trichloroacetonitrile, or to tribromonitromethane. Horizontal gene transfer can occur in microbes that have exposure to nonantibiotic ECs, namely, triclosan, carbamazepine, disinfectants (such as free chlorine, chloramine), hydrogen peroxide, metals (such as copper, silver, chromium, and zinc), etc.

1.2.2 Impact of ECs on animals and plants

The utilization of various products by the increasing population that release ECs, contamination of environments, followed by their bioaccumulation and biomagnification, are serious substantial issues. The bioaccumulation and biomagnification of chlorinated hydrocarbons, originated from the pesticide di-chlorodiphenyltrichloroethane (DDT), resulting in delicate egg shell and declination in bird population is well known. Penicillin, sulfonamides, and tetracylines affect higher trophic level through food chain modification, as they predominantly alter the microbial community structure in nature. Many of the ECs are characterized as endocrine-disrupting chemicals. They originate from natural and synthetic hormones and from other pharmaceuticals, pesticides, plasticizers, and organometallic compounds. Endocrine disrupting chemicals significantly influence the functioning of the endocrine system, with subsequent adverse effects in reproduction, development, and immune functions of an organism. Hence, incidents of EC effects, including alterations in reproductive behavior, aggression, boldness, activity, socialization and feeding rate in fish population, alteration in behavior of tadpoles (*Bufo arabicus*) and *European perch* during exposure to fluoxetine, and oxazepan, respectively, are recognized. ECs that originate from polymer resins and plastics, and released from flame retardants, furniture, and electronics can cause neurotoxicity and thyroid hormone. Estrogenic effects in rats are evidenced from exposure to bisphenol A (BPA) released from food packaging plastic or resin items. Feminization in fish occur from its contact with estrone and 17-β estradiol, a steroidal estrogen and 17-α ethynylestradiol (synthetic contraceptive), enabling imitating estrogen/hormone effect to nontarget organisms. Few other exclusive examples of effect from ECs exposure are disorganized reproductive tissues and abnormal ratios of estrogen and testosterone in juvenile alligators in Lake Apopka, Florida, and sexual abnormalities in fish living near WWTP outfalls (male fish producing female yolk precursor protein vitellogenin) [5].

1.2.3 Impact of ECs on human

The list of diseases occurring in current days is steadily expanding with the addition of many new or modern ones that were not present in earlier days. Majority of these diseases are attributed to ECs when considered in-depth. ECs are accountable for a wide range of dire as well as prolonged negative effects on health. Endocrine disruption, neurological disorders, immunotoxicity, and cancer in humans, are many times connected to exposure to ECs. As seen from the categories of ECs, they are mostly of synthetic origin but are reactive elements that can interact with many cellular or molecular functions in human cells/health [10]. For example, a few medicines are known to possess persistency in the human body, long after application, showing different modes of action. The pesticides used for pest control cause neurotoxic and mutagenic effects. The byproducts from disinfection, plasticizers, and fire retardants are widely established as sever carcinogenic materials, reactive with vital organs, resulting in several complex diseases of the systems related to respiratory, renal, cardio vascular, reproductive, as well as digestive organs of humans. The parabens alkyl-*p*-hydroxybenzoates, which are commonly used as preservatives in many cosmetics, including in skin creams, tanning lotions, hygiene products, etc., have estrogenic activity. ECs, for example, perfluorinated compounds (PFCs), influence thyroid hormone levels, body weight, and insulin resistance [11]. Perfluoroalkyl acids (PFAAs) are associated with the dysfunction of reproductive systems, such as infertility, fecundity, breastfeeding, low number as well as quality of semen, etc. in humans. At higher levels, perfluoroalkyl substances (PFASs) in serum negatively affect the thyroid hormone, with increased thyroid stimulating hormone (TSH), T3, and T4, and negatively related to T4 of pregnant women with higher TPOAb. The PFC, namely, perfluorooctanoic acid (PFOA), influences lipid mobilization genes, and is related to increased total lipid and low-density lipoprotein cholesterol (LDL-C); it can convert to hypercholesterolemic condition – higher rate of hyperuricemia – which is considered to be potential risk factor for hypertension, along with cardiovascular diseases. Higher levels of perfluorononanoic acid (PFNA) cause hyperglycemia and higher β-cell activity. Nanomaterials, such as amorphous silicon dioxide (SiO_2), carbon nanotubes (CNTs), titanium dioxide (TiO_2), etc., are also considered to be carcinogenic. The waterborne enteric pathogens (bacteria, viruses, and protozoa) are responsible for dermal as well as other vital organ infections and severe health disorders [5]. Heavy metals, such as mercury, lead, chromium, cadmium, and arsenic, are often associated with ECs that originated from industrial and agricultural chemical, causing several severe health concerns and are beyond the scope of explanation in this chapter. Detail appropriate information concerning ECs and their affects on human health are still in progress in the present scenario.

2 Treatment of ECs

The vast number of diverse ECs often cause challenges in determination, remediation, and regulation. Advancing science and technology has contributed to the development of important gears and tools, for example, gas chromatography-mass spectrometry (GC-MS), Liquid chromatography-mass spectrometry (LC-MS), nuclear magnetic resonance spectroscopy (NMR), and transmission electron microscopy (TEM) that highly favor the detection as well as analysis of many frequently occurring, persistent as well as previously undetected or unconsidered pollutants. Detection of pathogens, and antibiotic resistance bacteria (ARB)s and antibiotic-resistant gene (ARGs) are advanced with the utilization of either culture-based/genetics-based or molecular-based, DNA probes, PCR methods, etc. Therefore, the rate of investigation on ECs is highly influenced by adequate analytical prospects. Various research groups are continuously working on identified ECs through modified conventional as well advanced nonconventional physical–chemical–biological treatment process, and a brief assessment is presented in this section.

2.1 Conventional treatment

Considering the properties of ECs, a few of them are removed by the conventional units structured with activated sludge process, microalgae reactor, coagulation, flocculation and sedimentation, filtration/membrane reactors, activated carbon filtration, etc. that are used for treatment of surface water source for drinking purposes.

Bioremediation is the application of biological origin agents such as unicellular tiny microbes, algae, and plants in the form of either live or dead biomass, for the treatment of pollutants. The organisms are equipped with various cellular-molecular and metabolic activities that enable them to break down the high molecular weight molecules into smaller molecules or immobilize the pollutants, while extracting energy and nutrient for sustaining life. The activated sludge discussed earlier is a widely utilized biological process for the removal of a wide range of organic pollutants.

2.2.1 Activated sludge process

Aeration in activated sludge prompt redox potential, favorable for property changes of cations/anions present in fluid as well as of the metabolic activity of the microbes that are present. Bioremediation occurs following mechanisms of bio-adsorption, bio-uptake, and biodegradation, according to the prevalent microbes as well as the properties of pollutants. These can further assist in the removal of the present ECs along with other major pollutants. The EPS, cellular, and molecular functioning govern the bioremediation through secretion of various proteins/enzymes or chelation agents required for bioremediation. ECs, namely trimethoprim, sulfamethoxazole, carbamazepine, diclofenac, triclosan, and methyl 3,5-dichloro-4-hydroxybenzoate, present in the effluents of conventional wastewater treatment plant can be significantly removed in acclimated activated sludge reactor where the ECs were distributed in water and sludge phases [12]. A High-Rate Activated Sludge System treated endocrine disruptors, namely, nonylphenol, bisphenol A, and triclosan, and the pharmaceuticals, ibuprofen, naproxen, diclofenac, and ketoprofen in poor-to-moderate extent through biotransformation, where sorption of the contaminants onto the sludges dominated removal efficiency [13].

2.2.2 Microalgae reactor

Microalgae are widely acknowledged for their concomitant bioremediation efficiency toward nutrients as well as heavy metals from wastewater, even at low concentrations, through bio-adsorption, bio-uptake, and biodegradation mechanisms. Similar activities are followed to accomplish treatment of ECs. Functional groups, such as amines, carboxyls, hydroxyls, carboxyl, sulphates, and other charged groups, in microalgal cell surfaces enhance ECs removal through adsorption or uptake following various mechanisms, for example, passive diffusion, passive-facilitated diffusion, energy-dependent/active uptake, and active transport of the contaminant into the cell, where it binds to intracellular proteins and other compounds across the cell membrane through adsorption, micro precipitation, chelation etc. [14, 15] On the other hand, biodegradation is carried out mostly for organic ECs by the enzymatic reaction at favorable conditions. Being an autotrophic phenomenon in microalgae reactor, there is no necessity of external oxygen supply; contrarily, addition of natural oxygen resulted from photosynthesis, favorable for other microbes. Microalgae reactors also present the facility of resource recovery through their biomass. In a critical review on ECs removal by microalgae, researchers observed the general trend to be pharmaceuticals > PCPs > EDCs > pesticides [16]. A few examples of ECs' removal through biouptake are antibiotics, triclosan, and triclocarban by filamentous green alga, *Cladophora* sp. and *Nannochloris* sp., antiepileptic drug carbamazepine removal by the green alga, *Pseudokirchneriella subcapitata*, ethinylestadiol by the green alga *Desmodesmus subspicatus*, etc. Microalgae reactors successfully decontaminant EDCs namely, E1, E2, and EE2, nitrophenols, polyfluoroalkyl substances, carbamazepine, ibuprofen, and pharmaceuticals (analgesic and anti-inflammatory drugs, including stimulant caffeine). Several species, such as *Chlorella, Cladophora, Oscillatoria, Scenedesmus* of the genera *Anabaena* are recognized to have "hyperaccumulator" and "hyperadsorbent" properties.

2.2.3 Constructed wetland (CWs)

Constructed wetlands are generally land-based treatment systems, structured with low-depth ponds rich in floating/emergent or rooted vegetations, typical to wetlands. The mode of operation follows mechanisms of bioremediation, including microbiological degradation, biofilm, root and plant uptake, and a few physicochemical processes like evaporation, photo degradation, oxidation, hydrolysis, retention, or root sorption into the soil surface, according to the bed property. The structure of CW is a sum of several environments with different micro environments, which enable them to show a high performance. CW configuration design, namely, pore water, upper layer exposed to the sunlight, plants, biofilm on the substrate, and roots' biofilm hydraulic mode, temperature and seasonality, pH, oxygen, and redox potential, etc. are major parameters governing performance of the same [17]. There have been successful implementations of

CW for emerging pollutants like antibiotics in livestock, other pharmaceuticals, sunscreen compounds, fragrances, antiseptics, fire retardants, surfactants, pesticides, plasticizers, etc. [18, 19]

2.2.4 Coagulation, flocculation, and sedimentation

Application of metal hydroxides for coagulation of low water-soluble hydrophobic chemicals and organics for floc formation, followed by their sedimentation, removes high amount of turbidity caused by suspended inorganic and organic particles. The additions of coagulants can influence ECs present in fluid, and favor floc formation that contribute to ECs' removal. In a study with Aluminum- (Al) and iron- (Fe) based chemically enhanced primary sedimentation of sewage, accompanied with acidogenic fermentation of sludge, expressive removal of retinoids (retinoic acids (RAs) and their metabolites), and oestrogenic endocrine disrupting chemicals, namely, 4-nonylphenol, bisphenol A occurred where sludge fermentation has played significant role in pollutants removal [20, 21]. Removal of microplastic polystyrene (PS) and polyethylene (PE) is recommended using powdered activated carbon and $FeCl_3$ coagulation in alkaline condition, followed by higher stirring, where charge neutralization agglomeration adsorption occurred in PS system, with promoting effects by carbonate ions; there is little effect by chloride and inhibitory effect by sulphate ions.

2.2.5 Filtration

Filtration process follows retaining of larger size molecules, based on pore size of the media, using both biological (e.g., specific organisms' usage and trickling filter) or physical (sand and/or coke filters) approaches. The physical and chemical properties of micro emerging contaminants can cause attachment with other large molecules, making a larger size in total, which enhance separation in the filtration process. Complete removal of emerging contaminants (caffeine, theobromine, theophylline, amoxicillin and penicillin G) is feasible in intermittent sand and/or coke filters [22].

The issue with separation techniques mentioned above is that the through adsorption, filtration, etc., the contaminants get concentrated in secondary products and demand further final disposal of these residue; otherwise, they may create severe negative impact on the environment.

2.2.6 Activated carbon filtration

Activated carbon (AC) is one of the most frequently used treatment agents derived from raw materials, including many waste materials, sludge, waste rubber tire, coconut shells, wood char, lignin, petroleum coke, bone char, peat, sawdust, carbon black, rice hulks, bagasse, peach pits, fish residues, fertilizer waste, etc. Its specific characteristics, such as high porosity, broad range of pore sizes (ranging from visible cracks to molecular level) and specific surface area with basal planes, heterogeneous superficial groups (functional groups, such as hydroxyl, carboxyl, phenol, lactone, lactol, and quinone), inorganic ash, etc., significantly favor adsorption of wide ranges of pollutants in AC. The parent material of AC largely governs the above mentioned parameters.

Granular/powdered AC has provided suitable results of adsorption of various target ECs, including pesticides (Atrazine, Propoxur, Methidathion bentazon, Phthalic acid (PA), diphenolic acid (DPA), 2,4-dichlorophenoxy-acetic acid, 2,4-D, 2-methyl-4-chlorophenoxyacetic acid (MCPA), chlorophenoxyacetic acid, isoproturon, and many more); personal care products (acetaminophen, caffeine, diclofenac-Na, naproxen, sulfamethoxazole, sulfamethazine, atenolol, carbamazepine, ciprofloxacin, diclofenac, erythromycin, ketoprofen, metronidazole, ibuprofen, bezafibrate, clofibric acid, carbamazepine, diclofenac, norfloxacin, etc.); microplastics (polystyrene, polyethylene, etc.); and EDCs (bisphenol A, triclocarban, nonylphenol, 17-β-estradiol, amitrol, etc.) [23–25].

2.2 Nonconventional treatment

2.2.1 Membrane system

Currently, membrane reactor plays a significant role in the treatment of emerging pollutants with microunits, and function on the basis of hydrostatic pressure utilization for the removal of larger molecules or suspended solids. Properties such as pressure, temperature, viscosity, transmembrane pressure, osmotic pressure, flux rate, concentration polarization, pore size of the membrane, fouling, etc. are to be carefully monitored for suitable outcomes. The prominent membrane systems are micro- or ultra-filtration or nanofiltration, forward osmosis, and reverse osmosis. They have been successfully established for many emerging contaminants from all categories, such as EDCs/pharmaceuticals/antibiotic resistant bacteria from water, and wastewaters. Few notable examples of ECs with successful removal are estradiol, ibuprofen, diclofenac, naproxen, carbamazepine, acetaminophen, ethynilestradiol, atrazine, sulfonamides, diclofenac, macrolides, bisphenol A, nonylphenol atenolol, and trimethoprim. Membrane systems are also sometime associated with other treatment systems that provide excellent final treated effluent. For example, carbon nanotubes, with their large specific surface area and developed pore structure, excellent photocatalytic activity, and high mechanical strength, are capable of adsorbing and remediating PPCPs and EDCs such as triclosan, acetaminophen, ibuprofen, etc. Graphene adsorption reactor (GAR), coupled with conventional sand filtration, is also highly fruitful for removal of pharmaceutical pollutants Caffeine, carbamazepine, ibuprofen, and diclofenac from urban wastewater, whereas porous graphene material (PG), as a filter medium, can satisfactorily remove different ECs, namely, atenolol (ATL), ciprofloxacin (CIP), carbamazepine (CBZ), Diclofenac (DCF), Gemfibrozil (GEM), and Ibuprofen (IBP), from different water bodies.

2.2.2 Advanced oxidation processes (AOPs)

Presently, there are various successful AOPs, such as hydroxyl-based, ozone-based, UV-based, sulfate-radical-based, Fenton-based, and few other advanced radical-based. Peroxymonosulphate, persulphate, catalytic ozonation, ultrasonication and hydrodynamic cavitation, gamma radiation, electrochemical oxidation, modified Fenton, plasma-assisted AOPs, etc. are continuously growing with more exploration [5, 26–28]. Application of highly reactive hydroxyls, hydroperoxyls, superoxides, and sulphate radicals for partial or complete mineralization of contaminants provide high removal capability of ECs from influents. The AOPs-modified electrochemical oxidation, gamma radiation, and plasma-assisted systems are highly appreciated due to their high potential for effective and sustainable operation during contaminant removal, with recommended environmental safety, compatibility, efficient transformation, etc.

2.2.3 Ozonation

Ozonation is a dark oxidation method with strong oxidant ozone application, which is frequently used for the treatment of ECs, with high efficiency. ECs such as carbamazepine, diclofenac, indomethacin, sulpiride, atrazine, meprobamate, ibuprofen, and trimethoprim can be treated almost completely by the application of low ozone dose of 5 mg/L, whereas some ECs like bezafibrate show resistance. Lower pH supported suitable ozonation of EDCs, corrosion inhibitors, biocides/pesticides, anti-inflammatories, antiepileptics, antibiotics, natural and synthetic estrogens, and removal of pharmaceuticals from treated effluents. Ozonation is reported to be successful for the treatment of ketoprofen, diclofenac and antipyrine, diclofenac, ketoprofen, mefenamic acid anddiuron, gabapentin, trimethoprim, metformin, primidone, azithromycin and clarithromycin, clofibric acid, ibuprofen, diclorfenac, carbamazepine, paracetamol, clofibric acid, bisphenol A, 17-estradiol antibiotic (clarithromycin), alachlor, atrazine, chlorfenvinphos, carbamazepine isobroturum, diuron) antibiotic (metronidazol), benzafibrate (lipid regulator) ibuprofen, bezafibrate, amoxicillin, sulfamethoxazole, etc. [16, 28]

2.2.4 Fenton-related AOPs: Fenton and photo-Fenton

In the Fenton process, metal ion Fe^{2+}/ferryl ions reacts with active H_2O_2 to produce hydroxyl radicals, which are strong reactive species. Complete elimination or partial decontamination of ECs, such as norfloxacin, atrazine, some phenylurea, herbicides, estrone, 17-estradiol, 17-ethinylestradiol, diethylstilbestrol, antibiotic (amoxiline), antibiotic (metronidazol), etc., is achieved through the Fenton process [28].

2.2.5 UV-based AOPs, photocatalysis (TiO_2)

In the presence of catalysts or oxidants, for example, titanium dioxide (TiO2), gets excited to produce positive holes in the valence band (hv + vb) with an oxidative capacity, and negative electrons at the conduction band (e-cb) with a reductive capacity, and induces hydroxyl radicals for the decontamination of ECs like EDCs and PPCPs. High destruction of estrogens (17-estradiol, estrone, and 17-ethinylestradiol), bisphenols and antiepileptics, herbicide of metsulfuron-methyl (MM), sulfonylurea herbicide, diclofenac, 17-estradiol, estrone and 17-ethinylestradiol, bisphenol A, 17-estradiol, and 17-ethinylestradiol, 17-estradiol, 17-ethinylestradiol, and estriol, clofibric acid, carbamazepine, and iomeprol is achieved through photocatalysis [29, 30].

3 Bioremediation

The great adaptability, cellular simplicity, genetic diversity, and functionality of microbial populations offer enormous potential to react and remediate large number of pollutants in the environment. The pollutant removal potential is also present in higher life, such as algae and plants, along with bacteria, archaea, and fungi. Bioremediation of pollutants occurs through four major phenomena, namely, **biodegradation**, by pool of enzymes suitable for breakdown of organic pollutants and inorganics; **biosorption**, caused by various organic acids and functional groups suitable for binding inorganics, especially heavy metals, through processes like ion exchange, chelation, adsorption, diffusion through the cell wall and cell membrane, etc., **bioaccumulation** of organic pollutants and inorganics within or outside the cell by redox changes or molecular binding, and **bioleaching** of pollutants at low concentrations, and of metals by the enzymes present in the cells. Groups of protein perform efflux of metal ions or few other refractory contaminants. Mechanism, such as intracellular bioaccumulation, is carried out by microbes with metallothioneins and phytochelatins (PCs, polypeptide with high number of gammaPCs, dipeptide residues) for the accumulation, confiscation, and immobilization of the contaminants within the cell. Few microbes carry out extracellular sequestration through exopolysaccharides (EPS) with functional groups, namely, hydroxyls, carboxyls, amides, phosphoryls, sulfhydryls, etc., as well as high molecular weight polyanionic polymers) for the protection and resistance to contaminants. Enzyme detoxification is another successful mode of decontamination followed by microbes, which enables alteration of the chemical compositions in pollutants, following precipitation and confiscation in the cell mass.

Modern technology, such as bioremediation of pollutants, has also geared up with the identification of efficient treatment agents, and the consequent suitable working conditions for better efficiency. Therefore, reactor configurations are modified, and we currently have various biological reactors for the treatment of complex waste matter. Few successful biological reactors of the current days, especially with the application of microbe for removal of ECs, are described in this section.

3.1 Biological membrane reactors

Being critical contaminants, either in terms of concentration or properties, ECs demand special arrangements for their treatment. In biological membrane reactors, microbes are arranged in the form of a membrane, provided with some substratum or completely on their adhesion properties to each other for their membrane structure. When waste influents with ECs are added to reactors, the active microbes come into contact with the pollutants and work upon them, which is mostly sorption and biodegradation, according to microbial specialization and contaminant properties. Many times, sludge retention favors high removal through promotion in biomass growth and bioremediation. Other factors influencing removal of ECs in MBR are sludge age, sludge EC concentration, characteristics of wastewater, micro-environment (according to oxygen), redox condition, and operating conditions such as hydraulic retention time, pH, temperature, and conductivity. For example, anaerobic membrane bioreactor (AnMBR) and anaerobic fluidized membrane bioreactor, based on anerobic digestion with high system stability and microbial community abundance, are highly recognized for destroying a number of ECs, such as diclofenac, 17β-estradiol, 17a-ethinylestradio, sulfamethoxazole, ciprofloxacin, amoxicillin, ceftriaxone, cefoperazone, ampicillin, triclosan tetrahydrofuran, sul, intl1, tet C, bisphenol, androstenedione, androsterone, diazinnon, butylparaben, and Linuron [31]. Biological MBR provide suitable outcomes for the treatment of pharmaceuticals belonging to the categories of anticonvulsant, Carbamazepine, salicylic acid and propyl parabens analgesic, NSAID, H2 blocker, antibiotic, β-blocker, hormones, many PCPs, and pesticides.

3.2 Biofilm reactor

Bioremediation is often accelerated in biofilm reactor, provided with the microbial cells present in high biomass grown in fixed or moving substratum, where mixing can be through aeration (aerobic reactor) or other mixing systems (anaerobic/anoxic system).

The biofilm retaining microbial cells are generally robust with higher tolerance against stressful environmental conditions, pathogens, anti-microbials, and harmful chemicals, and are prominent in the sorption and metabolization of contaminants, executed by extracellular enzymes, which involve a well-regulated cell communication signaling system, known as quorum sensing (QS). The common biofilm reactor for various waste treatment are moving bed biofilm reactor (MBBR), fluidized bed biofilm reactor (FBBR), upflow anerobic-packed bed biofilm reactor (UAnPBBR), rotating bed biofilm reactor (RBC), trickling filter (TF), etc. High treatment efficiency, low energy consumption, prevention of secondary pollution, feasibility of simultaneous aerobic and anoxic conditions, and less space requirement are the significant advantages of biofilm reactors [1, 32–34]. Biofilm reactors successfully decontaminate caffeine, carbamazepine, and three estrogens (estrone, 17β-estradiol, and 17α-ethynylestradiol), as well as *E. coli* and F + specific coliphage, in aerobic process. MBBR exhibits complete removal of certain pharmaceuticals (ibuprofen and ofloxacin) from hospital wastewater, pesticides (organophosphates, carbamates, and pyrethroids), pathogens, and PCPs [32].

3.3 Hybrid system

A good success rate in treatment efficiency is most often reported from hybrid techniques such as chemical oxidation-based treatment, for example, ozonation/Fenton/AOPs, combined with a biological process; the combination of a physical process accompanied by the next stage advanced biological process, such as activated carbon, etc.; MBR- reverse osmosis/filtration/ozonation/microfiltration/ultrafiltration/activated sludge, followed by ultrafiltration/microfiltration; and MBR-reverse osmosis, followed by biological activated carbon, etc. [16, 35]

4 Systems biology for the advanced management of pollutants

The systems biology approach has offered great inputs in biological studies, which applies "omics" techniques (genomics, proteomics, transcriptomics, and metabolomics) and metagenomic studies, along with latest technologies necessary for modern biological research as shown in Figure 2. With time, several new contaminants are entering the environment, causing severe issues to human and other organisms; hence, researchers are continuously working to solve the same, including taking knowledge from multiple disciplines and tools, in addition to understanding the interaction of pollutants and studying the organisms' response analysis to understand the function of biological systems. It is also helpful to understand how biological systems work, as also as the underlying mechanisms that control these processes.

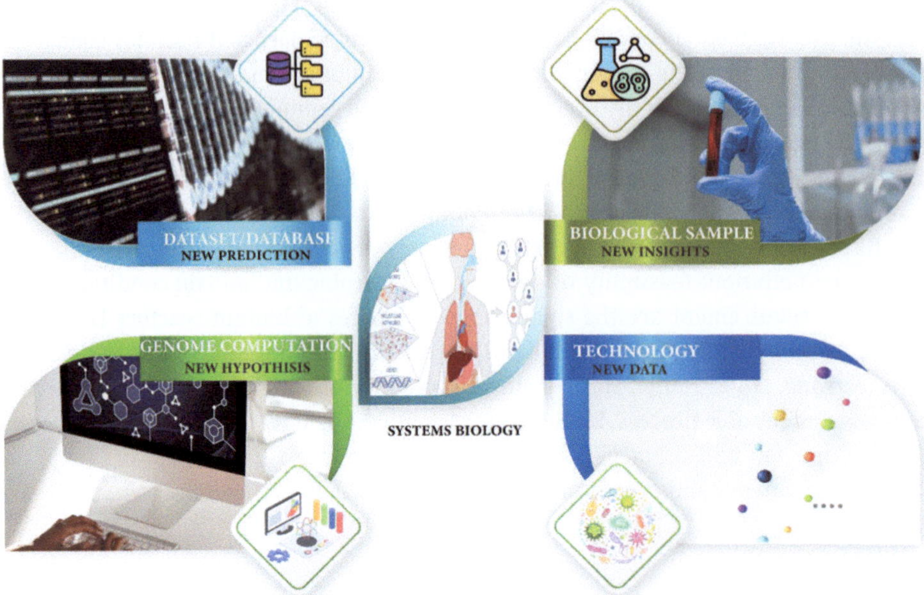

Figure 2: Systems biology for ecotoxicology and metagenome analysis.

As a discipline, systems biology encompasses both experimental and theoretical questions related to the practical nature of biological systems: how cells function, how do they respond to external stimuli, etc. The primary goal is to unveil the myriad intricacies involved in large-scale biological processes such as development and metabolism. It is also manly focused on the analysis of large-scale data sets generated from experimental cells or organismal models. It combines traditional methods used in molecular biology with new techniques, to create computer programs capable of simulating dynamic networks of interactions, at the molecular scale. Tools of system biology include various databases and computational/bioinformatic tools, and software. Recognized databases, for example, the publicly available "Comparative Toxicogenomics Database" (CTD Database), database from National Center for Biotechnology Information (NCBI), and Gene Expression Omnibus (GEO) database provide information on the molecular toxicology of chemicals, drugs and other agents on health issues, ecotoxicology, expression profiles from a diverse collection of human tissues and cell lines, and responsible pollutants derived from advanced techniques such as next-generation sequencing (NGS), liquid chromatography-mass spectrometry (LC-MS), gas chromatography-mass spectrometry (GC-MS), nuclear magnetic resonance (NMR), and in-silico analyses [36] Tools highly recommended for toxicogenomic studies are rich in numerous plugins that enhance the utility, for example, MEGAN (Metagenome Analyzer), MG-RAST (metagenomics RAST), MOTHUR, UPARSE-High-accuracy, high-throughput OTU clustering, UNIX and OBITOOL, PRINSEQ tool, CLC Microbial Genomics Module, MicrobiomeAnalyst

(Microbiome data analysis), EBI-Metagenomics, PICRUSt, BURRITO, Visualization and Analysis of Microbial Population Structures (VAMPS), etc. [37]. Systems biology incorporates information about the interactions of contaminants with bioremediating agents, in terms of gene expression, enzymes, biosynthetic pathways, secondary metabolites in microbes, alteration of existing pathways under conditions of stress, etc., and helps in structuring conceptual models. The major tools used for monitoring are functional genomic microarray, phylogenetic analysis, and metabolomics, proteomics, and quantitative PCR. DNA-based genomics tools, such as 16S rRNA clone library, PhyloChip, or sequencing are recommended for identifying the microbial community structure, whereas RNAseq, GeoChip (for RNA), and several mass spectroscopy methods are recommended for proteins. These tools can be used for the analysis of metagenomes, identification of the functional genes involved, understand the associated cellular pathways with microbial bioremediation, and for the identification of suitable agents for treatment [38, 39].

5 Advances and limitations of microbial bioremediations of ECs

The significant advantage of bioremediation lies in its natural activities such as transforming large molecules to small molecules through metabolisms, and assimilation with none or negligible secondary waste. The organisms can modify their cellular and molecular properties to adapt to the pollutants and sustain the same while bioremediation is proceeding. In the case of bioremediation of emerging pollutants, major limitations are raised by uncertain pollutants as well as the upcoming intermediates, resulting from the breakdown of the parent ECs, as there is evidence that, sometimes, the intermediates can be more toxic than the parent compounds. Therefore, prior knowledge of the bioremediation pathway of the considered pollutants is necessary while implying bioremediation. Another key limitation is that in initial conditions, bioremediation is often very slow as microbes/biological agents require time for adapting their functional properties according to the exposed conditions. For few persistent pollutants, co-metabolisms and addition of external substrates might be necessary to maintain the necessary bioremediation rate.

6 Current status and future scope

As the term suggests, ECs are continuously getting added to the environment, and they is continuously under monitoring and research. However, as stated above, many treatment systems are investigated for a number of ECs with successful treatment rates. In the case of bioremediation also, many works have been done till date, but still there is scope for the application of a variety of recognized efficient microbes in different reactor configurations for further efficiency improvement for variety of ECs in different categories.

7 Conclusion

ECs are associated with various challenges, including poor recognition/identification, unexpected intermediates or final product, etc. The group of each ECs is increasing, including their type with complex properties of bioaccumulation, biomagnification, as well as treatment challenges because of formation of their non-targeted complexes. The expounding researchers are also tirelessly way working on the remediation of the recognized one with various treatment modes. Bioremediation has also proved a lot successful for various ECs, and more suitable results are possible from the application of diverse microbes for specific ECs. The identification and operation of a suitable microbial team can be recognized by systems biology approaches, following proper analysis of databases and the application of suitable tools for the same.

References

[1] Ortúzar M, Esterhuizen M, Olicón-Hernández DR, González-López J, Aranda E. Pharmaceutical pollution in aquatic environments: A concise review of environmental impacts and bioremediation systems. Front Microbiol 2022, 13, 1393.

[2] Poynton HC, Vulpe CD. Ecotoxicogenomics: Emerging technologies for emerging contaminants1. JAWRA J Am Water Resour Assoc 2009, 45, 83–96.

[3] Noguera-Oviedo K, Aga DS. Lessons learned from more than two decades of research on emerging contaminants in the environment. J Hazard Mater 2016, 316, 242–251.

[4] Caliman FA, Gavrilescu M. Pharmaceuticals, personal care products and endocrine disrupting agents in the environment – a review. CLEAN – Soil Air Water 2009, 37, 277–303.

[5] Bolong N, Ismail AF, Salim MR, Matsuura T. A review of the effects of emerging contaminants in wastewater and options for their removal. Desalination 2009, 239, 229–246.

[6] Rzymski P, Drewek A, Klimaszyk P. Pharmaceutical pollution of aquatic environment: An emerging and enormous challenge. Limnol Rev 2017, 17, 97–107.

[7] Bilal M, Adeel M, Rasheed T, Zhao Y, Iqbal HMN. Emerging contaminants of high concern and their enzyme-assisted biodegradation – A review. Environ Int 2019, 124, 336–353.

[8] Stefanakis AI, Becker JA A review of emerging contaminants in water: Classification, sources, and potential risks. https://servicesigi-Glob-1-4666-9559-7ch003. 1AD, 55–80.

[9] Gomes IB, Maillard JY, Simões LC, Simões M. Emerging contaminants affect the microbiome of water systems – strategies for their mitigation. Npj Clean Water 2020, 3, 1, 1–11.

[10] Pereira LC, de Souza AO, Bernardes MFF, Pazin M, Tasso MJ, Pereira PH, et al. A perspective on the potential risks of emerging contaminants to human and environmental health. Environ Sci Pollut Res 2015, 22, 18, 13800–23.

[11] Lei M, Zhang L, Lei J, Zong L, Li J, Wu Z, et al. Overview of emerging contaminants and associated human health effects. BioMed Res Int 2015, 2015.

[12] Wang S, Wang J. Degradation of emerging contaminants by acclimated activated sludge. Environ Technol 2018, 39, 1985–1993.

[13] Koumaki E, Noutsopoulos C, Mamais D, Fragkiskatos G, Andreadakis A. Fate of emerging contaminants in high-rate activated sludge systems. Int J Environ Res Public Health 2021, 18, 1–16.

[14] Yu X, Lee K, Ulrich AC. Model naphthenic acids removal by microalgae and Base Mine Lake cap water microbial inoculum. Chemosphere 2019, 234, 796–805.

[15] Sutherland DL, Ralph PJ. Microalgal bioremediation of emerging contaminants – Opportunities and challenges. Water Res 2019, 164, 114921.

[16] Ahmed MB, Zhou JL, Ngo HH, Guo W, Thomaidis NS, Xu J. Progress in the biological and chemical treatment technologies for emerging contaminant removal from wastewater: A critical review. J Hazard Mater 2017, 323, 274–298.

[17] Ravikumar Y, Yun J, Zhang G, Zabed HM, Qi X. A review on constructed wetlands-based removal of pharmaceutical contaminants derived from non-point source pollution. Environ Technol Innov 2022, 26, 102504.

[18] Younas F, Khan N, Bibi I, Afzal M, Hussain K, Shahid M, et al. Constructed wetlands as a sustainable technology for wastewater treatment with emphasis on chromium-rich tannery wastewater. 2022, 422, April, 2021.

[19] Matamoros V, Salvadó V. Evaluation of the seasonal performance of a water reclamation pond-constructed wetland system for removing emerging contaminants. Chemosphere 2012, 86, 111–117.

[20] Zhou GJ, Lin L, Li XY, Leung KMY. Removal of emerging contaminants from wastewater during chemically enhanced primary sedimentation and acidogenic sludge fermentation. Water Res 2020, 175, 115646.

[21] Zhou G, Wang Q, Li J, Li Q, Xu H, Ye Q, et al. Removal of polystyrene and polyethylene microplastics using PAC and FeCl3 coagulation: Performance and mechanism. Sci Total Environ 2021, 752, 141837.

[22] Egea-Corbacho A, Gutiérrez S, Quiroga JM. Removal of emerging contaminants from wastewater through pilot plants using intermittent sand/coke filters for its subsequent reuse. Sci Total Environ 2019, 646, 1232–1240.

[23] Luo Y, Guo W, Ngo HH, Nghiem LD, Hai FI, Zhang J, et al. A review on the occurrence of micropollutants in the aquatic environment and their fate and removal during wastewater treatment. Sci Total Environ 2014, 473–474, 619–641.

[24] Jeirani Z, Niu CH, Soltan J. Adsorption of emerging pollutants on activated carbon. Rev Chem Eng 2017, 33, 491–522.

[25] Katsigiannis AS, Κοψιδας O, Katsigiannis A, Noutsopoulos C, Mantziaras J, Gioldasi M. Removal of emerging pollutants through Granular Activated Carbon SCALE UP OF ADSORPT ION IN FIXED-BED COLUMN SYST EMS Removal of emerging pollutants through Granular Activated Carbon. https://doi.org/10.1016/j.cej.2015.05.109.

[26] Esplugas S, Bila DM, Krause LGT, Dezotti M. Ozonation and advanced oxidation technologies to remove endocrine disrupting chemicals (EDCs) and pharmaceuticals and personal care products (PPCPs) in water effluents. J Hazard Mater 2007, 149, 631–642.

[27] Kar P, Shukla K, Jain P, Sathiyan G, Gupta RK. Semiconductor based photocatalysts for detoxification of emerging pharmaceutical pollutants from aquatic systems: A critical review. Nano Mater Sci 2021, 3, 25–46.

[28] Priyadarshini M, Das I, Ghangrekar MM, Blaney L. Advanced oxidation processes: Performance, advantages, and scale-up of emerging technologies. J Environ Manage 2022, 316, 115295.

[29] Prieto-Rodriguez L, Miralles-Cuevas S, Oller I, Agüera A, Puma GL, Malato S. Treatment of emerging contaminants in wastewater treatment plants (WWTP) effluents by solar photocatalysis using low TiO2 concentrations. J Hazard Mater 2012, 211–212, 131–137.

[30] Gude VG, Barrera K, SA F, Veera, Gude G, Truax DD, et al. A Critical Evaluation of Advanced Oxidation Processes for Emerging Contaminants Removal Related papers 3.0 Advanced Oxidat ion Processes Lit erat ure Review A Critical Evaluation of Advanced Oxidation Processes for Emerging Contaminants Removal. https://doi.org/10.1007/s40710-017-0207-1.

[31] Ji J, Kakade A, Yu Z, Khan A, Liu P, Li X. Anaerobic membrane bioreactors for treatment of emerging contaminants: A review. J Environ Manage 2020, 270, 110913.

[32] Husain Khan A, Abdul Aziz H, Khan NA, Ahmed S, Mehtab MS, Vambol S, et al. Pharmaceuticals of emerging concern in hospital wastewater: removal of Ibuprofen and Ofloxacin drugs using MBBR method. https://doi.org/101080/0306731920201855333. 2020. https://doi.org/10.1080/03067319.2020.1855333.

[33] Sahariah BP, Chakraborty S. Kinetic analysis of phenol, thiocyanate and ammonia-nitrogen removals in an anaerobic-anoxic-aerobic moving bed bioreactor system. J Hazard Mater 2011, 190, 260–267.

[34] Sandhya M, Huang Y, Li J, Wu X, Zhou Z, Lei Q, et al. Biofilm-mediated bioremediation is a powerful tool for the removal of environmental pollutants. Chemosphere 2022, 133609.

[35] Saidulu D, Majumder A, Gupta AK. A systematic review of moving bed biofilm reactor, membrane bioreactor, and moving bed membrane bioreactor for wastewater treatment: Comparison of research trends, removal mechanisms, and performance. J Environ Chem Eng 2021, 9, 106112.

[36] Choudhari JK, Choubey J, Verma MK, Chatterjee T, Sahariah BP. Metagenomics: The boon for microbial world knowledge and current challenges. Bioinformatics 2022, 159–175.

[37] Choubey J, Choudhari JK, Verma MK, Chaterjee T, Sahariah BP. Systems biology aided functional analysis of microbes that have rich bioremediation potential for environmental pollutants. Microb Remediat Azo Dyes Prokaryotes 2022, 157–170.

[38] Choubey J, Choudhari JK, Sahariah BP, Verma MK, Banerjee A. Molecular tools: Advance approaches to analyze diversity of microbial community. Wastewater Treat Cut-Edge Mol Tools Tech Appl Asp 2021, 507–520.

[39] Anandkumar J, Choudhari JK, Choubey J, Chatterjee T, Verma MK, Sahariah BP. Microbial response to lead exposure. Microb Community Stud Ind Wastewater Treat 2022, 21–31.

Harlina Ahmad*, Mardiana Idayu Ahmad, Rekah Nadarajah,
Nishalini Ratha Pukallenthy, Norli Ismail

Bioremediation of Pesticides in the Environment

Abstract: The extensive application of pesticides in agriculture initially increased crop productivity and production. Since the agricultural revolution, pesticides have significantly benefited farmers in reaching this goal. Pesticides made for commercial use are mixtures with a single, declared, and regulated ingredient. Pesticides may be transported from their application sites to nearby streams and rivers through transport processes. Aquatic populations, which include fish, benthic invertebrates, and others, may suffer if pesticides are prevalent in rivers and streams. Bioremediation is extremely necessary due to the level of toxicity brought on by pesticides. Plants that uptake and transform pesticides are used in tandem with endophytes that can break pesticides. Phytoremediation technologies are perfectly suited for use on pesticide-containing soils. Plants for phytoremediation can be cultivated and used favorably. It is advantageous to add soil amendments that promote microbial communities' biological variation and food sources. The soils that have been contaminated by pesticides have a great chance of being restored through bioremediation. Recent studies have advanced our fundamental knowledge of how pesticides degrade at low concentrations in a variety of environmental conditions, and they are likely to aid in the creation of practical bioremediation techniques for soil and water resources that have been contaminated by pesticides.

Keywords: Bioremediation, pesticides, ex situ, in situ, fate of pesticides

1 Introduction

Bioremediation, by definition, is the use of microorganisms to eliminate or immobilize environmental pollutants. Incorporating microorganisms and their metabolic activity to destroy contaminants, bioremediation is seen as a cost-effective, safe, dependable, efficient, and environmentally friendly alternative to traditional remediation approaches and can be carried out in two ways: ex situ and in situ [1].

Bioremediation is a new area of research and practice that has developed in response to the shortcomings and negative consequences of conventional waste treatment

***Corresponding author: Harlina Ahmad,** Environmental Technology Division, School of Industrial Technology, Universiti Sains Malaysia, Penang 11800, Malaysia, e-mail: harlinaa@usm.my
Mardiana Idayu Ahmad, Rekah Nadarajah, Nishalini Ratha Pukallenthy, Norli Ismail, Environmental Technology Division, School of Industrial Technology, Universiti Sains Malaysia, Penang 11800, Malaysia

https://doi.org/10.1515/9783111016825-004

methods. The Exxon Valdez oil disaster of 1989 in the United States profoundly impacted the historical development of bioremediation. In this case, nitrogen and phosphate fertilizers were added to help microorganisms that naturally break down dangerous hydrocarbons in oil grow faster.

The development of bioremediation is highlighted as a model for new environmental technology. Methods and procedures for remediation play a crucial role in the comprehensive clean-up, containment, removal, reclamation, and restoration of polluted environments. The type and quantity of pollutants, the physical, chemical, and biological parameters of the contaminated environment, the cost, time constraints, and the microbial community present or requirement for augmentation all play a role in determining the remediation technique selected for each contaminated environment [2].

Pesticides are especially important since they are extensively used to minimize agricultural pest infestations, thereby safeguarding crops from potential productivity and quality losses [3]. Nonetheless, the excessive use of pesticides has not only caused significant difficulties for the environment and human health but also for the biodiversity of plants and animals [4]. One of these risks is the pesticide's inherent toxicity, which slows or stops the development of organisms (both plant and animal). As a consequence, efficient remediation solutions must be developed to mitigate the negative impact that pesticides have on the surrounding ecosystem. Since bioaccumulation and biomagnification of hazardous pesticides are the major causes of biodiversity loss, microbial bioremediation has been envisioned as a safe and sustainable approach to purifying the environment [5]. The purpose of this chapter is to provide a comprehensive overview of bioremediation and its use in the process of remediating pesticides, with a special emphasis on agricultural settings.

2 The fate of pesticides in the environment

Pesticides are chemical compounds used to kill or control pests. They can be classified as insecticides, herbicides, rodenticides, bactericides, fungicides, and larvicides. Plant extracts from pyrethrum and tobacco, along with organic compounds such as arsenic, sulfur, and copper, were used as pesticides in the early twentieth century. Organochlorines, organophosphates, and other insecticides, fungicides, and herbicides, however, were only produced in the 1940s and afterward [6].

Only a small portion of the pesticides applied to a field of crops or any other area where pests are present reach the target organisms. Fewer than 0.1% of applied pesticides were likely to reach their target pests [7]. The remainder of the pesticides may be distributed across the environmental compartments: air, water, and soil and may potentially reach nontarget organisms, including humans. The fate and behavior of pesticides in the environment are governed by pesticide properties, ambient conditions, and interactions involving chemical, physical, and biological processes.

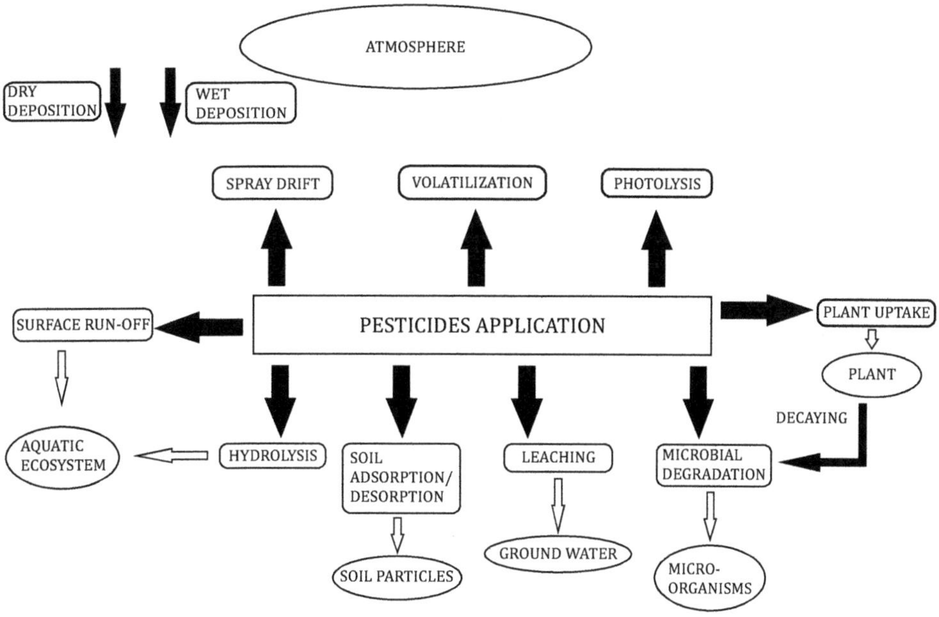

Figure 1: General fate of pesticides in the environment upon application in agricultural fields.

Figure 1 displays the processes that occur in the various environmental compartments after the application of pesticides. Spray drift (atmospheric dispersion helped by wind) and volatilization are the methods through which pesticides are released into the atmosphere. Volatilization occurs when liquid pesticide molecules escape into the atmosphere as a result of a comparatively higher vapor pressure. If precipitation occurs, pesticide residues will combine with rainfall and settle via wet deposition on the land and surface water; otherwise, they will settle by dry deposition due to gravity sedimentation.

The pollution of aquatic ecosystems with pesticides as a result of agricultural surface runoff has emerged as one of the most prominent sources of environmental degradation in aquatic systems. Due to the high affinity that pesticides have for soil, the soil component serves as a storage enclosure for the said chemicals. However, surface water resources such as streams, estuaries, and lakes, as well as groundwater, are susceptible to contamination from pesticides due to the soil's close interconnection with water bodies. Pesticides with low biodegradability have a prolonged half-life and are more likely to persist in the environment, potentially poisoning water supplies. In addition, the pesticide's mobility is determined by its solubility and absorption capacity. Strongly adsorbed pesticides are much less likely to infiltrate the soil layer; however, they can easily be transferred by eroding soil particles through runoff water to adjacent water bodies [8].

Precipitation and irrigation water frequently leach soluble pesticides from soils [9]. In horizontally moving water, pesticides are dispersed, dissolved, suspended, or emulsified, but they do not enter the soil. Infiltration, however, is another process that describes the downward movement of water into the soil, which may bring pesticide residues together and contaminate the water table, shallow groundwater, and deeper aquifers. Due to the presence of pesticide residues in groundwater, consumers of drinking water are exposed to substantial health risks. Due to this fear and the prospect of other catastrophic impacts of pesticides on fish and wildlife, regulatory agencies have been compelled to exert severe control over pesticides and aggressively enforce laws and regulations [10].

An important factor that plays a role in determining how long pesticides will remain in the environment is the rate at which pesticides degrade, are transformed into degradation products, or are partially or completely mineralized. This decomposition could be an abiotic process, such as hydrolysis or photolysis, or it could be a biotic process, namely biodegradation. Photolysis is the process by which pesticide molecules are broken by photons from the sunlight. The hydrolysis of pesticides is a chemical reaction that occurs when pesticide molecules contact water, resulting in the disintegration of the material into its constituent parts. On the other hand, the process of biodegradation refers to the transformation of materials into ecologically friendly by-products such as water, carbon dioxide, and biomass through the action of microbes under normal environmental conditions.

In plant uptake, most plant components are able to absorb pesticides that are soluble in soil due to the facilitation of absorption through transpiration [11]. Roots are responsible for the transmission of pesticides, followed by the circulatory system. The prevalence of their metabolites in the crop vascular system is determined by factors including their reactivity with soil and plants, dosages of sprayed pesticides, biochemical and physicochemical parameters, and the mode of said agrochemical absorption [12].

3 Bioremediation of pesticide contamination in water and soils

Bioremediation takes advantage of the natural biodegradation process that occurs when hazardous pollutants are broken down by organisms in the environment. This is achieved by greatly boosting the activity and growth of the microorganisms that convert harmful substances into environmentally safe products. Microorganisms have long been used for bioremediation. However, numerous findings on bioremediation utilizing plants, fungi, algae, or enzymes have expanded the scope of bioremediation [13].

Pesticides often have a very sluggish rate of biodegradation in comparison to other methods that have been described. Selecting the appropriate microbes is necessary to maintain an adequate rate of degradation. An essential step in the process of bioremediation is the collection, from a polluted area, of naturally occurring microorganisms that are capable of breaking down the contaminants. Isolated from polluted soils and sludge created in agricultural and industrial locations, the strains of *Acinetobacter johnsonii*, *Lysini bacillus*, *Bacillus* sp., and *Pseudomonas* sp. are being exploited for the breakdown of pesticides [14].

The involvement of microorganisms, fungi, and enzymes in pesticide bioremediation has been elaborated in detail [15]. Essential to the process is the oxidation of the parent molecule, which releases carbon dioxide and water and offers a source of energy for the microorganisms responsible for the biodegradation of the pesticide. When a soil's native microbial population is unable to efficiently manage pesticides, it is recommended to introduce additional microflora that can degrade chemicals. In addition to the enzyme system, the capacity of microbes to break down pesticides depends on environmental conditions such as temperature, pH, and nutrition availability. Some pesticides degrade easily; while others, due to the presence of anionic species within the molecule, are resistant to breakdown. The minute structural changes that fungi make to pesticides in order to degrade them, thereby turning them into nontoxic compounds and releasing them into the soil, where they are susceptible to further degradation. Enzymes play a crucial role in the degradation of pesticide chemicals, both within the organism being targeted (through intrinsic detoxification systems and developed metabolic resistance) and in the broader ecosystem (via biodegradation by soil and water microorganisms).

4 Bioremediation technology

Remediation technologies have been divided into four categories based on the method by which contaminants are eliminated [4]: (i) removal, which occurs when a technology physically removes contaminants or contaminated media from the site without separating them from the host medium; (ii) separation, which involves removing the contaminant from the host medium (soil or water); (iii) destruction, in which contaminants are eliminated chemically, biologically, or neutralized to generate less hazardous substances; and (iv) containment, where the surface and subsurface movement of the contaminant is hindered or immobilized.

There are two primary treatment approaches available: in situ bioremediation and ex situ bioremediation, both of which are dependent on the location where the pesticide treatment will take place. In situ, or on-site, bioremediation involves the treatment of toxic chemicals on-site, while ex situ bioremediation involves the treatment of toxic pollutants off-site.

4.1 In situ bioremediation techniques

4.1.1 Bioremediation of pesticides in contaminated rivers and lakes

Utilizing in situ bioremediation technology in contaminated rivers instead of excavating and transporting pollutants is known as in situ bioremediation. Native bacteria are employed in this procedure, sometimes in combination with farmed microbes. Additionally, it typically requires several steps to increase the impact of technology. Microbial remediation, aquatic plants, and aquatic animals are examples of in situ bioremediation. The strategic way to reduce surface water pollution should be bioremediation. These techniques must be explored on contaminated surface streams in addition to rivers and lakes. Furthermore, bioremediation processes need to be tuned depending on flow conditions and nutrient availability.

a. Bioremediation by using microorganisms

Microbes are utilized to degrade, transform, and absorb pollutants in water. Current research indicates that the appropriate microbial functional groups exist and can eradicate certain pollutants from wastewater. Two microorganism-based approaches are used for in situ surface water treatment. The first method employs microbial dosing, whereas the second method utilizes biofilms [16, 17].

i. Microbial dosage

Microbial dosing removes contaminants from water using efficient bacteria. For water purification, a bio-energizer and coupled water mixing increased microbial breakdown.

ii. Biofilm

Under the conditions of artificial aeration or dissolved oxygen, pollutant transfer in the river occurs through adsorption, degradation, and filtration. The bio-film method employs a microcarrier and a bio-membrane connected to the natural riverbed. There is an incorporation of thin layer flow, subsurface stream purification, gravel contact oxidation, and artificial packing contact oxidation. Some countries are employing biofilm technology for river purification. In Japan, indirect river treatment entailed constructing purification facilities on the river's bank and utilizing the river's flow to guide water into the facilities for purification before discharge.

b. Bioremediation by using aquatic plants

Plants, particularly aquatic plants, can purify water. Plants with high pollution absorption and tolerance could be grown in contaminated water. Water was purified by plants through adsorption, absorption, accumulation, and decomposition, which removed or remedied pollutants in the water. Commonly employed restoration plants include reed, *Eichhornia crassipes* (water hyacinth) and cattail, *Alternanthera philoxeroides.*

c. Bioremediation by using aquatic organism
The aquatic organisms were utilized to remediate eutrophied water and alter the nutritional structure of the water by modifying the species distribution and fish density. To efficiently manage phytoplankton-induced eutrophication, silver carp, common carp, and other filter-feeding fish may be utilized (algae). In lakes and streams, silver carp can live for anything between 6 and 10 years, and in extreme cases for over 20. The common silver carp is a filter-feeding omnivore that can filter particles as fine as 10 μm, including zooplankton and phytoplankton.

4.1.2 Bioremediation in contaminated soil

Numerous highly effective chemical pesticides have been developed since the introduction of several synthetic organic pesticides in the 1940s, such as dichloro-diphenyl-trichloroethane and hexachlorocyclohexane. These chemical pesticides are frequently used to control field weeds, plant infections, and agricultural pests [11]. In order to maintain crop yields, pesticides are unavoidable; yet the incorrect and prolonged use of these chemicals has put both the environment and the public's health in serious jeopardy. In addition, crops that are amenable to rotation may face considerable phytotoxicity from pesticide residues [18]. Consequently, the public and scientific communities have focused only on the elimination of residual pesticides from contaminated soils.

The cleanup work for the in situ approach is typically conducted using aerobic processes because it is performed directly in the contaminated region. The most common types of in situ mitigation strategies are natural attenuation, bioaugmentation, biostimulation, bioventing, and biosparging [19] have elaborated on the in situ technique as follows:

a. Natural attenuation, which utilizes the microorganisms found in the contaminated soil. The pollutant absorbs into the soil's organic matter and clay minerals, where it then degrades biologically, volatilizes, disperses, dilutes, decays radioactively, and degrades through sorption.

b. Bioaugmentation is known as the addition of microbial strains or enzymes to contaminated soils. In this way, microbes with metabolic abilities support the biodegradation pathways. The utilization of additional microbial cultures in a land treatment unit is constrained by two factors: Most soils exposed to biodegradable wastes include indigenous microorganisms that are efficient degraders if the land treatment unit is managed. Non-native cultures seldom compete with an indigenous population sufficiently to expand and maintain useful population levels.

c. Biostimulation method involves optimizing the types and quantities of nutrients delivered to stimulate and enhance the growth of indigenous microorganisms. To be able to decompose the contaminants in their environment, local microorganisms must have access to nitrogen, phosphorus, carbon, and oxygen.

d. Bioventing promotes the growth of microorganisms capable of degrading contaminants by delivering oxygen to unsaturated soil regions. Bioventing can be conducted either actively or passively with regard to aeration. Active bioventing requires the use of a blower to inject air into the soil, whereas passive bioventing depends only on atmospheric pressure to exchange gases through the vent wells. Bioventing treatment can last anywhere between six months and five years, depending on the type and quantity of contamination, the rate of biodegradation, and the soil's characteristics, such as permeability and moisture content.

e. Biosparging increases the oxygen level and promotes microorganisms to degrade the contamination by pumping pressurized air into the saturated soil zone. Comparable to bioventing, however, the air is introduced directly into the saturated zone during biosparging. Due to the unsaturated zone, volatile organic compounds may ascend to the surface, aiding biodegradation. The infiltration rate, which influences the bioavailability of the pollutant to microorganisms, and the biodegradability of the pollutant are two elements that influence the performance of the process. Small-diameter air injection points are simple and affordable to install, thereby providing system designers and builders with a significant lot of design flexibility.

4.2 Ex situ bioremediation of pesticides

4.2.1 Bioremediation of the slurry

These methods involve the remediation of excavated soil in a bioreactor's-controlled environment. The soil is cleaned of stones and debris before being mixed with water to a predetermined concentration based on the pollutant content, rate of biodegradation, and physical characteristics of the soil. Typical slurries may have 10–40% solids. In order to maximize biological activity, electron acceptors and nutrients are added to the reactor, and variables such as pH and temperature are modified. Specialized species may be put into the reactor if no suitable population exists. For heterogeneous soils, soils with limited permeability, sites where it would be difficult to remove underlying groundwater, or when speedier treatments are required, bioreactors are recommended over in situ biological techniques.

4.2.2 Bioremediation of pesticides in soils

Ex situ bioremediation is a biological procedure where excavated soil is put in a lined above-ground treatment area and aerated after processing to improve the breakdown of organic pollutants by the local microbial population [19]. Ex situ treatments include bioreactors, composting, land farming, and biopiles.

a. Bioreactors

The bioreactor is utilized to physically separate stones and rubble, the excavated earth is first treated. Prewashing is often done as well to concentrate the pollutants into a smaller amount of dirt. The quantity relies on changing the concentration for an appropriate rate of biodegradation to occur. Contaminated soil, silt, or sludge is mixed with water and nutrients to generate aqueous slurry. The slurry typically has a solid content of 10–30 wt%. The above-depicted bioreactor is where this is then put. To maintain suspended solids and maintain microbial interaction with the soil pollutants, the slurry is stirred. After the procedure is completed, the slurry is dewatered, and the treated soil may be repositioned. Only the polluted fines and the collected wastewater need further treatment.

b. Composting

Involves the combination of polluted soil with additives to encourage the pesticides' aerobic breakdown. This method includes biopiles and land cultivation. This strategy is especially recommended when the pesticide level is low. The pollutant's microbial bioavailability is essential for composting. Controlling the water content, soil composition, and characteristics of the additional amendment is crucial for this reason. For example, biochar can be added to polluted soils to speed up the decomposition processes. Black carbon known as biochar is created when biomass is thermally converted under oxygen-restricted circumstances (gasification) or without oxygen (pyrolysis). High porosity and a broad surface area are two characteristics that make it conducive to pesticide adsorption. In addition, biochar is a source of carbon that encourages microbial activity, which aids in biodegradation. It has been shown that applying biochar enhances soil aeration and boosts the soil's ability to hold water.

i. Land farming

In land farming, the contaminated substance is applied to the soil's surface, mixed and aerated. In rare instances of highly superficial pollution, it may be possible to just till the top layer of the site without needing to dig anything up. The degrading process has slow kinetics and can take an exceedingly long time. This procedure primarily addresses fuel, nonhalogenated VOC, SVOC, pesticide, and herbicide-related soil pollution. Halogenated organics can be treated using this method, but it is less successful. Although the method is straightforward and affordable, it requires a large area, and occasionally volatilization help in the reduction of pollutant concentrations compared to microbial degradation. Toxic substances (both the parent compound and its metabolites) must not leak or volatilize throughout the process. Before the process is started, the soil must have a waterproof cover in place to prevent any chance of penetration.

ii. Biophiles

Biopiles are a combination of land cultivation and composting. To promote rapid biodegradation, it is designed with optimal temperature, moisture content, aeration, and nutrient concentrations. Typically, local microbes are responsible for degradation. Compost heaps with ventilation are used to construct engineered cells. They are a more advanced form of land farming and are widely used for the remediation of petroleum hydrocarbon surface contamination. Leaching and volatilization tend to control the physical losses of pollutants. This strategy is aimed to address soil contamination caused by herbicides. This approach can be used to treat halogenated organic compounds; however, it is less effective. Although the procedure is simple and inexpensive, it requires a significant amount of space, and sometimes the reduction in pollutant concentrations is due more to volatilization than to biodegradation. Treatment durations vary from 6 to 16 weeks based on the level of pollution, concentration targets, soil properties, and the season. The highest rates of biodegradation occur during the warmer summer months; frigid winter conditions can significantly slow or stop biodegradation. The temperature has a significant effect on bioremediation.

Table 1 depicts a few applications of bioremediation techniques to remove pesticides in soils.

Table 1: Bioremediation application in treating pesticide contamination in soils.

Bioremediation technique	Pesticides	Microorganism	Application	References
Land farming	Endosulfan insecticide	*Bordetella* sp., *Bordetella petrii*, and *Achromobactery xylosoxidans* form the consortium	99% of endosulfan was removed in just 457.75 h	[20]
Bioaugmentation	s-Triazine herbicide	*Arthrobacter* sp.	Completely degrading 25 mg/L atrazine	[21]
Bioaugmentation using fungi	Lindane	*Aspergillus niger*, *Talaromyces atroroseus*, *Alaromyces purpurogenus*, *Yarrowia lipolytica*, and *Aspergillus flavus*	The synergistic interaction of the found rhizospheric fungi and *Megathyrsus maximus* roots might be exploited to eliminate lindane in the soil.	[22]
Biostimulation	Bentazone, mecoprop, and dichlorprop	Indigenous microorganism	Oxygen at 1 µg/L in anaerobic aquifer sediment material degraded bentazone, mecoprop, and dichlorprop.	[23]

4.3 Advantages and disadvantages

Bioremediation can be used to remove a wide range of contaminants. Many dangerous compounds can be turned into harmless goods. The treatment's residues, which include carbon dioxide, water, and cell biomass, are frequently safe by-products. When compared to other cleaning processes, the benefits of bioremediation include lower costs and less disruption of the contaminated environment. Bioremediation has proven to be a promising technique for pesticide breakdown, while its long-term viability in the field is still in doubt. Complete pesticide destruction or detoxification in nature would be difficult because microbial species' metabolic paths rely on soil physicochemical properties. Bioremediation is limited to biodegradable substances for success, such as the availability of metabolically competent microbial populations, optimal environmental growth conditions, and the proper concentrations of toxins and nutrients. It is possible that there are contaminants in the form of solids, liquids, or even gases. However, the use of biological methods is restricted by requirements such as the need to be compatible with the environment, uncomplicated access of the microbial population to the pesticide molecules, and procurement of suitable pesticide-degrading microorganisms. These requirements must be met for the use of biological methods to be considered viable [24]. In order for one to make use of biological techniques, it is necessary to fulfill certain prerequisites.

References

[1] Tyagi B, Kumar N. Bioremediation: Principles and applications in environmental management. In: Bioremediation for environmental sustainability. Elsevier, 2021, 3–28.

[2] Ossai IC, Ahmed A, Hassan A, Hamid FS. Remediation of soil and water contaminated with petroleum hydrocarbon: A review. Environ Technol Innov 2020, 17, 100526.

[3] Damalas CA. Understanding benefits and risks of pesticide use. Sci Res Essay 2009, 4, 10, 945–949.

[4] Gavrilescu M. Fate of pesticides in the environment and its bioremediation. Eng Life Sci 2005, 5, 6, 497–526.

[5] Boudh S, Singh JS. Pesticide contamination: Environmental problems and remediation strategies. In: Emerging and eco-friendly approaches for waste management. Springer, Singapore, 2019, 245–269.

[6] Matthews GA. A history of pesticides. Cabi, 2018.

[7] Pimentel D. Amounts of pesticides reaching target pests: Environmental impacts and ethics. J Agri Environ Ethics 1995, 8, 1, 17–29.

[8] Syafrudin M, Kristanti RA, Yuniarto A, Hadibarata T, Rhee J, Al-Onazi WA, Algarni TS, Almarri AH, Al-Mohaimeed AM. Pesticides in drinking water – a review. Int J Environ Res Public Health 2021, 18, 2, 468.

[9] Pérez-Lucas G, Vela N, El Aatik A, Navarro S. Environmental risk of groundwater pollution by pesticide leaching through the soil profile. Pest Use Misuse Impact Environ 2019, 1–28.

[10] Demir AEA, Dilek FB, Yetis U. A new screening index for pesticides leachability to groundwater. J Environ Manage 2019, 231, 1193–1202.

[11] Sharma A, Kumar V, Shahzad B, Tanveer M, Sidhu GPS, Handa N, Kohli SK, Yadav P, Bali AS, Parihar RD, Dar OI. Worldwide pesticide usage and its impacts on ecosystem. SN Appl Sci 2019, 1, 11, 1–16.
[12] Lushchak VI, Matviishyn TM, Husak VV, Storey JM, Storey KB. Pesticide toxicity: A mechanistic approach. EXCLI J 2018, 17, 1101.
[13] Bernardino J, Bernardette A, Ramrez-Sandoval M, Domnguez-Oje D. Biodegradation and bioremediation of organic pesticides. Pesticides – recent trends in pesticide residue assay. InTech, 2012.
[14] Geed SR, Shrirame BS, Singh RS, Rai BN. Assessment of pesticides removal using two-stage Integrated Aerobic Treatment Plant (IATP) by Bacillus sp. isolated from agricultural field. Bioresour Technol 2017, 242, 45–54.
[15] Uqab B, Mudasir S, Nazir R. Review on bioremediation of pesticides. J Bioremed Biodegr 2016, 7, 3, 343.
[16] Wang J, Liu XD, Lu J. Urban river pollution control and remediation. Proc Environ Sci 2012, 13, 1856–1862.
[17] Ateia M, Yoshimura C, Nasr M. In-situ biological water treatment technologies for environmental remediation: A review. J Bioremed Biodeg 2016, 7, 348, 2.
[18] Huang X, Wang L, Ma F. Arbuscular mycorrhizal fungus modulates the phytotoxicity of Cd via combined responses of enzymes, thiolic compounds, and essential elements in the roots of Phragmites australis. Chemosphere 2017, 187, 221–229.
[19] Raffa CM, Chiampo F. Bioremediation of agricultural soils polluted with pesticides: A review. Bioengineering 2021, 8, 7, 92.
[20] Pipit A, Rinanti A. Endosulfan insecticide removal planning with bioaugmentation-landfarming bioremediation method. IOP Conf Ser Earth Environ Sci 2021, November, 894, 1, 012042, IOP Publishing.
[21] Wang H, Liu Y, Li J, et al. Biodegradation of atrazine by Arthrobacter sp. C3, isolated from the herbicide-contaminated corn field. Int J Environ Sci Technol 2016, 13, 257–262.
[22] Asemoloye MD, Ahmad R, Jonathan SG. Synergistic rhizosphere degradation of γ-hexachlorocyclohexane (lindane) through the combinatorial plant-fungal action. PLoS One 2017, 12, 8, e0183373.
[23] Levi S, Hybel AM, Bjerg PL, Albrechtsen HJ. Stimulation of aerobic degradation of bentazone, mecoprop and dichlorprop by oxygen addition to aquifer sediment. Sci Total Environ 2014, 473, 667–675.
[24] Vikrant K, Giri BS, Raza N, Roy K, Kim KH, Rai BN, Singh RS. Recent advancements in bioremediation of dye: Current status and challenges. Bioresour Technol 2018, 253, 355–367.

Mardiana Idayu Ahmad*, Nur Kamila Ramli, Nur Anis Ahmad, Harlina Ahmad

Pesticides and Risk Assessment in Agriculture

Abstract: With the advent of the industrial revolution and the production of numerous pesticides such as organophosphates (OPs), organochlorine (OC), carbamates, and pyrethroids, our agricultural productivity has undoubtedly increased, and the bulk of our crops are now protected against pests. Due to their very persistent nature, the majority of pesticides and their residues tend to accumulate in the environment, where they can have detrimental impacts on human health and diverse ecosystems. Even in low concentrations, the resulting pesticide mixtures can be hazardous to human health. The considerable increase in agricultural pesticide use has aroused concerns about potential toxicity, hazards, and negative consequences on the environment and human health, especially in countries where rules are not properly implemented, and farmers' understanding of safe handling methods is often lacking. The continuous exposure to pesticides in farming activities poses a threat to the health of farmers, making it imperative to monitor and evaluate occupational exposure to pesticides. Risk assessments must be conducted in order to prevent the problems from getting even more severe. This chapter is intended to focus on the following topics: history and classification of pesticide, harmful effects of pesticide on humans and the environment, and risk assessment of pesticide in agriculture. In addition, it covers the classification of pesticides in terms of their target pests, chemical group, and active components, as well as agricultural risk assessment studies. It is vital to conduct risk assessments in order to avoid the problems from becoming even more serious. This chapter is structured to focus on the following topics: the history and categories of pesticides, the adverse effects that pesticides have on both human and the environment, and the risk assessment of pesticide use in agricultural settings. It also discusses pesticide classification in terms of target pests, chemical groups, and active components, as well as studies of risk assessment in agriculture.

Keywords: Pesticides, agriculture, risk assessment

1 Introduction

The term "pesticide" refers to a class of toxic and hazardous substances that are formulated to either eliminate a wide variety of undesirable organisms or bring under

***Corresponding author: Mardiana Idayu Ahmad,** Environmental Technology Division, School of Industrial Technology, Universiti Sains Malaysia, Penang 11800, Malaysia, e-mail: mardianaidayu@usm.my
Nur Kamila Ramli, Nur Anis Ahmad, Harlina Ahmad, Environmental Technology Division, School of Industrial Technology, Universiti Sains Malaysia, Penang 11800, Malaysia

https://doi.org/10.1515/9783111016825-005

control the numbers of pests that are native to the affected regions. Since 1950, pesticides have been the preeminent and most well-known method used around the globe for the elimination of certain species that are considered to be pests [1]. It is projected that 5.2 billion pounds of pesticides were used around the globe in 2006, and the same amount was used the following year [2]. There are approximately 10,000 distinct formulations of pesticides, and these formulations comprise over 800 different active compounds [3, 4]. Pesticides may be broken down into many classes of chemicals. Farmers in developing nations have been exposed to widespread advertising for pesticides and have used them extensively in agriculture. Additionally, it may be utilized for the management of product quality as well as the elimination or reduction of yield loss [5]. Insecticides, fungicides, herbicides, rodenticides, molluscicides, and nematicides are the typical categories into which pesticides are placed according to the species that they are designed to kill [6].

Although pesticides are effective crop protection agents that help ensure a broad diversity of food sources, there is the possibility that they may have harmful impacts on both the environment and the health of humans [7]. Farmers in poor nations face a number of significant challenges, and some of the most significant of them include inadequate sprayer maintenance, incorrect spraying practices, unsuitable sprayers, and the pesticide usage that is prohibited or outlawed in other countries [8]. A few pesticides that are not taken up by plants are passed on to the surrounding environment through the soil, water, and air. It has the potential to taint human and animal blood, as well as milk, meat, and plant tissue. In addition, due to the physicochemical features they possess, they are able to be transmitted to locations located hundreds of kilometers distance.

In addition, there is a substantial body of information that can be gleaned from research conducted in the past that suggests that pesticides not only provide a significant potential danger to humans but also have unfavorable and detrimental impacts on the environment and other living things [9, 10]. Extensive research has shown that a number of accidents and cases of poisoning have occurred as a direct result of exposure to environmental dangers [11]. The number of accidents was affected by a variety of different variables. For instance, the vast majority of small farmers lack awareness regarding the fundamental safety measures that should be taken while applying and handling pesticides, as well as the hazards associated with the improper use of pesticides [12].

Humans can be exposed to pesticides in a variety of ways, including proximity to agricultural regions, consumption of polluted water and food, and occupational exposures. Skin contact, inhalation, and ingestion are distinct routes of pesticide exposure for pesticide users in the workplace [13]. Exposure to some of the pesticides that are suspected to be endocrine disruptor chemicals (EDCs) can have harmful effects once they enter the body. This is due to the fact that these pesticides can disrupt the body's hormones. For example, impairment of reproductive function and embryonic development in human and wildlife is one of the adverse effects of endocrine disruptors [14]. Humans can be exposed to pesticides through a variety of different channels, such as being in the vicinity of plantation areas, drinking polluted water and food, or being exposed to pesticides on the job. When it comes to occupational exposure to pesticides, there are a few distinct sorts of routes that might occur. These pathways include skin contact, inhalation, and ingestion. In addition, the total pesticide exposure is necessary for determining the dangers posed by pesticides when all and multiple exposure pathways are taken into consideration [13]. When within the body, certain pesticides that are thought to be EDCs may have adverse consequences if the person is exposed to them. This is due to the fact that these pesticides may wreak havoc on the body's hormonal balance. Some of the negative impacts of endocrine disruptors include, for instance, a reduction in the ability of humans and other animals to reproduce and an interruption in the development of embryos [14].

Regulatory agencies for marketing pesticides are actively collecting and distributing data on the harmful effects of active ingredients on human health [5, 15], despite the fact that estimating the hazards of pesticide use is challenging for several reasons. Since the 1940s, the concept of risk assessment has been advocated for use all over the globe. Scientific data analysis is utilized in risk assessment, an instructive technique that describes the size, shape, and features of potential dangers to people and the environment [16]. To ascertain the extent to which a chemical poses a threat to human health, an evaluation was carried out using a set of established principles and methods for gathering and analyzing relevant information [17]. There have been some risk assessment studies, which have mostly focused on the effects of pesticides on humans. Therefore, it is essential to conduct a risk assessment of potential exposure to pesticides, taking into account end-user knowledge, attitudes, and habits in order to identify potential consequences and circumstances. It is necessary and is of utmost importance that comprehensive pesticide risk mitigation plans for good agricultural practices need to be developed. A risk assessment is a process that involves identifying and characterizing various hazards and risks which can be carried out in four phases: (i) identification of hazards; (ii) characterization of hazards; (iii) exposure assessment; and (iv) risk characterization [5, 18]. Thus, the interaction between people and the environment is a vital part of risk assessment programs. It involves the use of knowledge and practices to minimize the risks associated with human health.

Farmers who are exposed to pesticides for extended periods of time put themselves at risk for a variety of unfavorable health consequences, including those related to the respiratory system, reproductive and developmental damage, neurological illness, and cancer [19–21]. As a result, many scientists study pesticides in these settings, focusing on issues including risk assessment [22], environmental health hazard and exposure [23], toxicology [24], and soil [25] and water [26] contamination by pesticides. The research to explore and analyze knowledge, attitudes, and practices related to the use and handling of pesticides worldwide has also been done, the world to better understand the working environment and working conditions of agricultural workers [27]. Many severe environmental and human health impacts of pesticides have been recorded globally, but only several studies have targeted the possible hazards of pesticides by establishing risk assessment methodologies with reference to farmers' knowledge, attitudes, and actions

[28]. In order to prevent the escalation of these issues, risk assessment is essential. This chapter has been structured to focus on the following topics: history and pesticide categories; adverse effects of pesticides on human and environment; and risk assessment of pesticides in agriculture.

2 History and pesticide categories

Traditional approaches to agricultural production have largely been rendered obsolete by the widespread use of chemical pesticides since the 1950s. There are various different categories of pesticides, and these types span 10,000 different formulations, each of which contains more than 800 active chemicals [3, 29, 30]. More than 500 distinct pesticide formulations are utilized in agricultural practices around the globe [31]. Synthetic insecticides first appeared on the global market between 1960 and 1980. For instance, throughout the 1940s and 1950s, OC insecticides saw widespread use. This is because of its significantly increased contribution to agricultural productivity as well as insect management. Pesticides containing OC are effective against illnesses such as malaria and typhus. Unfortunately, OC pesticides were restricted or banned in 1960, particularly in countries that were technologically advanced [6] and developing countries [32], after the long-term side effects of these chemicals were studied and understood. After that, in the year 1960, OPs were first made available [6]. As a result of their mechanism of action, which involves the inhibition of acetylcholinesterase in the nervous systems of pests [33], OP pesticides have flourished and have now become one of the most widely used pesticides in the entire world [33]. The following year, 1970, saw the introduction of carbamate pesticides, which were then followed by pyrethroids in the year 1980. Insecticides known as carbamates destroy the nervous system of their targets by interfering with the activity of the enzyme responsible for regulating acetylcholine levels. The manufacture of pyrethroids, which target the nerve systems of pests, came about in the realm of agrochemical research and development [34]. Between 1970 and 1980, publications of fungicides and herbicides were made [6, 35]. When applied to plantation, fungicides inhibit or kill the growth of spores and fungi by disrupting the membranes of fungal cells and interfering with the production of energy by fungal cells [34, 36]. This allows the fungicide to kill or inhibit the growth of spores and fungi. On the other hand, herbicides have been utilized to get rid of undesired vegetation and weeds that have developed into rivals for crops. Because of the drug's existence, it quickly spread across the area. Since then, companies that make pesticides have endeavored to develop herbicides that are able to eradicate and kill unwanted weeds without having a negative impact on crops. During this time period, up to the turn of the twentieth century, an enormous number of selective herbicides were found, developed, and brought to market across the world [36, 37].

In general, pesticides may be divided into two categories: (i) the pests that they are intended to control, and (ii) the primary groupings and the chemical structures that they share [38]. First, the target pest categorization organizes pesticides according to the kinds of pests that they are effective against. The meaning of the suffix "-cide" can be simplified to "kill." Today, many different types of pesticides are utilized, including fungicides, defoliants, desiccants, bactericides, herbicides, insecticides, miticides, molluscicides, nematicides, plant growth regulators, rodenticides, and wood preservatives [38]. There are a few different categories that pesticides fall into. For instance, bactericides and fungicides are both chemicals that are used to eliminate the presence of bacteria and fungi, respectively [39]. Herbicides are most commonly and extensively utilized in agriculture for the purpose of weed control and elimination. At the same time, pesticides are utilized all over the world in an effort to regulate insect populations.

Miticides or acaricides, in addition to herbicides and insecticides, are used to control mites and ticks. On plantations, molluscicides are widely used to control snails and slugs. Nematicides are used to control nematodes. Rodenticides have been utilized for rodent control [29]. Defoliants and desiccants are both used in agriculture to reduce plant foliage. Wood preservatives are also used to counteract wood-destroying organisms. As a plant growth process, plant growth regulators are employed to manage pests [40]. When classifying pesticides into major groups with comparable chemical structures, pesticides belonging to the same major group typically contain active chemicals with identical chemical structures [40]. However, there are differences across the categories in terms of application volume, substance, and usage. Organophosphorus compounds, for instance, have approximately 80 distinct active components, including phosphonate, phosphate, phosphorodithioate, and phosphorothiolate. Nonetheless, certain pesticides, such as imidazolones, have few active components [41]. Pesticide compounds containing pyrethroids, organophosphorus, and OC are frequently utilized as acaricides, insecticides, or nematicides, respectively. In addition, triazole and conazole are subgroups of the azole group of pesticides, which are typically used as fungicides. These pesticides fall within the azole category. Carbamates are versatile chemicals that may be used in a variety of applications, including as insecticides and herbicides. In addition, carbamates, amides, and triazines are classes of pesticides that are frequently utilized as herbicides. These pesticides are typically applied to crops such as rice, soybeans, and maize [41]. Classifications of pesticides based on their chemical structures typically have the same features and modes of action that are responsible for their lethal effect [40]. The categorization of pesticides may be found in Table 1, which is organized according to the primary classes and subgroups of pesticides.

The World Health Organization (WHO) has made a list of pesticide categories that can be used as a guide.

Table 1: Classification of pesticides based on main pesticide groups and subgroups [41].

Groups	Related subgroups
Amide	Chloroacetanilide, acylanaline, dichloroanilide
Aryloxyalkanoic acid	Phenoxacetic acid and salts
Azole	Conazole, triazole
Carbamate	Carbamate, dithiocarbamate, dimethyldithiocarbamate, thiocarbamate
Diazine	
Dinitroaniline	
Diphenyl ether	
Imidazolinone	
Organophosphorus	Phosphonate, phosphate, phosphorodithionate, phosphorothiolate, phosphoroamidate
Organochlorine	
Pyrimidine	
Pyrethroid	
Triazine	1,3,5-Triazinone, 1,3,5-triazine, 1,2,4-triazine
Sulfonylurea	
Urea	Phenylurea
Various	Single structure, no specific structure (benzoic acid derivatives)

The classification is based on the acute toxicity hazard categories of the Globally Harmonized System of Classification and Labeling of Chemicals [42]. Additionally, the WHO classification of pesticides is being extensively disseminated through discussions with regional and international organizations as well as governments all over the world [43]. This is done since the classification undergoes periodic updates from time to time. The categorization of pesticides according to the WHO recommendations is given in Table 2. This classification is based on several classifications.

Table 2: Classification of pesticides [43].

World Health Organization (WHO) classification		LD 50 for rats (mg/kg body weight)	
		Oral	**Dermal**
Ia	Extremely hazardous	<5	<50
Ib	Highly hazardous	5–50	50–200
II	Moderately hazardous	50–2,000	200–2,000
III	Slightly hazardous	Over 2,000	Over 2,000
U	Unlikely to present acute hazard	5,000 higher	

3 Adverse effects of pesticides on humans and environment

The use of pesticides is necessary in order to accomplish the benefits that are desired without causing harm to ecosystems or the environment [23, 44]. Unfortunately, when pesticides are used in activities, they not only adhere to things but also indirectly harm groundwater, rivers, lakes, and the surrounding surroundings [45–48]. In addition, there is evidence from the literature which mentioned that humans, organisms, and the environment are at potential risk of adverse effects of pesticides [6]. Continued use of pesticides can also lead to significant environmental damage in the long term. Pesticides adversely affect biodiversity and animal species both within and outside the areas to which they are applied [23].

It has been shown that pesticides used in agriculture over the past 30–40 years have caused chronic difficulties among agricultural workers [49]. Exposure to pesticides in the workplace, for instance, can be destructive to the respiratory system, eyes, and can cause death [50, 51]. It can also result in cancer, endocrine disturbance, reproductive dysfunction, and neurotoxicity [24]. Farmers exposed to pesticides experienced abdominal discomfort, trouble breathing, rashes, numbness, diarrhea, red eyes, fatigue, sneezing, and muscular cramps, according to a study conducted in the central Peninsular of Malaysia [52].

Pesticides have detrimental environmental effects, including pollution of the air, water, and soil, as well as nontarget organisms [53, 54]. In general, pesticide misuse can result in species population reduction, water pollution from drifting or migrating chemicals, agricultural damage from excessive pesticide usage, air pollution from volatile pesticides, and so on. [5]. It has been shown that the use of pesticides has harmful effects on animals and plants as well as insects [23].

4 Risk assessment of pesticide

Pesticide risk is defined as the likelihood of an adverse event (loss, damage, or death) occurring as a result of exposure to a physical or chemical agent under certain conditions [55]. Furthermore, the risk is defined as a mix of hazard and exposure [56]. The risk equation is as follows:

$$\text{Risk} = \text{hazard} \times \text{exposure}$$

A hazard is a collection of features contained in a substance, a mixture of materials, or a process involving a material that, under the conditions of manufacturing, use, or disposal, has the potential to influence the environment or organisms negatively dependent on the amount of exposure [57]. Some researchers have underlined that a danger is an occurrence that might negatively affect terminal receptors, such as people [56]. Exposure, on the other hand, can be described as the amount, concentration, or intensity with which a specific environmental, physical, or chemical agent impacts an intended target organism, population, organ, tissue, or cell, based on the duration, frequency, and concentration of the substance [55].

Risk assessment may be defined as assessing and defining the risk repercussions of a certain occurrence or application, whether caused by physical or chemical causes. For instance, physical agents can be recognized as risks in the workplace, whereas dangerous substances can be identified as chemical agents [55]. Moreover, risk assessment has attracted considerable international attention since 1970, both from the perspective of regulatory agencies and the scientific method [58]. A risk assessment involving any or all these elements essentially involves four processes: (i) hazard identification, (ii) effect assessment, (iii) exposure assessment, and (iv) risk characterization [59, 60]. There are three important factors involved in risk assessment: (i) identified contaminant (source), (ii) pathway, and (iii) receptor. When these three factors interact with one another, there is potential for risk [61]. It has been stated that sources may be described as a material that has been deposited beneath, in, or on land, and that has the potential to have a negative impact on water resources, the environment, and human health. Pathway, on the other hand, might be interpreted as the route that pollutants take as they migrate or travel [62]. These procedures could have an effect on the receptors or expose them to pollution. In a nutshell, things that are subjected to the negative impacts of pollutants are referred to as receptors. People, bodies of water, the ecosystem, or even property may all act as receptors. A risk assessment element can be written as shown in Figure 1.

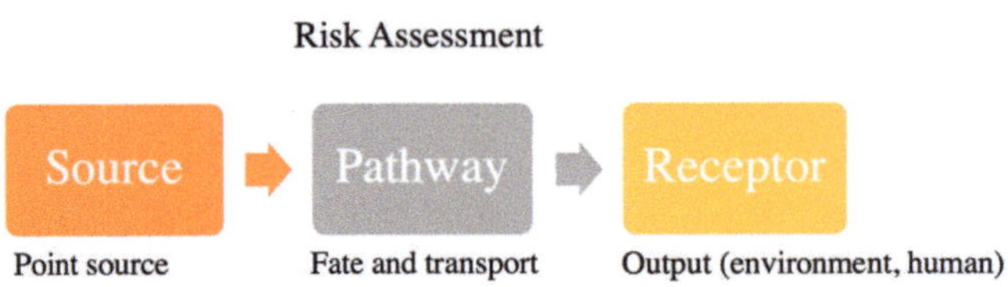

Figure 1: The relationship between the source, pathway, and the receptor in the context of risk assessment.

4.1 General steps in risk assessment

Before doing a risk assessment, it is necessary to first analyze the problem formulation. Problem formulation includes the phases of planning and defining the scope. In this stage, the risk assessor is responsible for establishing the scope, objective, and technical components to be utilized [59]. The risk is then determined. Hazard identification is the process of identifying the undesirable consequences of substances. Typically, risk and hazard are differentiated by the probability of harm resulting from exposure [63]. At this level, evaluators decide if the stressor may have negative impacts on the ecosystem or on humans under certain conditions [60].

After identifying hazards, the dose–response evaluation is the following step. A dose–response evaluation evaluates the severity and likelihood of adverse health consequences (responses) in proportion to the exposure conditions and levels of a particular dosage [60]. It is possible for the dose–response connection to be as simple as a linear relationship, or it might be rather complicated. In addition, this stage focuses on the regulation of harmful substances and their effects on species, including humans [59].

The risk assessment technique is then applied to the exposure evaluation. The purpose of exposure assessment is to determine the extent to which animals, people, and other species are exposed to dangerous chemicals. In order to estimate chemical exposure levels, the duration, frequency, and concentration of compounds present in the environment can be quantified [59]. Moreover, exposure information was required for two primary situations: accidents and anticipated exposures [55]. In exposure evaluations, the frequency, intensity, and duration of pathogen exposure are also quantified and estimated.

Risk identification is the final phase of the risk assessment process. Risk identification is the process of identifying the level of threat to the environment, human health, and other kinds of life in the wild. Complete the risk identification procedure with an exposure evaluation and effect result analysis. Risk identification incorporates both ecological and human health risk assessments. The risk identification includes the risk assessors' thoughts on the nature of the risk and the existence or absence of dangers, as well as information on how the risk was evaluated [60]. Figure 2 depicts the four risk assessment phases.

4.2 Significance of risk assessment

Risk assessment is one of the necessary items for handing chemical substances. Both the pure material and the active components in chemical products can cause immediate and chronic health harm. In general, a risk assessment should be conducted to establish the threat that chemical users pose to human health. According to previous studies, chemicals used in commercial home items and the industrial sector lead to various human health hazards, including respiratory impacts and central nervous system toxicity [64]. In addition, all critical parts of risk assessment and risk identification are required to handle human health risk concerns. This is because risk assessment may be utilized as a technique for assessing public health [65].

In addition to human health hazards, risk assessment must also consider ecological threats. The ecological assessment is an essential evaluation to guarantee the protection of biodiversity [66]. Assessments of ecological risk are performed to determine the balance between environmental sustainability and economic development. In

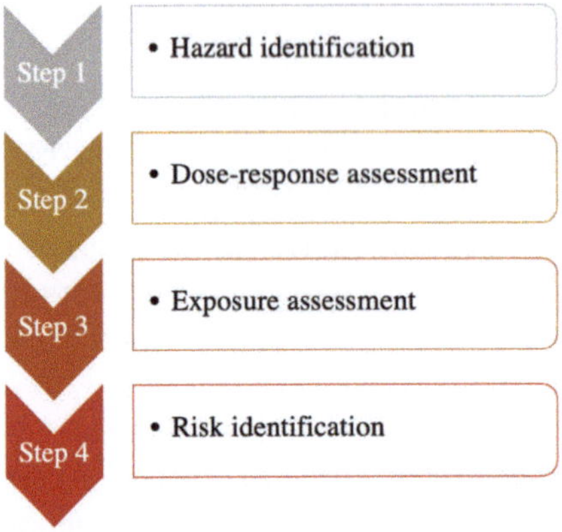

Figure 2: Risk assessment steps.

addition, chosen parties, such as ecological stressors, the public, and chemical users, require the data generated from ecological risk assessments in order to preserve the ecological integrity of the environment [67].

5 Risk assessment studies of pesticides in agriculture

In today's day and age, there are a plethora of risk assessments and occupational exposure studies being carried out among employees all over the world. Accidents at work with pesticides, such as leaks, spills, or damaged application equipment, might result in employee exposure to chemicals. Negligence of basic hygiene practices such as washing hands before eating or after handling pesticides, failure to emphasize personal protective equipment (PPE) in basic safety guidelines, and overlooking the correct methods of pesticide usage and handling all contribute to an increase in worker exposure [68]. The intake of food and water is often how the general population is exposed to pesticides. Owing to the small distance between their office and their houses, people have a significant risk of being exposed to pesticides because they bring contaminated objects home with them from their jobs [69].

For instance, a risk assessment study was carried out in Vietnam with the purpose of determining occupational exposure to pesticides in relation to safety practices such as correct application techniques, proper pesticides disposal techniques, correct use of PPE, and adherence to label instructions. According to the findings of these investigations that were carried out, the health of farmers was at a significant threat when they were exposed to pesticides [70]. Furthermore, a health risk evaluation of organophosphorus (OP) pesticides in Thai vegetative agriculture revealed that the agricultural populations were more likely to be exposed to OP pesticides than nonagricultural groups [71]. Pesticides like chlorpyrifos, dicrotophos, and profenofos are among them.

5.1 Fruits and vegetables

Risk assessment and occupational exposure to agricultural pesticides are extensively done around the globe, particularly in emerging nations such as Asian countries, with the goal of reducing the potential for adverse health effects and protecting the natural environment. For instance, in Thailand, farmers were subjected to an occupational exposure and health risk assessment in order to investigate the impacts of OP pesticides such as dicrotophos, chlorpyrifos, profenofos, and parathion methyl [72]. In order to keep up output throughout this trial, growers sprayed tropical fruits with pesticides when they were still in the planting stage. Acute symptoms such as impaired vision, respiratory arrest, lung edema, muscular weakness, and difficulty breathing were documented in health risk evaluations [7, 73]. A research on the efficacy of pyrethroid and carbamates was conducted and it was found that during the process of applying pesticides in the field, assessments were carried out, and the results revealed a number of aspects that may contribute to possible health hazards [74].

5.2 Stable crops

Around the world, there are around 50,000 plants that have been planted. Rice, maize, and wheat are the only three plants that are regarded to be staple foods, and this is out of a total of 50,000 crops. According to the Food and Agriculture Organization of the United Nations, staple foods are defined as those items that are consumed on a consistent basis and form the basis of the diet for particular communities all over the world. Approximately 60% of the daily energy intake required for human nutrition comes from staple foods [75].

Rice (*Oryza sativa* L.) is the world's most common grain, feeding almost half the population [76]. Rice is a significant staple item that fulfills the dietary requirements of over 70% of the population in India, Southeast Asia, East Asia, Latin America, and Africa [77]. Rice is widely consumed in these regions. In order to keep up with and improve rice production, there is a need for a diverse selection of pesticides, which are used to eliminate rice-destroying organisms such as insects, diseases, fungal, bacterial, and weeds [76]. In light of this, it is essential to carry out a risk assessment and investigate the occupational exposure to pesticides in food staples. A study on risk assessment was carried out among paddy smallholder farmers in the Cagayan Valley of the Philippines in order to evaluate 20 various kinds of pesticides [23]. This study was carried out in order to corroborate the data that was presented earlier. The possible threat was evaluated using a guideline for conducting potential risk assessments that was developed by the Environmental Protection Agency in the United States.

In addition to rice, maize, also known as *Zea mays* L., is another important cereal grain that is utilized all over the world. The year 2008 saw the production of more than 750 million metric tons of maize, with the United States being the biggest provider in the world, followed by the European Union, China, Brazil, Mexico, and India [78]. As a result of the rising demand, enormous volumes of pesticides were utilized in the fight against illnesses and vermin [79]. Approximately 87% of farmers reported using pesticides in order to protect their maize crops from various insects and illnesses [80]. During the process of maize cultivation, farmers were found to be significantly exposed to systemically hazardous pesticides such as endrin, dieldrin, aldrin, chlorfenyinp, and c-chlordane, according to the findings of a health risk assessment that was carried out [79].

Wheat is an additional key and vital staple food for the population all over the world. Rice and maize are the other two primary staple foods. Wheat has contributed to 30% of the world's average plant use with a total quantity of 680 million in the year 2009. In Germany, an investigation of the effects of pesticides on human health was carried out among those who apply chemicals. It was established how much exposure to pesticides a person had based on the pathways of exposure, which included ingesting, breathing in, and emitting chemicals when pesticides were being sprayed. During the course of this evaluation, a number of different pesticides, including prochloraz, tebuconazole, chlorothalonil, cyproconazole, deltamethrin, and pirimicrab [81], were utilized.

The aforementioned research suggests that pesticides are widely utilized around the world, particularly in agricultural areas. Even though pesticides are commonly used, they may have dangerous and damaging effects on the environment and human health [82]. Movement of pesticides into surface water and soil, which can pollute the food chain, creates environmental problems. In addition, there are dangers to human health when agricultural workers are directly exposed to pesticides during pesticide application. Moreover, people are indirectly exposed to pesticide residues on food [6]. Consequently, risk assessment is required to control risks, decrease workplace incidents, and establish risk management standards in accordance with acceptable legal requirements and safe practices.

6 Pesticide risk minimization strategies

6.1 Safety measures

Priority should be placed on safety when exposed to pesticides. Pesticides, such as vapor drift and concentrates, pose a considerable danger of harming the environment and persons, such as bystanders and employees. Consequently, it is essential to limit risks by implementing suitable security measures. Approximately 20,000–40,000 occupational poisonings per year are attributable to chemicals, including pesticides [83]. Poisoning by chemicals can be caused by a number of circumstances, including inadequate equipment and a lack of PPE [84]. Therefore, appropriate PPE and adequate pesticide handling procedures are crucial safety precautions. According to a previous study, applicators are required to provide decontamination stations, training, and the proper PPE to mitigate chemical risks [85]. Important safety measures include the use of appropriate PPE (e.g., face scarves, coveralls, appropriate gloves, and shoes) and best practices (e.g., bathing after work, not eating during work, washing clothes separately, and properly disposing of pesticide containers) [86]. PPE is utilized to minimize dangers and reduce risks, particularly in agricultural operations worldwide. The Worker Protection Standards mandate the use of specified PPE by applicators while handling, mixing, or applying chemicals [87].

Coverall protection is one of the proper PPEs. During agricultural fieldwork, chemicals can enter the body without protective equipment. Transdermal absorption through the skin, for instance, is one of the primary mechanisms by which pesticides enter the human body [86]. Therefore, it is crucial to utilize coveralls as safety precautions. Several studies involving pesticide exposure demonstrated that coveralls are much more protective than pants and shirts alone [88]. Risk assessments completed by additional writers validated this statement. Coverall protection reduces cutaneous exposure among pesticide workers compared to those who simply wear T-shirts while handling, according to Cattani et al. [89]. In addition, improper usage of the face increases the danger of exposure to the face, particularly the eyes, which can result in instant harm [90]. Respiratory diseases such as airway inflammation, coughing, and wheezing [91] and chronic respiratory diseases such as lung cancer, asthma, and chronic obstructive pulmonary disease [92, 93] are typically caused by occupational exposures to pesticides through inhalation. To avoid respiratory infections in agriculture, respiratory protection must be used. Respiratory masks are used to decrease respiratory distress caused by human inhalation and are indirectly a significant agricultural safety precaution [94].

When working with pesticides, one of the safety considerations is to comprehend pesticide labels. The majority of information on critical chemicals is derived from pesticide labels. Labels on pesticides give vital information on their safe usage, therefore, reducing health risks and preventing massive environmental damage [87]. Also of issue is pesticide-handling safety precautions. Risk exposure reduction is influenced by the formulation of the pesticide used. Using water-soluble chemical packages to combine chemicals is a safer alternative to injecting concentrates directly into employees [88] compared to open injection of concentrates.

6.2 Personal protective equipment (PPE)

PPE is a crucial factor in limiting pesticide's harmful effects to farmers. Goggles, respirators, and protective overclothes have been identified as PPEs that can aid in risk reduction [95]. Different protective equipment uses and advantages vary. Chemical-resistant clothes, for example, can be used for skin protection to keep pesticides from moving through fabrics during pesticide contact. Multiple studies in Ghana discovered that farmers who did not use protective gear were more vulnerable to pesticide exposure

[96]. Furthermore, waterproof gloves can limit the possibility of increased pesticide exposure. Inadequate protection, including using leather or canvas footwear when working with pesticides, contributes to increased exposure levels. As a result, wearing chemical-resistant footwear during pesticide applications is strongly encouraged. Furthermore, to protect the eyes from irritants, protective eyewear can be used. A respirator is another important piece of PPE that can help farmers avoid inhaling harmful air.

Using a mask may limit pesticide inhalation exposure [97]. Also, a study conducted in remote Indonesia revealed that farmers had increased dermal exposure to pesticides when no respirator was used while they wore short-sleeved, wet clothing while being exposed to pesticides [98]. In addition, PPEs should be maintained according to the labels' instructions, and their usage should be monitored. Another study demonstrated that farmers who wear right PPEs tend to practice greater cleanliness when handling pesticides [99].

6.3 Integrated pest management (IPM)

In addition to wearing the proper PPE, risk reduction strategies including integrated pest management (IPM) offer cost-effective practices that employ a number of practices and approaches to prevent pest infestation. IPM aids in the reduction of pesticide applications. For example, the IPM method is to select pesticides with lower concentrations and volatilities of active components to limit occupational pesticide exposure through the respiratory system [100]. When an IPM system is implemented, the impacts of pesticides on the environment and human health will automatically decrease, and plant yield will increase [53, 101, 102]. In accordance with IPM, pesticides are chosen based on a predetermined set of criteria. In addition, guidelines must be followed while applying insecticides to plantations. First, IPM pesticides are chosen for their biological effectiveness, which includes low resistance risk, enhanced crop tolerance, the most ideal residual effects, and quick plantation impact. Insecticides must also be easy to use. Some user-friendly pesticide selection criteria include optimum formulation, simple administration, low chronic and acute toxicity to people, prolonged storage stability, and ease of use.

The IPM method calls for the use of pesticides that are not only user-friendly but also safe for the surrounding ecosystem. Pesticides are considered environmentally friendly if they degrade quickly in the environment, do not leave residues in food or fodder that are above the maximum residue levels, have low mobility in soil, are applied infrequently, and have a low level of toxicity to organisms that are not the intended target. In the last step of the IPM method, pesticides are chosen according to how profitable they are or how efficient they are economically. Pesticides that are economically efficient and lucrative have a broad range of action, novel product qualities, deliver better profits to farmers, are patented, and compete well in terms of market value [103].

6.4 Monitoring and awareness training program

A monitoring program is another strategy that ought to be explored for the purpose of reducing the potential dangers posed by improper use of agricultural or chemical techniques. For instance, organizations that are accountable for their actions have to play their part in the periodic monitoring of how pesticides are used and handled by farmers. This is because consistent monitoring has the potential to protect agricultural workers. Aside from that, manufacturers of pesticides and the organization responsible for their regulation are obligated to provide farmers with pertinent data and information on precautions, safe use of pesticides, and pesticide hazards. This is done with the goal of reducing the likelihood that farmers will be exposed to hazardous pesticides.

In addition, training on the appropriate use and management of pesticides should be stressed as a means of preserving and ensuring good agricultural practices among farmers. Furthermore, these safety trainings need to have an emphasis on the detrimental health consequences of pesticide exposure, the safe handling of pesticides, and the interpretation of pesticide labels, as the latter is the primary reference source that farmers use to gain information about pesticides [96]. Farmers should be provided with awareness training on safety measures as an alternate technique for reducing the potential for risk to excellent agricultural practices. This training should be performed. For instance, responsible organizations can run campaigns among farmers to increase their understanding of the dangers that pesticides pose to human health and the environment. These efforts should focus on the risks and hazards that pesticides pose.

It is essential for the responsible organization to run educational campaigns on safety attitudes and practices during the use and handling of pesticides in order to guarantee that farmers have a safe working environment when applying pesticides in their spraying operations. This will ensure that farmers have a safe working environment during the application of pesticides. Positive attitudes and healthy practices, such as wearing a hat while spraying, changing clothes immediately after applying, and not smoking while applying pesticides, have been proven to have a significant impact on minimizing the symptoms of acute pesticide exposure, particularly among male farmers.

7 Conclusions

Most pesticides and their residues accumulate in the environment, where they can harm the human health and ecosystems. When pesticides are combined, even little dosages might be dangerous. Significant development in pesticide usage in agriculture has raised concerns about toxicity, hazards, and bad impacts on the environment and human health, especially in countries where farmers' awareness of safe handling practices is limited, and laws are not effectively enforced. Due to the ongoing exposure of pesticides in farming, monitoring and evaluating occupational exposure is crucial. Risk assessment and occupational exposure to agricultural pesticides are done globally, especially in developing nations of Asia, to reduce harmful health consequences and safeguard the environment. Neglecting fundamental hygienic procedures and failing to stress personal protection equipment in basic safety rules all enhance workers' or farmers' exposure to harmful effects of pesticides. Besides, as part of risk minimization strategies, before being allowed on the market, new pesticides need to be tested and reviewed. Existing pesticides need to be reassessed on a regular basis in order to guarantee that they continue to satisfy the necessary level of safety requirements.

References

[1] Abhilash PC, Singh N. Pesticide use and application: An Indian scenario. J Hazard Mater 2009, 165, 1–12.

[2] U.S. EPA. Pesticide market estimate: 2006–2007 [Online]. Washington, DC. Available: http://www.epa.gov/opp00001/pestsales/07pestsales/table 2011 [Accessed February 14th, 2014].

[3] Hernández AF, Parrón T, Tsatsakis AM, Requena M, Alarcón R, López-Guarnido O. Toxic effects of pesticide mixtures at a molecular level: Their relevance to human health. Toxicology 2013, 307, 136–145.

[4] Shahid M, Khan MS. Ecotoxicological implications of residual pesticides to beneficial soil bacteria: A review. Pestic Biochem Physiol 2022, 188, 105272.

[5] Damalas CA, Eleftherohorinos IG. Pesticide exposure, safety issues, and risk assessment indicators. Int J Environ Res Public Health 2011, 8, 1402–1419.

[6] Aktar MW, Sengupta D, Chowdhury A. Impact of pesticides use in agriculture: Their benefits and hazards. Interdiscip Toxicol 2009, 2, 1–12.

[7] Panuwet P, Prapamontol T, Chantara S, Thavornyuthikarn P, Montesano MA, Whitehead RD Jr, Barr DB. Concentrations of urinary pesticide metabolites in small-scale farmers in Chiang Mai Province, Thailand. Sci Total Environ 2008, 407, 655–668.

[8] Asogwa EU, Dongo LN. Problems associated with pesticide usage and application in Nigerian cocoa production: A review. Afr J Agric Res 2009, 4, 675–683.

[9] Lozowicka B, Kaczynski P, Paritova AE, Kuzembekova GB, Abzhalieva AB, Sarsembayeva NB, Alihan K. Pesticide residues in grain from Kazakhstan and potential health risks associated with exposure to detected pesticides. Food Chem Toxicol 2014, 64, 238–248.

[10] Penagos-Tabares F, Sulyok M, Faas J, Krska R, Khiaosa-ard R, Zebeli Q. Residues of pesticides and veterinary drugs in diets of dairy cattle from conventional and organic farms in Austria. Environ Pollut 2022, 316, Part 2, 120626.

[11] Schreinemachers P. Prasnee Tipraqsa Agricultural pesticides and land use intensification in high-, middle- and low-income countries. Food Policy 2012, 37, 616–626.

[12] Recena PCM, Caldas DE, Pires XD, Pontes CJE. Pesticides exposure in Culturama, Brazil – knowledge, attitude, and practices. Environ Res 2006, 102, 230–236.

[13] EFSA. Scientific opinion of the panel on plant protection products and their residues (PPR Panel) on a request from the EFSA to evaluate the suitability of existing methodologies and, if appropriate,

the identification of new approaches to assess cumulative and synergistic risks from pesticides to human health with a view to set MRLs for those pesticides in the frame of Regulation (EC) 396/2005. EFSA J 2008, 704, 1–85.
[14] Wojciech H, Joanna J. The risk of adverse reproductive and developmental disorders due to occupational pesticide exposure: An overview of current epidemiological evidence. Int J Occup Med Environ Health 2004, 17, 223–243.
[15] Temkin AM, Uche UI, Evans S, Anderson KM, Perrone-Gray S, Campbell C, Naidenko OV. Racial and social disparities in Ventura County, California related to agricultural pesticide applications and toxicity. Sci Total Environ 2022, 853, 158399.
[16] Leeuwen CJV, Vermeire TG. Risk assessment of chemicals: An introduction. Springer Science & Business Media, The Netherlands, 2007.
[17] Van Engelen JGM, Hakkinen PJ, Money C, Rikken MGJ, Vermeire TG. Human exposure assessment. In: Van Leeuwen CJ, VERMEIRE TG (eds.). Risk assessment of chemicals: An introduction. Springer, The Netherlands, 2007.
[18] Maino JL, Thia J, Hoffmann AA, Umina PA. Estimating rates of pesticide usage from trends in herbicide, insecticide, and fungicide product registrations. Crop Prot 2022, 163, 106125.
[19] Bassil KL, Vakil C, Sanborn M, Cole DC, Kaur JS, Kerr KJ. Cancer health effects of pesticides. Can Fam Phys 2007, 53, 1704–1711.
[20] Kanavouras K, Tzatzarakis MN, Mastorodemos V, Plaitakis A, Tsatsakis AM. A case report of motor neuron disease in a patient showing significant level of DDTs, HCHs and organophosphate metabolites in hair as well as levels of hexane and toluene in blood. Toxicol Appl Pharmacol 2011, 256, 399–404.
[21] Parrón T, Requena M, Hernández AF, Alarcón R. Association between environmental exposure to pesticides and neurodegenerative diseases. Toxicol Appl Pharmacol 2011, 256, 379–385.
[22] Matthew KC. Effectiveness of interventions in reducing pesticide overexposure and poisonings. Am J Prev Med 2000, 18, 4, Supplement 1, 80–89.
[23] Snelder DJ, Masipiqueña MD, De Snoo GR. Risk assessment of pesticide usage by smallholder farmers in the Cagayan Valley (Philippines). Crop Prot 2008, 27, 747–762.
[24] Curwin BD, Hein MJ, Sanderson WT, Striley C, Heederik D, Kromhout H, Reynolds SJ, Alavanja MC. Urinary pesticide concentrations among children, mothers and fathers living in farm and non-farm households in Iowa. Ann Occup Hyg 2007, 51, 53–65.
[25] Đurović-Pejčev R, Đorđević T, Radivojevic L, Šantrić L, Gajić-Umiljendić J. Multiresidue Analysis of Pesticides in Soil by Liquid-Solid Extraction Procedure. Pestic Phytomed 2012, 27, 239–244. 10.2298/PIF1203239D.
[26] Vega BA, Frenich GA, Martinez-Vidal JL. Monitoring of pesticides in agricultural water and soil samples from Andalusia by liquid chromatography coupled to mass spectrometry. Anal Chim Acta 2005, 538, 1–2, 117–127.
[27] Karunamoorthi K, Mohammed A, Jemal Z. Peasant association member's knowledge, attitudes, and practices towards safe use of pesticide management. Am J Ind Med 2011, 54, 965–970.
[28] Yassin MM, Abu-Mourad TA, Safi JM. Knowledge, attitude, practice, and toxicity symptoms associated with pesticide use among farm workers in the Gaza Strip. Occup Environ Med 2002, 59, 387–393.
[29] Mardiana IA, Nur-Anis A, Syahidah-Akmal M, Norizan E. A survey on use, hazards and potential risks of rice farming pesticides in Permatang Keriang, Pulau Pinang (Malaysia). Int J Sci Res Publ 2014, 4.
[30] Zhang J, Sheng X, Cao J, Fang S, Liu X, Liu X, Peng L, Sun H, Weng R. Occurrence and risk exposure assessment of multiple pesticide residues in edible mint in China. J Food Compos Anal 2022, 116, 105071.
[31] Azevedo ASON. Assessment and simulation of atrazine as influenced by drainage and irrigation. An interface between RZWQM and ArcView GIS. Iowa State University, 1998.

[32] Çok I, Bilgili A, Özdemir M, Özbek H, Bilgili N, Burgaz S. Organochlorine pesticide residues in human breast milk from agricultural regions of Turkey, 1995–1996. Bull Environ Contamin Toxicol 1997, 59, 577–582.

[33] Waddell B, Zahm S, Baris D, Weisenburger D, Holmes F, Burmeister L, Cantor K, Blair A. Agricultural use of organophosphate pesticides and the risk of non-Hodgkin's lymphoma among male farmers (United States). Cancer Causes Control 2001, 12, 509–517.

[34] U.S. EPA. Types of pesticides [Online]. Available: http://www.epa.gov/pesticides/about/types.htm 2012b, [Accessed May 12, 2014].

[35] USGC NWHC. Chemical toxins: Organophosphorus and carbamate pesticides. In: Ciganovich EA, FRIEND M, Franson JC (eds.). Field manual of wildlife diseases: General field procedures and diseases of birds 2013. Washington, D.C.

[36] National Pesticide Information Center (NPIC). Fungicides [Online]. Available: http://npic.orst.edu/ingred/ptype/fungicide.html 2013, [Accessed May 24, 2014].

[37] Duke SO. Taking stock of herbicide-resistant crops ten years after introduction. Pest Manage Sci 2005, 61, 211–218.

[38] U.S.EPA. What is Pesticide? [Online]. Available: http://www.epa.gov/pesticides/about/#what_pesticide 2014, [Accessed May 21, 2014].

[39] Khalidah AL, Nor-Kartini AB, Nurzawani MI. Preliminary study of difenoconazole residues in rice paddy watersheds. Malays J Sci 2010, 29.

[40] Arias-Estévez M, López-Periago E, Martínez-Carballo E, Simal-Gándara J, Mejuto J-C, García-Río L. The mobility and degradation of pesticides in soils and the pollution of groundwater resources. Agric Ecosyst Environ 2008, 123, 247–260.

[41] Barcelo D, Hennion MC. Trace determination of pesticides and their degradation products in water. Elsevier Science, Amsterdam, The Netherlands, 1997.

[42] Nations Union. Globally harmonized system of classification and labelling of chemicals (GHS). United Nations, New York and Geneva, 2011.

[43] WHO. The WHO recommended classification of pesticides by hazard and guidelines to classification 2009. In: World Health Organization, W. (ed.) 2011. Geneva, Switzerland.

[44] Sangchan W, Bannwarth M, Ingwersen J, Hugenschmidt C, Schwadorf K, Thavornyutikarn P, Pansombat K, Streck T. Monitoring and risk assessment of pesticides in a tropical river of an agricultural watershed in northern Thailand. Environ Monit Assess 2014, 186, 1083–1099.

[45] Schulz R. Rainfall-induced sediment and pesticide input from orchards into the Lourens river, Western Cape, South Africa: Importance of a single event. Water Res 2001, 35, 1869–1876.

[46] Müller K, Stenger R, Rahman A. Herbicide loss in surface runoff from a pastoral hill slope in the Pukemanga catchment (New Zealand): Role of pre-event soil water content. Agric Ecosyst Environ 2006, 112, 381–390.

[47] Vryzas Z, Alexoudis C, Vassiliou G, Galanis K, Papadopoulou-Mourkidou E. Determination and aquatic risk assessment of pesticide residues in riparian drainage canals in North-Eastern Greece. Ecotoxicol Environ Saf 2011, 74, 174–181.

[48] Vryzas Z, Papadakis EN, Vassiliou G, Papadopoulou-Mourkidou E. Occurrence of pesticides in transboundary aquifers of North-Eastern Greece. Sci Total Environ 2012, 441, 41–48.

[49] Dalkvist T, Topping CJ, Forbes VE. Population-level impacts of pesticide-induced chronic effects on individuals depend more on ecology than toxicology. Ecotoxicol Environ Saf 2009, 72, 1663–1672.

[50] Davis DL, Blair A, Hoel DG. Agricultural exposures and cancer trends in developed countries. Environ Health Perspect 1992, 100, 39–44.

[51] Woodruff TJ, Kyle AD, Bois FY. Evaluating health risks from occupational exposure to pesticides and the regulatory response. Environ Health Perspect 1994, 102, 1088–1096.

[52] Rozita H, Azimatun NA, Shamsul AS, Mohd-Rohaizat H, Nazarudin S, Mohd-Hasni J. Chlorpyrifos blood level and exposure symptoms among paddy farmers in Sabak Bernam, Malaysia. Public Health Res 2011, 1, 1–6.

[53] Bürger J, De Mol F, Gerowitt B. The “necessary extent” of pesticide use – Thoughts about a key term in German pesticide policy. Crop Protect 2008, 27, 343–351.

[54] Isman MB. Botanical insecticides, deterrents, and repellents in modern agriculture and in increasingly regulated world. Ann Rev Entomol 2006, 51, 45–66.

[55] Illing HPA. Risk assessment. In: Duffus J, Worth H (eds.). Fundamental toxicology for chemists. UK The Royal Society of Chemistry, Cambridge CB4 4WF, 1996.

[56] Robson M, Toscano W. Risk assessment for environment health. Introduction to risk assessment in public health. 1st edition. Wiley Imprint, USA, 2007.

[57] IUPAC, International Union of Pure and Applied Chemistry. Glossary for chemists of terms used in toxicology. Pure Appl Chem 1993, 65, 2003–2122.

[58] Paustenbach. Human and ecological risk assessment: Theory and practice. John Wiley and Sons, New York, NY, 2002.

[59] Mark GR, William AT. Risk assessment for environmental health. Jossey-Bass, Market Street, San Francisco, United Stated of America, 2007.

[60] U.S. EPA. Human health risk assessment [Online]. United States: Environmental Protection Agency. Available: http://www.epa.gov/risk_assessment/health-risk.htm 2012a, [Accessed 9/6/2015 2015].

[61] El-Ghonemy H, Watts L, Fowler L. Treatment of uncertainty and developing conceptual models for environmental risk assessments and radioactive waste disposal safety cases. Environ Int 2005, 31, 89–97.

[62] ESG (Environment Scientific Group). Risk assessment: Sources, pathways and receptors[Online]. Burton, United Kingdom: Environmental Scientific Group Ltd. Available: http://www.esg.co.uk/services/risk-assessment-sources-pathways-and-receptors/ 2015, [Accessed 20/4/2015 2015].

[63] Van Der Oost R, Beyer J, Vermeulen NPE. Fish bioaccumulation and biomarkers in environmental risk assessment: A review. Environ Toxicol Pharmacol 2003, 13, 57–149.

[64] Maazza-Yousufabdulah M. Risk assessment of hazardous chemicals – A case study in chemistry research laboratories of USM. Master of Science, Universiti Sains Malaysia, 2008.

[65] Fowles J, Dybing E. Application of toxicological risk assessment principles to the chemical constituents of cigarette smoke. Tobacco Control 2003, 12, 424–430.

[66] Rahimah W, Latiff AM, Pauzi MA. Input Ekologi Dalam Penilaian Impak Alam Sekitar (EIA) Bagi Langkah Penebatan Untuk Aktiviti Kuari di Malaysia. Malays J Environ Manage 2010, 11, 45–56.

[67] U.S. EPA. Ecological risk assessment. In: AGENCY USEP (ed). 2013.

[68] Fenske RA, Day EW Jr. Assessment of exposure for pesticide handlers in agricultural, residential, and institutional environments. In: Franklin CA, Worgan JP (eds.). Occupational and residential exposure assessment for pesticides. John Wiley & Sons, Chichester, UK, 2005b.

[69] Jaga K, Dharmani C. Sources of exposure to and public health implications of organophosphate pesticides. Pan Am J Public Health 2003, 14, 171.

[70] Phung DT, Connell D, Miller G, Rutherford S, Chu C. Pesticide regulations and farm worker safety: The need to improve pesticide regulations in Viet Nam. Bull World Health Organ 2012b, 90, 468–473.

[71] Jaipieam S, Visuthismajarn P, Sutheravut P, Siriwong W, Thoumsang S, Borjan M, Robson M. Organophosphate pesticide residues in drinking water from Artesian wells and health risk assessment of agricultural communities, Thailand. Hum Ecol Risk Assess 2009, 15, 1304–1316.

[72] Sematong S, Zapuang K, Kitana N. Pesticide use, farmer knowledge and awareness in Thong Pha Phum region, Kanchanaburi province. J Health Res 2008, 22, 15–20.

[73] Gönner JV, Bowler DE, Gröning J, Klauer A-K, Liess M, Neuer L, Bonn A. Citizen science for assessing pesticide impacts in agricultural streams. Sci Total Environ 2022, 857, Part 3, 159607.

[74] Jinky LL. Gender analysis of women in the Philippine agriculture and their occupational issues. J Int Women's Stud 2010, 11, 73–82.
[75] FAO. 2015. *Staple foods: What do people eat?* [Online]. United Kingdom. [Accessed April 2015].
[76] Abdullah AR, Bajet CM, Matin MA, Nhan DD, Sulaiman AH. Ecotoxicology of pesticides in the tropical paddy field ecosystem. Environ Toxicol Chem 1997, 16, 59–70.
[77] Prasanna-Kumar MK, Sidde-Gowda DK, Rishikant-Moudgal, Kumar KN, Pandurange-Gowda KT, Vishwanath K. Impact of fungicides on rice production in India. Intech Publ 2013.
[78] USDA/FAS. 2008. *Corn 2008: Production, supply, demand database*. [Online]. United Stated. Available: http://www.pecad.fas.usda.gov/cropexplorer/. [Accessed May 12, 2015].
[79] Akoto O, Andoh H, Darko G, Eshun K, Osei-Fosu P. Health risk assessment of pesticides residue in maize and cowpea from Ejura, Ghana. Chemosphere 2013, 92, 67–73.
[80] Dinham B. Growing vegetables in developing countries for local urban populations and export markets: Problems confronting small-scale producers. Pest Manage Sci 2003, 59, 575–582.
[81] Fantke P, Charles R, Alencastro LFD, Friedrich R, Jolliet O. Plant uptake of pesticides and human health: Dynamic modeling of residues in wheat and ingestion intake. Chemosphere 2011, 85, 1639–1647.
[82] Heinemann JA, Massaro CDS, Agapito-Tenfen SZ, Wen JD. Sustainability and innovation in staple crop production in the US Midwest. Int J Agric Sustain 2013, 12, 71–88.
[83] Blondell J. Epidemiology of pesticide poisonings in the United States, with special reference to occupational cases. Occup Med 1997, 12, 209–220.
[84] Palis FG, Flor RJ, Warburton H, Mahabub-Hossain. Our farmers at risk: Behaviour and belief system in pesticide safety. J Public Health 2006, 28, 43–48.
[85] U.S. EPA. 1993. The Worker Protection Standard for agricultural pesticides. How to comply. U.S. Government Printing Office, Washington, DC.
[86] Drexler H. Skin protection and percutaneous absorption of chemical hazards. Int Arch Occup Environ Health 2003, 76, 359–361.
[87] Carpenter WS, Lee BC, Gunderson PD, Stueland DT. Assessment of personal protective equipment use among Midwestern farmers. Am J Indust Med 2002, 42, 236–247.
[88] Keifer MC. Effectiveness of interventions in reducing pesticide overexposure and poisonings. Am J Prevent Med 2000, 18, 80–89.
[89] Cattani M, Cena K, Edwards J, Pisaniello D. Potential dermal and inhalation exposure to chlorpyrifos in Australian pesticide workers. Ann Occup Hyg 2001, 45, 299–308.
[90] Mokhele TA. Potential health effects of pesticide use on farmworkers in Lesotho. South Afr J Sci 2011, 107, 29–35.
[91] Sanborn MD, Cole D, Abelsohn A, Weir E. Identifying and managing adverse environmental health effects: 4.Pesticides. Can Med Assoc J 2002a, 166, 1431–1436.
[92] Hoppin JA, Valcin M, Henneberger PK, Kullman GJ, Umbach DM, London SJ, Alavanja MCR, Sandler DP. Pesticide use and chronic bronchitis among farmers in the agricultural health study. Am J Indust Med 2007, 50, 969–979.
[93] Hoppin JA, Umbach DM, London SJ, Henneberger PK, Kullman GJ, Coble J, Alavanja MCR, Beane Freeman LE, Sandler DP. Pesticide use and adult-onset asthma among male farmers in the Agricultural Health Study. Eur Respir J 2009, 34, 1296–1303.
[94] Kirkhorn SR, Garry VF. Agricultural lung diseases. Environ Health Perspect 2000, 108, 705–712.
[95] Ming Y, Jeremy B, Jonathan WM, Ambikaipakan S. Occupational pesticide exposures and respiratory health. Int J Environ Res Public Health 2013, 10, 6442–6471.
[96] Ntow WJ, Gijzen HJ, Kelderman P, Drechsel P. Farmer perceptions and pesticide use practices in vegetable production in Ghana. Pest Manag Sci 2006, 62, 356–365.
[97] Baharuddin MRB, Sahid IB, Noor Ma BM, Sulaiman N, Othman F. Pesticide risk assessment: A study on inhalation and dermal exposure to 2,4-D and paraquat among Malaysian paddy farmers. J Environ Sci Health 2011, Part B, 46, 600–607.

[98] Sekiyama M, Tanaka M, Gunawan B, Abdoellah O, Watanabe C. Pesticide usage and its association with health symptoms among farmers in rural villages in west Java, Indonesia. Environ Sci 2007, 14, 23–33.
[99] Lekei E, Ngowi A, London L. Farmers' knowledge, practices and injuries associated with pesticide exposure in rural farming villages in Tanzania. BMC Public Health 2014, 14, 389.
[100] Atreya K. Pesticide use knowledge and practices: A gender differences in Nepal. Environ Res 2007, 104, 305–311.
[101] Mariyono J. Direct and indirect impacts of integrated pest management on pesticide use: A case of rice agriculture in Java, Indonesia. Pest Manage Sci 2008, 64, 1069–1073.
[102] Nwilene FE, Nwanze KF, Youdeowei A. Impact of integrated pest management on food and horticultural crops in Africa. Entomol Exp Appl 2008, 128, 355–363.
[103] Palacios Xutuc CN Manual to train trainers on safe and correct use of plant protection products and integrated pest management (IPM). Crop Life Latin America 2010. Guatemala City, Guatemala: Crop Life International.
[104] Nordin RB, Araki S, Sato H, Yokoyama K, Muda W, Kyi DWIN. Effects of safety behaviours with pesticide use on occurrence of acute symptoms in male and female tobacco-growing Malaysian farmers. Indust Health 2002, 40, 182–190.
[105] Mancini F, Jiggins JLS, O'Malley M. Reducing the incidence of acute pesticide poisoning by educating farmers on integrated pest management in South India. Int J Occup Environ Health 2009, 15, 143–151.

Masoom Raza Siddiqui*, Moonis Ali Khan, Mohd Rafatullah,
Riti Thapar Kapoor

Emerging Environmental Pollutants: Sources, Consequences, and Method of Analysis and Removal

Abstract: This chapter deals with various aspects of the emerging contaminants. A brief introduction and classification of the target pollutants is mentioned. The ill effects and the consequences of such contaminants are also mentioned. One of the most required actions on such pollutants is their monitoring. Several analytical procedures ranging from the traditional spectrophotometric method to hyphenated liquid chromatography-mass spectrometry finds a place in this chapter. Sample preparation step is equally important for the accurate analysis and it has been mentioned in detail. Further, a few suggestions on the future challenges related to emerging contaminants have been given.

Keywords: emerging contaminants, analytical methods, extraction techniques, pollutants

1 Introduction

Environmental pollution is one of the major concerns of the modern society; of the different pollutants, a few do not have proper regulation or published health standard. These pollutants can be chemicals as well as microorganisms. Since a few classes of chemicals were not studied clearly, the researchers have little information about these classes of chemicals including their effect on human health, their impact on environment, and their method of analysis and removal. Such chemicals are categorized as the emerging contaminants (ECs). These chemicals are also named so because of new sources or new pathways in human. These chemical needs to be discussed on a wider platform, so that more information about these pollutants are available to the society, which will help in containing such pollutants.

***Corresponding author: Masoom Raza Siddiqui,** Advanced Materials Research Chair, Chemistry Department, College of Science, King Saud University, Riyadh, Saudi Arabia, e-mail: mrsiddiqui@ksu.edu.sa
Moonis Ali Khan, Chemistry Department, College of Science, King Saud University, Riyadh, Saudi Arabia
Mohd Rafatullah, Environmental Technology Division, School of Industrial Technology, Universiti Sains Malaysia, Penang 11800, Malaysia; Renewable Biomass Transformation Cluster, School of Industrial Technology, Universiti Sains Malaysia, Penang 11800, Malaysia
Riti Thapar Kapoor, Amity Institute of Biotechnology, Amity University Uttar Pradesh, Noida 201 313, Uttar Pradesh, India

https://doi.org/10.1515/9783111016825-006

In this chapter, we discuss selected emerging pollutants, their sources, effect on human health, methods of their analysis, and possible removal methods. Future challenges of these pollutants will also be discussed in this chapter.

1.1 Classification of emerging pollutants

There are several classes of chemicals that fall under this category; these may be pharmaceutical compounds either prescribed or illicit, industrial chemicals, personal care products, disinfection by-products, biocides and their metabolites, and algal toxins [1]. Table 1 contains the list of suspected emerging pollutants.

Table 1: Classification of emerging pollutants [2].

S. no.	Class of emerging pollutants	Common examples
1.	Pharmaceutical (both illicit and prescribed) and veterinary products	**Antibiotics**: ofloxacin, ciprofloxacin, ampicillin **NSAIDs**: aspirin, diclofenac, ibuprofen **Hormones**: 17-b estradoil, estriol, estrone, mestranol **Antihypertensive**: metoprolol, atenolol, enalapril, diltiazem **Lipid regulators**: simvastatin, clofibric acid, atorvastatin **Antidepressent**: fluoxetine, paroxetine, diazepam **Illicit drugs and metabolites**: amphetamines, cocaine and its urinary metabolite benzoylecgonine and BE, cannabinoids, ketamine, ephedrine, lysergic acid diethylamide **Veterinary products**: tetracyclin, salinomycin, Fenbendazol, and feed additives such as tylosin and ractopamine
2.	Personal care products	Disinfectants; conservation agents such as musk ketone, musk xylol, tonalide; UV protection agents such as benzophenone-3, homosalate, 4-methyl-benzylidene; insect repellant such as *N*, *N*-diethyltoluamide
3.	Industrial chemicals	1,4-Dioxanes, perfluorooctane sulfonate (PFOS), perfluorooctanoic acid (PFOA), 1,2,3-benzotriazole, bisphenol A, di-2-ethylhexyl phthalate (DEHP), benzene, 1,2-dichloroethane, CCl_4
4.	Disinfectants and biocides	Triclosan, chlorophene, dichlorophene, parabens, quaternary ammonium compounds
5.	Waterborne pathogens	Cyanobacteria (cynotoxins and microcystin).
6.	Nanomaterials	Nanomaterials have environmental, scientific, medicinal and industrial applications, but there is lack of health and data sheet, which risks nanomaterials being classified as emerging contaminants
7.	Biological toxins	Ricin and pinnatoxins

While discussing the emerging pollutants, it is equally important to know the sources and the routes of these pollutants. Keeping aside natural causes, it is generally believed that pollution is related to human activities. Likewise, the spread of the emerging pollutants is also related to human activities. Agricultural, industrial, and domestic activities are some of the well-known reasons that help these chemicals spread.

1.2 Activities leading to the spread of emerging contaminants

Although hospitals play an important role in making the general masses healthy, yet they are major source of concern when it comes to the spread of the hospital and pharmaceutical pollutants, where the waste water discharged from the hospital has a high concentration of these unregulated pollutants [3]. Emerging pollutants such as pharmaceutical compounds not limited to analgesics, antibiotics, hormones, β-blockers, pathogens such as persistent viruses, antibiotic-resistant genes, and antibiotic-resistant bacteria are reported to be present in hospital waste water, which is a matter of concern for the environmentalists, as Antibiotic resistant gene (ARG) and Antibiotic resistant bacteria (ARB) sometimes survive the water treatment procedure and pose a threat to the ecosystem [4]. These waste waters are also characterized by the presence of disinfectants, radionuclides surfactants, and some solvents.

Like hospitals, agriculture also aids ecosystem imbalance due to the emerging contaminants. A few decades back, a series of events such as droughts, floods, other climate changes, and urbanization resulted in demand for higher food production to meet the ever-rising world food requirement. To fulfill the world food requirement, many measures were taken by agriculturalists, including reclaimed water for irrigation, use of fertilizers, and use of pesticides and herbicides. Although the use of the reclaimed water has its own advantage, such water is a source of safety concerns, too. These reclaimed waters may contain several organic contaminants that were identified as an emerging threat to the environments. These pollutants are transferred to the soil, which further aids in the propagation of these pollutants [5]. Veterinary pharmaceuticals are one of the major pollutants that create an imbalance in the agricultural ecosystem. These veterinary medicines are generally given as supplements in animal feeds, as hormones for milk increase and for the treatment of various infections. These medicines contaminate the soil or agricultural land when excreted through urine and feces (used as manure). These pollutants from the soil can easily be transferred to the waterbodies. The continuous presence and exposure to antibiotics may result in antimicrobial-resistant microorganisms [6]. Human medicines and personal care products are also transported to agricultural land through sewage sludge as biosolids and through irrigational water after waste water treatment. At several occasions, researchers have highlighted the ECs in hospital waste water [7].

The progress of a country is sometimes judged by its industrial growth. These industries release different types of chemicals into the atmosphere, among which ECs are in considerable numbers. Mining industries release a lot of ECs in the environment; to name a few, selenium, surfactants, $ClO4^-$, acesulfame, perfluorooctanoic acid, and perfluoro octane sulfonic acid [8]. Automobile industries too do not lag behind other industries in releasing the ECs, palladium nanoparticles, per- and polyfluoroalkyl substances (PFAS), polybrominated diphenyl ethers (PBDEs), polybrominated dibenzo-*p*-dioxins (PBDDs), and polybrominated dibenzofurans (PBDFs) [9] The pharmaceutical industries too are responsible for adding emerging and hazardous contaminants to the environments. The post production waste release of the pharmaceutical industries contains many solvents, unused APIs and their metabolites, and disinfectants used throughout the production areas, which release a lot of emerging contaminants. Some of them alter the normal human functioning resulting in various health ailments. After the release of the medicines, hormones, and other pharmaceutical products, they also contribute to ECs [10]. Nations throughout the world are tirelessly working to improve food security. The flourishing of the food industry is a consequence of such efforts. However, health risks and safety problems come along with it. Food industries are accountable for toxic residue from the food materials, on a wide scale. Perfluorinated compounds, because of the property to repel both oil and water, are used in the food packaging. Halogenated flame retardants are the contaminants commonly traced in meats and sea foods [11]. Nanomaterials are yet another EC in the food industry, where the nanotechnology products find their application in food processing, packaging, delivery of food, and nutritional supplements as well. Nanoclay is used as diffusion barrier in food packaging, while the nanosilver represents the antimicrobial agent. It is believed that the human exposure to nanomaterials is due to the migration of the nanoparticles from the food packaging to the food they consume. Biotoxins are another class of ECs to which humans are exposed through the food and food industries. These kinds of toxins are generally chemicals produced by bacteria; plants and marine algae approximately have 5,000 species, out of which about 70–80 are believed to produce toxins. These toxins are accumulated in fish or mollusks. Botulinum toxins, ciguatoxins, palytoxins, and brevetoxins are some of the common bio toxins categorized as ECs [12, 13]. To sum up, we can conclude that water is one the major sources of the emerging pollutants. This water includes ground water and the water released from the wastewater treatment plants, since wastewater from industries, hospitals, agriculture, and human settlements arrive at the water treatment plants. The treatment process at the wastewater treatment facility works on certain standards and operating procedures that normally remove the regulated contaminants and pathogens. However, certain chemicals including the emerging contaminants are not eliminated, either because of their nature or low concentration. Ground water is used by a vast majority of population, worldwide. It is generally believed that ground water is vulnerable to emerging pollutants exposure, thanks to the discharges from the wastewater treatment plants and infiltration from the septic systems. Similarly, sea water too is an important source of ECs – riverine outputs and the wastewater treatment plant's effluents are the major contributors to sea water contamination with emerging pollutants. These contaminants pose a great threat to the aquatic ecosystem as well as humans [14].

2 Consequences of the emerging pollutants

There are thousands of chemicals that are an integral part of our day-to-day life. There are many that either escaped the eyes of the environmentalist or have not been studied and regulated completely. ECs are in such a group of chemicals. Some of the known ECs are being studied extensively, and each and every aspect of such contaminants is available to the masses; one such aspect is the "consequences of the emerging pollutants." A large chunk of the emerging pollutants is found in and around water – it is most obvious that the aquatic environment is one of the most affected ecosystems. Classes of compounds such as herbicide, ultraviolet (UV) filters, insect repellents, organophosphorus flame retardants, bactericides, pharmaceuticals, metabolites, and pesticides are the most common pollutants in the quantic ecosystem. The life of aquatic inhabitants is at risk, but lives of human beings, who consume many aquatic animals as food, are simultaneously at risk. As we know, humans are exposed to these chemicals in one way or the other, and it affects our health in multiple ways. Immunotoxicity, endocrine disruption, cancer, and neurological disorders are some of the ailments known to be caused by these emerging contaminants. Perfluorinated compounds are related to reproductive disorders. There are many disinfectants used in the swimming pool and in water purification that prevent waterborne diseases in humans. These disinfectants are generally oxidizing agents that also react with many deoxidizers, resulting in production of many undesirable by-products that are not good for human health, e.g., chlorinated disinfection byproducts and iodinated trihalomethanes.

3 Detection and determination of emerging pollutants

Quantitative monitoring of the ECs is an uphill task of the environmentalist and the researchers. To make is easier, there are several techniques used to detect and determine different ECs from different matrices. Among the several available techniques, mass spectrometry is one of the widely accepted techniques for the screening of the ECs. Sometimes, the technique is used alone, while hyphenated forms are widely available. The screening of such contaminants is based on the mass-to-charge ratio of the analytes in question: ultra-performance liquid chromatography (UPLC)-quadrupole time-of-flight

(QTOF)- mass spectrometry (MS), liquid chromatography (LC)-high-resolution mass spectrometry (HRMS). The pollutants include pesticides, pharmaceutical pollutants, and illicit drugs [15, 16]. In addition to screening, there are several techniques used for the quantitative analysis of these emerging pollutants, extraction procedure being an important aspect in this analysis. Spectrophotometry is one of the comparatively older techniques used in current times, too. The determination of the techniques is based on the reflection or transmittance by an analyte as a function of wavelength. There are several ECs analyzed by spectrophotometry, such as a wide variety of pharmaceutical contaminants and hormones [17].

Thin-layer chromatography is an old technique for quantitative analysis of ECs. However, when combined with the liquid chromatographic (LC) technique, it has proved to be a useful tool. Thin-layer chromatography, coupled with high-performance liquid chromatography (HPLC)-diode array detector (DAD)/ultra-HPLC (UHPLC)-HRMS is the tool that can serve the purpose of determination of emerging halogenated organic contaminants. The technique is believed to be solvent-saving and sensitive and can determine a number of ECs such as polyhalogenated carbazoles, tetrabromobisphenol-A analogs, and halogenated phenols [18]. Electrochemical methods are another traditional technique used for determination of ECs. They are advantageous over other techniques in terms of cost effectiveness, simplicity, and speed. There are several techniques under combined electrochemical methods used for the determination of different analytes, including ECs. Potentiometry is one such technique, where the difference in the potential between the two electrodes is exploited for the purpose of quantitative analysis . This potential difference is achieved when the titrant is added, bringing about a change in ionic concentration. In the literature, potentiometry is reported for selective determination of residue of benzodiazepine in wastewater. Sodium tetraphenylborate (TPB) and phosphotungstic acid (PTA) were used as ion pairing agents. As usual, the electrochemical methods have the ability to determine analytes at lower concentration; in this case too, the contaminants were determined in the range of 1×10^{-6} M to 1×10^{-3} M. The results suggest detection limit of 0.25 µg/mL for the benzodiazepine-PTA and benzodiazepine-TPB membrane electrodes [19]. Another important electrochemical technique is voltammetry, where the quantitative information of the analyte is gathered by measuring the current as a function of the applied potential. Reported literature presents the determination of a series of endocrine disruptors, including bisphenol A, 4-nitrophenol (4-NP), and phenols using square wave voltammetry, where magnetic "needle-like" Fe_3O_4 nanorods (MNRs) along with a poly(3,4-methylenedioxy)aniline (PMDAN)-based nanocomposite-modified graphite paste electrode was used to explore the determination possibilities. The article suggests the calibration curves for bisphenol were linear up to 40.0 µM, for 4-nitrophenol (4-NP) it was up to 125.0 µM, while for phenol, it was 150 µM. Limits of detection for the three were found to be 4.3, 16.8, and 15.3 nM, respectively [20].

Amperometry is another popular technique among the electrochemical methods, where the current is measured at constant voltage applied by the dropping mercury electrode. The technique was used for quantitative monitoring of the marine pollutants categorized as pollutants of environmental concerns. The pollutants such as estradiol, el Irgarol 1051, chloramphenicol, and sulfapyridine were quantitated directly from the sea water. The applied potential of −10 V versus the silver reference electrode was required to acquire the chronoamperograms [21].

Polarographic techniques are reported to be applied for the determination of ECs; in this technique, there are two electrodes (one polarizable and the other nonpolarizable), where the voltage between them is gradually increased, and the current flowing between the two electrodes are recorded and used for the quantitative analysis of the analyte. In the literature, polarography was reported to determine benzotriazoles, a class of organic emerging pollutants that are used in antifreeze brake fluid, industrial cooling systems, lubricating oils, aircraft, and roadway deicing fluids. Nonchromatographic methods are particularly popular for the analysis of benzotriazoles when the extraction is not possible. One such procedure is based on polarography, where the analyte was determined without extraction, filtration, or centrifugation. Differential pulse polarography was applied for the said determination. Authors applied static mercury drop electrode and obtained a detection limit of 40 µg/L 5-methylbenzotriazole and 40 µg/L for 5-methylbenzotriazole [22]. Referring to the published article, it can be ascertained that when it comes to trace level determination and without extraction process, in certain conditions, electrochemical process is a good choice for analysis of ECs. Fluorescence data from the sample is yet another procedure that was explored in the past for quantitative information on ECs. The technique is also used for the quantitative analysis of the emerging pollutants. One of the determination processes is based on UV irradiation following which the excitation–emission photoinduced fluorescence matrices were acquired; this data was combined with the with second-order chemometric analysis. The combined procedure led to the determination of piroxicam, ofloxacin, and carbamazepine; it is important to note that it does not require the extraction step [23]. The qualitative analysis procedure reported is also based on the fluorescence spectra, where alternating trilinear decomposition (ATLD) was adopted to get the required information on the excitation emission of the sample [24].

Another analytical technique used for the determination of ECs is electrophoresis. This technique is helpful in determination of a number of organic pollutants simultaneously, as well as the detection of bacterial contamination [25]. Thirteen pharmaceutical emerging contaminants are reported to be determined using electrophoresis with capacitively coupled contactless conductivity detection, electrolyte being 9 mM Tris/5 mM lactic acid at pH 8.0 containing *n*-propanol, gamma-CD, hydroxyl-beta-CD, and dimethyl-beta-CD in different ratios [26].

Chromatographic methods are, in recent times, the most used technique for pollutants determination including ECs. The techniques involved are gas chromatography (GC), HPLC, and further modified versions of HPLC with different names such as rapid resolution LC and UPLC, and most importantly, the hyphenated techniques involving LC and MS.

GC is an important tool for analysts when it comes to quantitative analysis of volatile pollutants. This technique is more efficient when it is coupled with mass spectrometry. There are several reports available where ECs were determined by GC in combination with tandem mass spectrometry [27], time-of-flight mass spectrometry [28], and GC-HRMS coupled with atmospheric pressure chemical ionization source (APCI) [29].

Liquid chromatographic techniques have achieved wide recognition in terms of analysis of multiple components including pharmaceuticals, organic pollutants, and dyes; ECs too were reported to be analyzed using the technique. HPLC is, comparatively, an older technique, considering the recent developments in chromatographic analysis. The importance of the detector in the HPLC analysis cannot be ignored. HPLC-DAD was used to determine diclofenac, carbamazepine, and triclosan as ECs. The method involves the reversed phase chromatographic procedure [30]. Another set of 11 emerging pollutants, namely methylparaben (MeP), bisphenol A (BPA), ethylparaben (EtP), butylparaben(BuP), propylparaben (PrP), bisphenolF(BPF), ethinylestradiol (EE2), 17-β-estradiol (17-BE), triclocarban (TCC), 2-ethylhexyl 4-(dimethylamino) benzoate (OD-PABA), 4-methylbenzyllidene camphor (4-MBC), and triclocarban (TCC) were determined using the HPLC-DAD combination in water. The authors used natural deep eutectic solvent in membrane-based techniques. UHPLC is an advanced form of HPLC, where the column length is reduced, leading to reduction of plate height and ultimately, the reduction of the analysis time. UHPLC combined with the UV detector can be used to simultaneously determine 25 ECs in surface and wastewater. The contaminants belong to several classes such as central nervous system stimulants, veterinary antibiotics, steroids, preservatives, and nonsteroidal anti-inflammatory drugs [31]. Other LC techniques reported for the quantification of the emerging pollutants include LC-MS/MS [32], LC-QTOF-MS [33], and LC/TOF/MS [34]. Details of selected liquid chromatographic methods are mentioned in Table 2.

4 Pretreatment of the emerging contaminants prior to their quantitative analysis

There are a few reported methods in which direct sampling and determination processes are done. However, most of the reported analytical methods come up with a proper extraction process when it comes to determination of the ECs, although this may also be true for other analytes. Extraction processes are important in minimizing the interference from the matrix. There are several processes that are used for extraction of emerging contaminants from different matrices. Solvent extraction is one of the simpler techniques used for obtaining a pure analyte of concern from the complex matrix. Forty-four ECs were extracted by accelerated solid phase extraction, followed by application of solid phase cartridge for cleanup step. For extraction, Milli-Q water and methanol were used in equal ratio with other experimental conditions, such as temperature:

Table 2: Liquid chromatographic methods reported for the analysis of emerging contaminants.

S. no	Technique with detector	Samples	Analytical conditions	Result			Reference
1.	HPLC-DAD	Diclofenac, carbamazepine, triclosan	**Mobile Phase:** ethanol: water (85:15, v/v) **Column:** RP C18 column **Flow rate:** 0.8 mL/min	Emerging Contaminant	Linearity	LOD and LOQ	[30]
				Diclofenac	20–1,000 ng/mL	3.66 and 11.09	
				Carbamazepine	60–1,000 ng/mL	7.51 and 22.75	
				Triclosan	60–2,000 ng/mL	7.22 and 21.88	
2.	HPLC-DAD		**Mobile phase:** acetonitrile: ultrapure water **Column**: C-18 **Flow rate:** 1 mL/min	Emerging contaminant	Linearity µg/L	LOD and LOQ µg/L	[35]
				Methylparaben (MeP)	20–400	6.1 and 20	
				Bisphenol A (BPA)	5–400	1.5 and 5	
				Ethylparaben (EtP)	5–200	1.5 and 5	
				Butylparaben (BuP)	5–400	1.5 and 5	
				Propylparaben (PrP)	1–200	0.3 and 1	
				Bisphenol F (BPF)	10–300	3.0 and 10	
				Ethinylestradiol (EE2)	1–400	0.3 and 1	
				17-β-estradiol (17-BE)	20–300	6.1 and 20	
				Triclocarban (TCC)	1–300	0.3 and 1	

(continued)

Table 2 (continued)

S. no	Technique with detector	Samples	Analytical conditions	Result			Reference
				2-Ethylhexyl 4-(dimethylamino) benzoate (OD-PABA)	1–300	0.3 and 1	
				4-Methylbenzyllidene camphor (4-MBC)	1–200	0.3 and 1	
3.	HPLC-UV	Samples	**Mobile phase: a**cetonitrile (solvent B) and ultra-pure water containing 0.5% acetic acid (solvent A) Gradient elution **Column:** C-18 (4.6 mm × 150 mm; particle size 1.8 and 2.6 μm **Flow rate:** 1.0–1.3 mL/min	Name of emerging contaminant	Linearity μg/L	LOD and LOQ **μg/L**	[31]
				Sulfanilamide	5–1,000	10 and 33	
				Theophylline	5–1,000	12 and 40	
				Acetaminophen	5–1,000	9 and 30	
				Sulfacetamide	5–1,000	6 and 20	
				Caffeine	5–1,000	12 and 40	
				Sulfadiazine	5–1,000	7.5 and 25	
				Sulfathiazole	5–1,000	9 and 30	
				Sulfapyridine	5–1,000	3 and 10	
				Sulfamerazine	5–1,000	3 and 10	

Sulfamethazine	5–1,000	1.5 and 5
Sulfamethoxypyridazine	5–1,000	7 and 23
Sulfamonomethoxine	5–1,000	7 and 23
Acetyl salicylic acid Sulfamethoxazole	5–1,000	15 and 50
Methylparaben	5–1,000	7.5 and 25
Sulfadimethoxine	5–1,000	6 and 20
Sulfaphenazole	5–1,000	12 and 40
Ethylparaben	5–1,000	7.5 and 25
Propylparaben	5–1,000	7.5 and 25
Ketoprofen	5–1,000	8 and 26
17α-Ethinyl estradiol	5–1,000	15 and 50
Estrone	5–1,000	15 and 50
Fenoprofen	5–1,000	7 and 24
Flurbiprofen	5–1,000	3 and 10
Diclofenac	5–1,000	12 and 40

(continued)

Table 2 (continued)

S. no	Technique with detector	Samples	Analytical conditions	Result			Reference
4.	UHPLC		**Mobile phase:** Solvent A water with 0.1 mM NH_4Ac and 0.001% HCCOH; solvent B MeOH with 0.1 mM NH_4Ac and 0.001% HCCOH Column: C18. Flow rate: 300 µl/min	Name of contaminants	Linearity µg/L	LOQ aqueous phase (ng/L) and solid phase extraction (µg/kg)	[32]
		Thiabendazole		Thiabendazole	0.5–500	84 and 17	
		Acesulfame		Acesulfame	NA	NA	
		Fenofibric acid		Fenofibric acid	NA	NA	
		Valsartan		Valsartan	0.25–500	74 and 27	
		Irbesartan		Irbesartan	0.1–500	14 and 1.2	
		Salicylic acid		Salicylic acid	0.05–500	220 and 35	
		Diclofenac		Diclofenac	0.05–500	48 and 63	
		Carbamazepine		Carbamazepine	0.05–500	8.6 and 5.4	
		4-Aminoantipyrin		4-Aminoantipyrin	0.5–100	99 and 11	
		4-Acetyl aminoantipyrine		4-Acetyl aminoantipyrine	0.1–50	33 and 46	
		4-Formyl aminoantipyrine		4-Formyl aminoantipyrine	0.1–50	96 and 46	
		Venlafaxine		Venlafaxine	0.1–500	0.2 and 7.5	
		Benzoylecgonine		Benzoylecgonine	0.05–500	9.3 and 15	

5.	LC-Q-ToF-MS		**Mobile Phase:** water with 0.1% formic acid and acetonitrile; gradient elution **Column:** C18 analytical column of 2.1 × 50 mm and 1.7 μm particle size. Flow rate: 0.3 mL/min	Linearity μg/L	LOD and LOQ (ng^{-1})	[33]
		16-Hydroxyhexadecanoic acid		0.5–100	100 and 313	
		Benzophenone		0.5–100	8.7 and 26.4	
		Lignocaine		0.5–100	6.8 and 20.8	
		Metoprolol		0.5–100	4.4 and 13.4	
		2-Dodecyl benzene sulfonic acid		0.5–100	7.5 and 22.9	
		Mefenamic acid		0.5–100	50 and 169	
		β-Estradiol		0.5–100	25 and 75.9	
		Oxybenzone		0.5–100	7.4 and 200	

100 °C; preheating: 3 min; pressure: 1,500 psi; static: 5 min; flush volume: 80%; purge and cycle: 2 min and three cycles. The contaminants extracted from sewage sludge include selected antibiotics, anticancer agents, antihypertensive agents, and a few more classes of pharmaceuticals [36] Among the recent methodologies is salting-out assisted liquid–liquid extraction, which is used for the extraction of EC residue, prior to instrumental analysis. This technique used water-miscible organic solvent as extraction solvent. Gezahegn and coworkers used acetonitrile and $MgSO_4$ for extraction. The extraction process also involves centrifugation for phase separation, where the upper layer containing the analyte is taken out, dried, and reconstituted in water to prepare for analysis [37]. Dispersive Liquid–Liquid Microextraction is another extraction procedure in the series used for extraction of ECs. The extraction procedure requires the sample to be in acidic pH (6.0), eutectic mixtures formed by fenchol and acetic acid. The extraction procedure proceeds in such a way that solid droplets of the eutectic mixture are formed during extraction, which are removed and used quantitative analysis purpose [38]. Ultrasound-assisted extraction is also reported in the literature for the extraction of ECs. Personal care products, pharmaceuticals, and hormones were extracted using the techniques. To prepare the samples ready for analysis, extraction and cleanup procedure were adopted. The extraction procedure is a multistep procedure: in the first step, methanol is used as extraction solvent when the sample is vortexed and kept in an ultrasonic bath and centrifuged. The second step includes extraction with 2.5 mL each of methanol and water. The third extraction step uses 2 mL of acetonitrile and the fourth step uses methanol and water (3 mL; 1:2 v/v; pH 2 with formic acid). Finally, for the mixed sample in the cleanup step was dried, rinsed in methanol, pH adjusted to 9 with NH_4OH, and percolated through Oasis HLB cartridge [39]. Stir bar sorptive extraction is the technique that is used for the extraction of some classes of ECs. In this form of extraction, both polar and nonpolar pollutants were extracted with the help of two stir bars placed in the samples. The bars were coated with two different materials, polydimethylsiloxane (PDMS) and l ethylene glycol-modified silicone (EG-silicone). During the extraction procedure, these stir bars were placed in the sample solution and stirred at 600 rpm for 24 h. Following this procedure, desorption studies were conducted on the stir bars, which were placed in methanol and ultrasonicated. The desorbed samples were then dried, reconstituted in equivolume methanol water, and made ready for LC-MS analysis [40]. Prior to the gas chromatographic analysis mentioned earlier, authors used proper extraction (two-step) where the cereals were placed in a glass column filled with cellulose filter paper. The initial step involves the extraction with 7 mL of AcEt: MeOH (90:10 v/v) containing 3% of NH_4OH; the assembly was sonicated for 15 min. In the next step, 3% NH_4OH was replaced by 3% formic acid. The second extraction was collected and dried in the tube of first extraction, reconstituted in 1 mL acetonitrile, followed by dispersive solid phase extraction using primary secondary amine (PSA) and then by vortexing and centrifugation to make the sample ready for GC-MS/MS analysis [27]. The most important and widely used extraction procedure for ECs is solid phase extraction. The ECs were determined in various water-related matrices. Emerging

organic contaminants, including pharmaceutical pollutants, were extracted from the fish samples. Solid phase extraction and solid phase micro extraction were included in the sample preparation step. HBL cartridge was used in solid phase extraction, where MTBE and methanol were used for conditioning. The elution of the analyte was possible using 5 mL of solvent A (methanol) and 5 mL of solvent B (methanol:MTBE; 10:90 v/v) [41]. A modified solid phase was developed for pre-concentration of different ECs of a lagoon system. L Phenomenex Strata-X reverse phase cartridges were used as solid phase. Conditioning the cartridge was done using 6 mL each of methanol, ultrapure water, and ultrapure water acidified with HCl to pH values of 2 and 3 [42]. Several ECs including metolachlor, diclofenac, cybutryne (irgarol), perfluoroalkyl acids, and perfluorooctansulfonic acid were extracted from the water sample using automated solid phase extraction, Oasis HLB (200 mg) cartridge was used for the purpose. Prior to extraction, all the samples were spiked with internal surrogate standards. The conditioning of the cartridge was done by 10 mL each of ethyl acetate, methanol and Milli-Q water. Elution of the sample was performed with ethyl acetate (5 mL/min). Recovery ranged from 31% for diclofenac and 1H-benzotriazole to 170% (PFHpA (perfluoroheptanoic acid) [43].

5 Removal of emerging contaminants

In addition to the detection and determination process, removal of the ECs is equally important for the environmentalists. There are several procedures used for the removal of ECs. These processes include application of membrane technology — the process involves different types of filtration (micro, nano, ultra-filtration, sand filtration), forward and reverse osmosis, activated sludge process, coagulation / flocculation, and electrodialysis [44, 45]. Biological methods too have gained a lot of attention for the removal of pollutants, specially the pollutants of emerging concerns. There are several microorganisms, specially bacteria, algae, and fungi that are used for the degradation of such pollutants. Several ECs such as synthetic musks, UV filters, and pharmaceutical compounds were treated using the bacteria, including both bacillus and rhodococcus. The results suggest removal from 14–71% [46].

Advanced oxidation process (AOP) has also gained the attention of many researchers for the treatment of water containing ECs. This process is able to get a wide variety of contaminants removed from water. It is based on the generation of potent degradation of organic ECs; these organic pollutants are oxidized mineral salts, water, and CO_2. There are several oxidizing agents used in the AOPs; these agents include ozone, H_2O_2, TiO_2/O_3, UV/ H_2O_2, UV/HOCl, and UV/ClO_2. ECs such as bisphenol A, DDT, atrazine, testosterone, estrone, musk ketone, triclocarban, and triclosan can be removed using this procedure.

Adsorption is one of the most prominent reported methods for removal of ECs. There are many commercial adsorbents such as activated carbon available that are good for water treatment with respect to ECs; however, they get saturated very easily. There are many low-cost and different classes of adsorbents used for the removal of the ECs. Polymeric adsorbent, β-cyclodextrin-epichlorohydrin copolymer, and biopolymer composite are some of the many polymers reported for the removal of ECs [47, 48]. Other materials used for the removal process are nanocellulose, biofilms, metal organic frameworks, surfactant-modified Montmorillonite, hydrochars, and biochars – both magnetic and nonmagnetic, and many other kinds of adsorbents as reported in the literature. Some of the prominent techniques for removal of ECs are mentioned in Table 3.

Table 3: Selected techniques for the removal of emerging contaminants (ECs).

S. no.	Name of the removal process	Name of pollutants	Removal rate/ capacity	Reference
1.	Gravity-driven membrane filtration using acclimated activated sludge	Sulfamethoxazole (SMX)	74.88%	[49]
2.	Chemically enhanced primary sedimentation (CEPS) and acidogenic sludge fermentation	Retinoids and estrogenic endocrine-disrupting chemicals	Aluminum-based CEPS- retinoids 65%, EDCs 73% after acidogenic fermentation Iron-based CEPSretinoids 80%, EDCs 72%	[50]
3.	Coagulation/flocculation	Diclofenac and carbamazepine, ketoprofen, triclosan diclofenac, galaxolide, and tonalide	Diclofenac and ketoprofen >70% Triclosan and naproxen 40–70% Rest <30%	[51]
4.	Biological methods	Galaxolide, tonalide, padimate O, octocrylene, and carbamazepine	1471%	[46]
5.	Advanced oxidation process	Bisphenol A, DDT, atrazine, testosterone, estrone, musk ketone, triclocarban, triclosan, etc., can be removed using this procedure		[52]
6.	Adsorption methods/ adsorbent types			
	Biocomposite	Diclofenac	Adsorption capacity – 99.29 mg/g	[53]

Table 3 (continued)

S. no.	Name of the removal process	Name of pollutants	Removal rate/ capacity	Reference
	Biofilms	Diclofenac, miconazole, Atrazine, estrone, ofloxacin, butachlor, carbamazepine, carbendazim, fluconazole	NA	[54]
	Metal organic framework	Chloroquine phosphate	68–100%	[55]
	Humic acid (1), humasorb (2), montmorillonite (3)	Carbamazepine	In wastewater 87.0% (1), 87.3% (2), 78.2% (3)	[56]
	Nonmagnetic biochar (derived from soap nut seed)	Ciprofloxacin	33.44 mg/g	[57]
	Magnetic biochar (derived from agricultural waste and iron nanoparticle)	Perfluorooctane sulfonate	120.44 mg/gm	[58]
	Hyrochar (from olive oil waste)	Triclosan (TCS), ibuprofen (IBP), diclofenac (DCF)	TCS 98%, DCF 64%, IBP 43%	[59]

6 Future challenges related to emerging contaminants

There are many known chemicals which we are exposed to; however, there are many which are still not reported to be pollutants. More and more such chemicals should be tested for their toxicity and ill effects on human beings. New tools should be developed to monitor the contaminants; new innovation should be made to reduce the environmental effect of such pollutants. Efforts should be made to localize the ECs so that they do not travel the larger distance and affect wider population. Although required in some cases, the use of pesticides and insecticides should be limited. Wherever necessary, doctors should prescribe medicines for fewer days and they should be refilled from time to time, to prevent any sort of excessive and expired medicines. Industries should take measures to treat their wastewater to get rid of the some of the environmental pollutants; similarly, hospitals should also take measures to convert the pharmaceutical products into eco-friendly products. Use of antibiotics and hormones should be checked in food materials as many misuses of hormones are frequently reported in developing countries.

References

[1] Abdulrazaq Y, Abdulsalam A, Rotimi AL, Abdulbasit AA, Clifford O, Abdulsalam OA, Racheal ON, Joy AA, Victor FO, Johannes ZM, Bilal M, Salehdeen UM. Classification, potential routes and risk of emerging pollutants/contaminant. Emerg Contamin. 10.5772/intechopen.94447.

[2] Hoyett Z. Pharmaceuticals and personal care products: Risks, challenges, and solutions. In: Svalova V (ed.). Risk assessment. IntechOpen. 10.5772/intechopen.68673.

[3] Ajala OJ, Tijani JO, Salau RB, Abdulkareem AS, Aremu OS. A review of emerging micro-pollutants in hospital wastewater: Environmental fate and remediation options. Results Eng 2022, 16, 100671.

[4] Majumder A, Gupta AK, Ghosal PS, Varma M. A review on hospital wastewater treatment: A special emphasis on occurrence and removal of pharmaceutically active compounds, resistant microorganisms, and SARS-CoV-2. J Environ Chem Eng 2021, 2, 104812.

[5] Valverde MG, Bueno MJM, Gómez-Ramos MM, Aguilera A, García MDG, Fernández-Alba AR. Determination study of contaminants of emerging concern at trace levels in agricultural soil. A pilot study. Sci Total Environ 2021, 782, 146759.

[6] Kaczala F, Blum SE. The occurrence of veterinary pharmaceuticals in the environment: A review. Curr Anal Chem 2016, 3, 169–182.

[7] Ashfaq M, Noor N, Saif-Ur-Rehman M, Sun Q, Mustafa G, Nazar MF, Yu C-P. Determination of commonly used pharmaceuticals in hospital waste of Pakistan and evaluation of their ecological risk assessment. Clean Soil Air Water 2017, 45, 1500392.

[8] Etteieb S, Magdouli S, Zolfaghari M, Brar SK. Monitoring and analysis of selenium as an emerging contaminant in mining industry: A critical review. Sci Tot Environ 2020, 698, 134339.

[9] Mandalakis M, Stephanou EG, Horii Y, Kannan K. Emerging contaminants in car interiors: Evaluating the impact of airborne PBDEs and PBDD/Fs. Environ Sci Technol 2008, 42, 6431–6436.

[10] Samal K, Mahapatra S, Ali MH. Pharmaceutical wastewater as Emerging Contaminants (EC): Treatment technologies, impact on environment and human health. Energy Nexus 2022, 6, 100076.

[11] Aznar-Alemany O, Eljarrat E. Chapter five – food contamination on flame retardants. Compr Anal Chem 2020, 88, 141–189.

[12] Estevez P, Castro D, Pequeño-Valtierra A, Giraldez J, Gago-Martinez A. Emerging marine biotoxins in seafood from European coasts: Incidence and analytical challenges. Foods 2019, 8, 149.

[13] Kantiani L, Llorca M, Sanchís J, Farré M, Barceló D. Emerging food contaminants: A review. Anal Bioanal Chem 2010, 398, 2413–2427.

[14] Calvo-Flores FG, Isac-García J, Dobado JA. Emerging pollutants. Origin Struct Prop 2017, ISBN: 978-3-527-69121-0.

[15] Castro G, Ramil M, Cela R, Rodríguez I. Identification and determination of emerging pollutants in sewage sludge driven by UPLC-QTOF-MS data mining. Sci Total Environ 2021, 778, 146256.

[16] Nikolopoulou V, Ajibola AS, Aalizadeh R, Thomaidis NS. Wide-scope target and suspect screening of emerging contaminants in sewage sludge from Nigerian WWTPs by UPLC-qToF-MS. Sci Total Environ 2023, 857, 159529.

[17] Xu Z, Wu Q, Duan Y, Yang M, Ou M, Xu X. Development of a novel spectrophotometric method based on diazotization-coupling reaction for determination of bisphenol. J Braz Chem Soc 2017, 28, 1475–1482.

[18] Li Y, Wang L, Zheng M, Lin Y, Xu H, Liu A, Hua Y, Jiang Y, Ning K, Hu S. Thin-layer chromatography coupled with HPLC-DAD/UHPLC-HRMS for target and non-target determination of emerging halogenated organic contaminants in animal-derived foods. Food Chem 2023, 404, 134678.

[19] Abdel-Gawad SA, Arab HH. Potentiometric sensors for the selective determination of benzodiazepine drug residues in real wastewater effluents. Chemosensors 2022, 10, 74. https://doi.org/10.3390/chemosensors10020074].

[20] Patel BR, Hassan Q, Noroozifar M, Kerman K. Simultaneous square wave voltammetric detection of endocrine-disrupting agents using a nanocomposite of magnetic Fe3O4 Nanorods and Poly(3,4-Methylenedioxy)aniline. ACS Sustain Chem Eng 2020, 8, 40, 15108–15119.
[21] Salvador J-P, Kopper K, Miti A, Sanchis A, Marco M-P. Multiplexed immunosensor based on the amperometric transduction for monitoring of marine pollutants in sea water. Sensors 2020, 20, 5532. 10.3390/s20195532.
[22] Ababneh AN, Abu-Dalo MA, Horn C, Hernandez MT. Polarographic determination of benzotriazoles and their sorption behavior on granular activated carbon. Int J Environ Sci Technol 2019, 16, 833–842.
[23] Hurtado-Sánchez MDC, Lozano VA, Rodríguez-Cáceres MI, Durán-Merás I, Escandar GM. Green analytical determination of emerging pollutants in environmental waters using excitation–emission photoinduced fluorescence data and multivariate calibration. Talanta 2015, 134, 215–223.
[24] Yu J, Zhang X, Hou D, Chen F, Mao T, Huang P, Zhang G. Detection of water contamination events using fluorescence spectroscopy and alternating trilinear decomposition algorithm. J Spectroscopy 2017, 1485048.
[25] Rodriguez MA, Lantz AW, Armstrong DW. Capillary electrophoretic method for the detection of bacterial contamination. Anal Chem 2006, 78, 14, 4759–4767.
[26] Quek NM, Law WS, Lau HF, Zhao JH, Hauser PC, Li SFY. Determination of pharmaceuticals classified as emerging pollutants using capillary electrophoresis with capacitively coupled contactless conductivity detection. Electrophoresis 2008, 17, 3701–3709.
[27] Albero B, Tadeo JL, Pérez RA. Determination of emerging contaminants in cereals by gas chromatography-tandem mass spectrometry. Front Chem 2020. https://doi.org/10.3389/fchem.2020.57166
[28] Prebihalo S, Brockman A, Cochran J, Dorman FL. Determination of emerging contaminants in wastewater utilizing comprehensive two-dimensional gas-chromatography coupled with time-of-flight mass spectrometry. J Chromatogr A 2015, 1419, 109–115.
[29] Aalizadeh R, Nikolopoulou V, Alygizakis NA, Thomaidis NS. First novel workflow for semiquantification of emerging contaminants in environmental samples analyzed by gas chromatography–atmospheric pressure chemical ionization–quadrupole time of flight–mass spectrometry. Anal Chem 2022, 94, 27, 9766–9774.
[30] Safwat N, Ayad MF, Abdel-Ghany MF. Development of an eco-friendly reversed-phase HPLC method for the simultaneous determination of three emerging contaminants in water samples. Arch Pharmaceut Sci Ain Shams University 2020, 4, 1, 79–91.
[31] Shaaban H, Górecki T. Fast ultrahigh performance liquid chromatographic method for the simultaneous determination of 25 emerging contaminants in surface water and wastewater samples using superficially porous sub-3 μm particles as an alternative to fully porous sub-2 μm particles. Talanta 2012, 100, 80–89.
[32] Boix C, Ibáñez M, Fabregat-Safont D, Morales E, Pastor L, Sancho JV, Sánchez-Ramírez JE, Hernández F. Analytical methodologies based on LC–MS/MS for monitoring selected emerging compounds in liquid and solid phases of the sewage sludge. MethodsX 2016, 3, 333–342.
[33] Khalid NK, Devadasan D, Aravind UK, Aravindakumar CT. Screening and quantification of emerging contaminants in Periyar River, Kerala (India) by using high-resolution mass spectrometry (LC-Q-ToF-MS). Environ Monit Assess 2018, 190, 370.
[34] Ferrer I, Thurman EM. Liquid chromatography/time-of-flight/mass spectrometry (LC/TOF/MS) for the analysis of emerging contaminants. Trends Anal Chem 2003, 22, 10, 750–756.
[35] Morelli DC, Bernardi G, Morés L, Pierri ME, Carasek E. A green – high throughput –extraction method based on hydrophobic natural deep eutectic solvent for the determination of emerging contaminants in water by high performance liquid chromatography – diode array detection. J Chromatogr 2020, 1626, 461377.

[36] Riva F, Zuccato E, Pacciani C, Colombo A, Castiglioni S. A multi-residue analytical method for extraction and analysis of pharmaceuticals and other selected emerging contaminants in sewage sludge†. Anal Methods 2021, 13, 526–535.
[37] Gezahegn T, Tegegne B, Zewge F, Chandravanshi BS. Salting-out assisted liquid–liquid extraction for the determination of ciprofoxacin residues in water samples by high performance liquid chromatography–diode array detector. BMC Chem 2019, 13, 28.
[38] Ortega-Zamora C, Jiménez-Skrzypek G, González-Sálamo J, Mazzapioda L, Navarra MA, Gentili A, Hernández-Borges J. Extraction of emerging contaminants from environmental waters and urine by dispersive liquid–liquid microextraction with solidification of the floating organic droplet using fenchol: Acetic Acid Deep Eutectic Mixtures, 2022 Accepted article https://doi.org/10.1021/acssuschemeng.2c04044].
[39] De Sousa DNR, Grosseli GM, Mozeto AA, Carneiro RL, Fadini PS. Ultrasound-assisted extraction method for the simultaneous determination of emerging contaminants in freshwater sediments. J Sep Sci 2015, 38, 3454–3460.
[40] Aparicio I, Martín J, Santos JL, Malvar JL, Alonso E. Stir bar sorptive extraction and liquid chromatography–tandem mass spectrometry determination of polar and non-polar emerging and priority pollutants in environmental waters. J Chromatogr A 2017, 1500, 43–52.
[41] Wang S, Oakes KD, Bragg LM, Pawliszyn J, Dixon G, Servos MR. Validation and use of in vivo solid phase micro-extraction (SPME) for the detection of emerging contaminants in fish. Chemosphere 2011, 85, 1472–1480.
[42] Lopes VSA, Riente RR, Da Silva AA, Torquilho DF, Carreira RDS, Marques M. Development of a solid-phase extraction system modified for preconcentration of emerging contaminants in large sample volumes from rivers of the lagoon system in the city of Rio de Janeiro, Brazil. Marine Pollut Bull 2016, 110, 572–577.
[43] Loos R, Tavazzi S, Mariani G, Suurkuusk G, Paracchini B, Umlauf G. Analysis of emerging organic contaminants in water, fish and suspended particulate matter (SPM) in the Joint Danube Survey using solid-phase extraction followed by UHPLC-MS-MS and GC–MS analysis. Sci Tot Environ 2017, 607–608, 1201–1212.
[44] Dharupaneedi SP, Nataraj SK, Nadagouda M, Reddy KR, Shukla SK, Aminabhavi TM. Membrane-based separation of potential emerging pollutants. Sep Purif Technol 2019, 210, 850–866.
[45] Rai B, Shrivastav A. Chapter 26 – Removal of emerging contaminants in water treatment by nanofiltration and reverse osmosis. Emerg Contamin Wastewater Bio-nanotechnology 2022, 605–628.
[46] Azaroff A, Monperrus M, Miossec C, Gassie C, Guyoneaud R. Microbial degradation of hydrophobic emerging contaminants from marine sediment slurries (Capbreton Canyon) to pure bacterial strain. J Hazard Mat 2021, 402, 123477.
[47] Gubitos J, Mongiovi C, Romita R, Cosma P, Nuzzo S, Rizzi V, Fini P. Removal of emerging contaminants from water using cyclodextrin-based polymers and advanced oxidation processes: The case of carbamazepine. Processes 2022, 10, 1703. https://doi.org/10.3390/pr10091703.
[48] Liakos EV, Lazaridou M, Michailidou G, Koumentakou I, Lambropoulou DA, Bikiaris DN, Kyzas GZ. Chitosan adsorbent derivatives for pharmaceuticals removal from effluents: A review. Macromol 2021, 1, 130–154.
[49] Chen R, Hu L, Zhang H, Lin D, Wang J, Xu D, Gong W, Liang H. Toward emerging contaminants removal using acclimated activated sludge in the gravity-driven membrane filtration system. J Hazard Mat 2022, 438, 129541.
[50] Zhou G-J, Lin L, Li X-Y, Leung KMY. Removal of emerging contaminants from wastewater during chemically enhanced primary sedimentation and acidogenic sludge fermentation. Water Res 2020, 175, 115646.

[51] Matamoros V, Salvadó V. Evaluation of a coagulation/flocculation-lamellar clarifier and filtration-UV-chlorination reactor for removing emerging contaminants at full-scale wastewater treatment plants in Spain. J Environ Manag 2013, 117, 96–102.
[52] Fast SA, Gude VG, Truax DD, Martin J, Magbanua BS. A critical evaluation of advanced oxidation processes for emerging contaminants removal. Environ Process 2017, 4, 283–302.
[53] Kumar A, Parta C, Rajendran HK, Narayanasamy S. Activated carbon-chitosan based adsorbent for the efficient removal of the emerging contaminant diclofenac: Synthesis, characterization and phytotoxicity studies. Chemosphere 2022, 307, Pt 2, 135806.
[54] Zhang H-Y, Zhang C-Y, Rao W-L, Zhang H, Liang G-H, Deng X, Zhao J-L, Guan Y-F, Ying -G-G. Influence of biofilms on the adsorption behavior of nine organic emerging contaminants on microplastics in field-laboratory exposure experiments. J Hazard Mat 2022, 434, 128895.
[55] Wang F-X, Wang -C-C, Du X, Li Y, Wang F, Wang P. Efficient removal of emerging organic contaminants via photo-Fenton process over micron-sized Fe-MOF sheet. Chem Eng J 2022, 429, 132495.
[56] Alaghmand M, Alizadeh-Saei J, Barakat S. Adsorption and removal of a selected emerging contaminant, carbamazepine, using Humic acid, Humasorb and Montmorillonite. Equilibrium isotherms, kinetics and effect of the water matrix. J Environ Sci Health A Tox Hazard Subst Environ Eng 2020, 55, 534–1541.
[57] Velusamy K, Periyasamy S, Kumar PS, Jayaraj T, Krishnasamy R, Sindhu J, Sneka D, Subhashini B, Vo D-VN. Analysis on the removal of emerging contaminant from aqueous solution using biochar derived from soap nut seeds. Environ Pollut 2021, 287, 117632.
[58] Hassan M, Du J, Liu Y, Naidu R, Zhang J, Ahsan MAA, Qia F. Magnetic biochar for removal of perfluorooctane sulphonate (PFOS): Interfacial interaction and adsorption mechanism. Environ Technol Innov 2022, 28, 102593.
[59] Delgado-Moreno L, Gasco SBG, Méndez A, El Azzouzi M, Romero E. New insights into the efficient removal of emerging contaminants by biochars and hydrochars derived from olive oil wastes. Sci Total Environ 2021, 752, 141838.

N. M. Nurazzi*, M. N. F. Norrrahim, M. R. M. Asyraf, M. S. Samsudin

Biomass-Derived Nanocellulose for Heavy Metal Removal

Abstract: Heavy metal pollution caused by industrial operations has emerged as a primary alarm on a global scale as a result of its high toxicity, low biodegradability, pervasiveness in the food chain, and risk to human safety. These are the most dangerous pollutants as they can cause problems for people as well as for the environment as a whole. Copper, zinc, mercury, chromium, cadmium, lead, and nickel are some of the most frequently discharged toxic heavy metals. Other examples include cadmium and zinc. In the realm of sustainable advanced nanomaterials, nanocellulose, obtained from biomass sources such as plant fibers, is garnering a growing interest for its potential application in water treatment technology due to the wide variety of specific properties and functionalities that it possesses. Nanocellulose can be derived from various sources including plant fibers, animal cells, and bacteria. Through surface modification and functionalization, the adsorption of heavy metals from contaminated streams is significantly enhanced once the cellulose dimension is reduced to nanosize materials. This review provides the readers with the most recent studies (from the years 2018–2023) on nanocellulose from plant fiber classifications, characteristics, the significance of nanocellulose as a biosorbent, and its mechanism in numerous heavy metal bioremediation and in addition, delivers understanding on the challenges, future outlook, and updates.

Keywords: Adsorption, nanocellulose, heavy metal, biomass, biosorbent

1 Overview

Water pollution is escalating in severity as industrialization progresses. Water pollutants can be categorized into organic and inorganic contaminants. Organic water contaminants include dyes, insecticides, medicines, and solvents, while acids, alkalis, heavy

***Corresponding author: N. M. Nurazzi,** Bioresource Technology Division, School of Industrial Technology, Universiti Sains Malaysia, Penang 11800, Malaysia; Green Biopolymer, Coatings and Packaging Cluster, School of Industrial Technology, Universiti Sains Malaysia, Penang 11800, Malaysia, e-mail: mohd.nurazzi@usm.my

M. N. F. Norrrahim, Research Centre for Chemical Defence, Universiti Pertahanan Nasional Malaysia (UPNM), Kem Perdana Sungai Besi, Kuala Lumpur, 57000, Malaysia

M. R. M. Asyraf, Engineering Design Research Group (EDRG), Faculty of Mechanical Engineering, Universiti Teknologi Malaysia, Johor Bahru, Johor, Malaysia; Centre for Advanced Composite Materials (CACM), Universiti Teknologi Malaysia, Johor Bahru, Johor 81310, Malaysia

M. S. Samsudin, Environmental Technology Division, School of Industrial Technology, Universiti Sains Malaysia, Penang 11800, Malaysia

https://doi.org/10.1515/9783111016825-007

metals, fertilizers, acids, and salts are examples of inorganic water contaminants [1]. For heavy metals, Pb(II), Cd(II), Cu(II), Fe(III), Cr(VI), Hg(II), and Co(II) can be considered as the main heavy metals ion. Chemical precipitation, coagulation or flocculation, ion exchange, reverse osmosis, electrochemical treatment, adsorption, and other removal techniques are frequently used to remove heavy metal ions from solutions or water [2].

However, the treatment of low-concentration wastewater can be accomplished quickly, simply, and affordably by using an adsorption mechanism [3]. Although it is currently a common adsorbent, activated carbon has a high manufacturing cost and difficult regeneration conditions [4]. Sustainable advanced nanomaterials like nanocellulose derived from plant fibers have gained popularity due to their favorable environmental effects, low cost, repeatability, and high adsorption efficiency [5]. Cellulose consists of long single chains made up of D-glucopyranosyl units connected by 1,4-glycosidic linkages that are used to build cellulose and have a high amount of hydroxyl groups that are amenable to further modification. A glucose unit of cellulose contains one primary (C^6) and two secondary (C^2, C^3) hydroxyl groups and the specific area of cellulose is limited (1–4 m^2/g) due to the creation of many hydrogen bonds, which results in reduced hydroxyl exposure and poor hydrophilicity and adsorption [6]. Typically, cellulose nanoparticles have a specific surface area of 50–200 m^2/g, while nanocellulose aerogels have a specific surface area of 250–350 m^2/g [7]. Membranes made of cellulose nanofibers that have undergone the supercritical drying process have a specific surface area that can reach 482 m^2/g [8].

Nanocellulose can be separated into three groups based on the variations in raw materials, processing techniques, and morphology: cellulose nanocrystals (CNC), cellulose nanofibrils, or cellulose microfibrils (CNF). As nanocellulose's surface area increases, more hydroxyl groups become exposed, and the presence of many hydroxyl groups allows nanocellulose to undergo various surface modifications [9]. The high specific surface area of nanocellulose together with the characteristics and density of surface functional groups are critical factors in determining how well they adsorb substances. Because modified nanocellulose has more active binding sites and excellent ion exchange performance, they can increase the adsorption capacity of adsorbents to contaminants in an aqueous solution and decide their selectivity. Besides that, in order to improve the adsorption capacity of the nanocellulose it is necessary to directly or indirectly introduce key functional groups on the nanocellulose surface for the heavy metal ion adsorption sites [10]. Hence, this review summarizes the functional group-based modification of nanocellulose adsorbents, associated adsorption mechanisms, and application of nanocellulose as adsorbents.

2 Nanocellulose from plant fiber properties

Cellulose is recognized as the polymer with the highest levels of abundance on earth, in addition to being natural, renewable, and capable of biodegradation. Most plants, algae, animals, fungi, and bacteria can contain and produce cellulose [11]. Plant fibers and wood are the primary sources of cellulose used in the biocomposite production of microfibril cellulose, pectin, lignin, and hemicellulose [12]. In addition to being the biopolymer with the most significant market share on the planet, it also possesses many interesting properties. Researchers discovered that cellulose in plants is formed from millimeter-sized filaments to nanometer-sized microfibrils. These microfibrils signify the basic structure of cellulose in all sources [13]. Generally, nanocellulose can be extracted from all plants, whether they are hardwood (like eucalyptus or aspen or balsa or oak or elm or maple or birch or rubber wood), softwood (like pine or spruce or juniper or hemlock or yew or larch or cedar), or natural fibers (like cotton, jute, bamboo, or kenaf) [14]. The agroindustrial biomass has the most potential for extracting nanocellulose compared to all other sources. This is because it is readily available in large quantities, which means that it has the potential to contribute to a high level of nanocellulose production on a large scale. It has been reported that these characteristics are linked to their type, origin, height, and age. Cellulose, hemicellulose, and lignin are the three plant fiber types that comprise most of the plants [15].

The surface functionalization and surface modification process transform nanocellulose into a material with valuable properties. The existence of hydroxyl groups (–OH) on the cellulose backbone of nanocellulose makes it possible to functionalize nanocellulose straightforwardly. This facilitates adding the required functionality and may be advantageous for various applications [16]. Figure 1 provides a visual representation of the chemical structure of cellulose, which reveals that it has a plethora of hydroxyl groups.

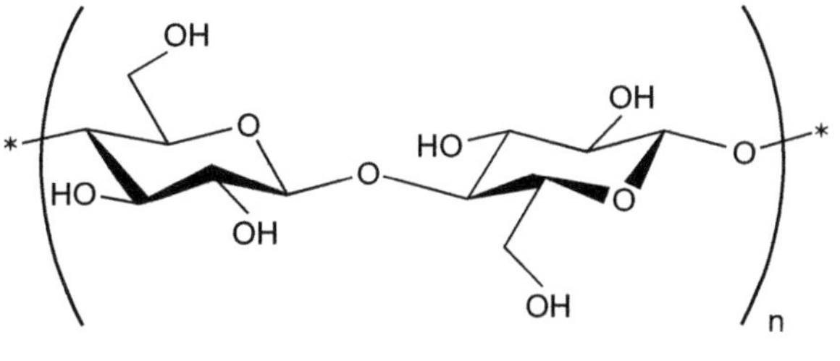

Figure 1: Chemical structure of cellulose.

3 Functionalization of nanocellulose for heavy metal detection

The strong intermolecular hydrogen bonding interaction and high surface energy of nanocellulose caused it to be aggregated in various conditions. Therefore, the surface properties and interfacial compatibility of nanocellulose are crucial for its functionalization and practical application [17]. The nanocellulose fiber has two primary drawbacks such as a high number of hydroxyl groups, which gives the product a gel-like structure, and high hydrophilicity, which restricts its applications. Thus, the modification aims to overcome these two limitations to increase the performance of nanocellulose and broaden its applications [18].

Stable dispersion is necessary for reproducible study results and for all projected nanocellulose products to meet quality control criteria. Nanocellulose can disperse in some strong polar solvents due to the intense interaction between the surface hydroxyls and the solvent molecules (particularly water). Hydrogen bonding between nanofibers produces aggregation at the microscale. In addition, the hydrophilicity of nanocellulose caused it hard to disperse in hydrophobic environments like low polar and nonpolar chemical solvents, and the majority of polymer matrices utilized for composite synthesis. Figure 2 presents an example of several common types of physical, chemical, and biological surface functionalizations of nanocellulose.

Surface functionalization of nanocellulose also improves adsorption performance since it increases the surface area available for adsorbate binding. Therefore, surface functionalization is essential in nanocellulose-based adsorbent materials to allow the adsorption of a specific class of pollutants and enhance adsorption capacity. To boost the adsorption capacity, it is required to increase or introduce ionizable or complexing sites like carboxylic sulfate and amine groups on the surface of the nanocellulose on which the metal is adsorbed. Modifying the nanocellulose surface with phosphate groups increases the adsorption of silver, copper, and ferric ions compared to the unmodified material. The removal efficiency was primarily due to the surface area, type, and density of functional groups on nanocellulose. The examples of functionalized and unfunctionalized nanocellulose as adsorbents for removing heavy metals are tabulated in Table 1.

4 Heavy metals in water

Heavy metals like arsenic, cadmium, chromium, lead, and mercury which typically linked to hazardous health problems and water pollution can be divided into two groups: anion (–) and cation (+). These are primary contaminants that are still under the subject of intensive scientific research. The human body can tolerate little metal under normal circumstances without developing major health problems. However,

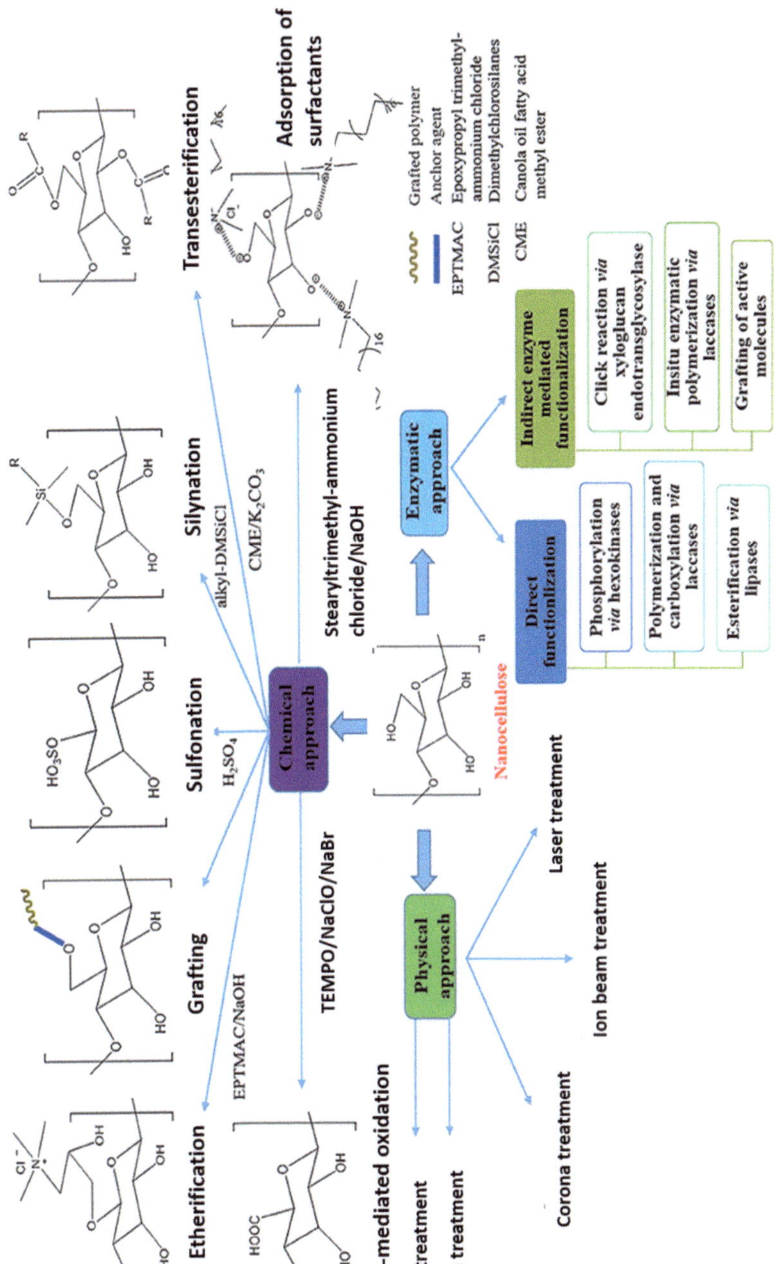

Figure 2: Certain biological, physical, chemical, and biological surface functionalizations of nanocellulose (adapted from ref. [19]).

Table 1: Certain functionalization process involves on nanocellulose as adsorbents toward the heavy metal removal.

Source of nanocellulose adsorbent	Functionalization	Heavy metals	Ref.
Rice husk	Highly charged carboxyl CNF	Lead(II) and lanthanum (III) (Pb(II) and La(III)) ions	[20]
Bamboo	Aminopropyltrimethoxysilane (APTMs) via a facile freeze-drying process	Cu^{2+}, Cd^{2+}, and Hg^{2+}	[21]
Douglas fir-pulp fibers	Modified by ethylene diamine tetraacetic acid (named UiO-66-EDTA) into CNF	Cr^{3+},	[22]
Amorpha fruticosa wood powder	Hybrid aerogel by integrating nanocellulose and ferroferric oxide (Fe_3O_4)	Cr(VI)	[23]
Cotton fibers	Hydrogel composites based on chitosan-*g*-poly(-acrylic acid) matrices filled with cellulose nanowhiskers	Pb(II) and Cu(II)	[24]
Bleached eucalyptus kraft pulp	Carboxylated CNF hybrid with carbon dots	Cu^{2+}	[25]
Peanut shells	Sulfonation to promote charged sulfate esters groups onto the surface	$Cu_2(BTC)_3$	[26]
Kapok fiber	Functionalized with polyethyleneimine andglutaraldehyde to produce bridge effect of iron ions	As(V)	[27]
–	Amino functionalized nanocellulose	Cu(II)	[28]
–	Thiourea-functionalized magnetic ZnO/ nanocellulose	Pb(II)	[29]

heavy metal poisoning of water has grown to be a significant global concern. According to Fu and Wang (2011), heavy metals are metals with a specific gravity greater than 5.0 and an atomic weight between 63.5 and 200.6 [30]. This is especially true when these metals are present in dissolved form. Extreme levels of any of these heavy metals pose a threat to people, the environment, and other living things. The strength of heavy metals and their toxicity depends on several variables that comprise the dosage, exposure route, chemical species, age, gender, genomes, and dietary condition of the affected person. In aquatic environments, heavy metals are highly soluble and readily absorbed by living things. Such metallic substances are considered cumulative toxins and have adverse health effects even at low concentrations. Even with lower occupational exposure levels, these metallic elements are cumulative toxicants with adverse health effects. These heavy metals are classified as human carcinogens by the International Cancer Research Agency and the US Environmental Protection Agency [31]. The primary producers of heavy metal pollution from the mining, metal alloy, paint and dye, leather, and electroplating businesses as well as those that produce fertilizer, cement, and other building materials. Even in low quantities, these heavy metals negatively impact aquatic habitats and can cause natural degradation that is harmful to everyone. The quality of the water is threatened by industrial activities and the direct discharge of municipal waste into water streams. The scarcity of potable water will quickly become a primary concern worldwide, given present patterns in population growth [32]. Table 2 presents the limitations and standard from EPA, WHO, EU, and Department of Environment Malaysia for heavy metals in drinking water.

Table 2: EPA, WHO, EU, and Department of Environment Malaysia for heavy metals for limitations and standards of selected heavy metal in drinking water.

Heavy metals	EPA limitations		WHO provisional guideline value (mg/L)	EU standard (mg/L)	Department of Environment, Malaysia (mg/L)
	Maximum contamination level (MCL) (mg/L)	Maximum contamination limit goal (MCLG) (mg/L)			
Lead	0.015	0	0.01	0.01	0.05
Chromium	0.1	0.1	0.05	0.05	0.05
Cadmium	0.005	0.005	0.003	0.005	0.003
Arsenic	0.010	0	0.01	0.01	0.01
Mercury	0.002	0.002	0.006	0.001	0.001
Copper	1.3	1.3	2.0	2.0	1.0
Zinc	5	–	3.0	–	3.0
Nickel	–	–	0.07	–	–
Ref.	[33, 34]			[35]	[36]

5 Performance and mechanism of nanocellulose as adsorbent for heavy metals

The term "adsorption" refers to a process in which a liquid solute or adsorbate is transferred to the surface of an adsorbent [37]. Physical adsorption, also referred to as physisorption, and chemical adsorption are the two types of adsorption that can be distinguished (chemisorption). The natural interaction between molecules and weak van der Waals forces are involved in physisorption. In contrast, the chemical bonding that results in chemisorption is responsible for the intense attraction between molecules. When it comes to removing heavy metals from wastewater with a low pollutant concentration, adsorption has been recognized as a method that is appealing, straightforward, and economical. Adsorption has been deemed an excellent operational method due to its simple operating procedure and low cost compared to other approaches [38]. A group of researchers reached this conclusion after conducting an in-depth analysis of the suitability of the adsorption method in the removal of heavy metals. Ion exchange, chemical adsorption, and physical adsorption are the three primary types of adsorption mechanisms. Chemical adsorption occurs as a consequence of the formation of chemical bonds, whereas the van der Waals force causes physical adsorption. In addition, the appropriate desorption can recover adsorbent using this method, and the process is considered to be the one that can be reversed. Due to the high surface area, a significant number of the adsorptive can be captured or retained by the adsorbate. In addition, the high porosity makes it possible for materials to quickly transfer mass within the pores, resulting in faster kinetics for eliminating pollutants. In order to eliminate excess metals productively, it is first necessary to identify the dominant charges carried by the pollutant in the wastewater. After this has been accomplished, appropriate functionalization can be done to the surface of the nanocellulose. The fundamental adsorption mechanism of the adsorbate onto the adsorbed phase is illustrated in Figure 3.

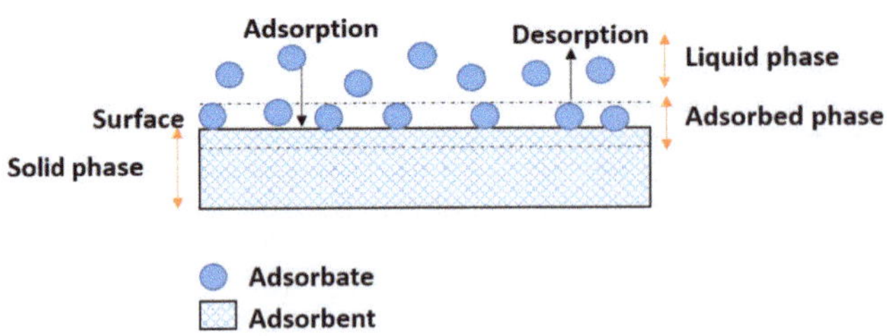

Figure 3: Fundamental adsorption mechanism of adsorbate (adapted from ref. [39]).

Adsorption can be affected by various factors including the pH of the solution, the concentration of metal ions, and the amount of adsorbent present. When optimized, the conditions of the variables offer a higher adsorption capacity, resulting in the development of an efficient method for eliminating metal contaminants [40]. The majority of the heavy metals that can be found in wastewater are in a cationic state. When the pH is low, positively charged heavy metal ions compete with hydronium ions, which have the charge H_3O^+, to be bound to the surface of the adsorbent. As a result, the amount of adsorption capacity that the adsorbate has for the adsorbent is significantly reduced. When the adsorption reaction occurs in an environment with a higher pH, a situation very similar to the one described above occurs. Because fewer active sites are available for the adsorbate binding, the adsorption process performance will become worse. The initial concentration of metal ions is yet another essential factor that plays a role in the uptake of metal ions by the cell. Since the concentration gradient drives adsorption capacity, having a higher initial concentration of heavy metals in solution results in greater adsorption capacity. When the initial concentration is higher than the value determined to be optimal for adsorption, the adsorbent's performance suffers because there are not enough binding sites available on the surface of the adsorbent.

If the concentration is increased beyond the optimal level, the binding sites will become full, and the surplus of metal ions will not be adsorbed. This results in a decrease in the percentage of ions that are removed from the solution. When the dosage of the adsorbent is increased to a specific limit, more adsorbing sites are available for the metal contaminants to bind to, which results in an increase in the efficiency with which ions can be removed from the solution [41]. On the other hand, increasing the dosage of the adsorbent results in an increase in the number of unsaturated metal ions on the surface of the adsorbent. This causes the adsorption capacity to decrease. Once the optimal dosage has been reached, there is no significant increase in the reported removal percentage; further increases in the dosage amount result in a reduction in the removal efficiency.

The concept of adsorption isotherm describes the connection amongst the concentration of heavy metal ions interacting with one another and the amount of adsorbate that is present on the surface of the adsorbent when the pH and temperature are held constant. Both the Langmuir and Freundlich models are examples of isotherm models frequently utilized in adsorption research. The Langmuir model assumes that the process is a monolayer type of adsorption, in which the adsorbates bind on the surface of the adsorbent only in one layer. In contrast, the Freundlich isotherm defines the process as a multilayer type of adsorption while the Langmuir model is based on the fact that the adsorbates bind on the surface of the adsorbent only in one layer [42]. It has been demonstrated that nanocellulose is a highly effective adsorbent for eliminating heavy metals. Figure 4 depicts the increasing trends in the number of publications related to the field of nanocellulose as adsorbents over the past five years, with a total of 14,460 publications. This can be seen by the research that was conducted (search on January 8, 2023, using the keyword "nanocellulose as adsorbent of heavy metal ions" on Google Scholar).

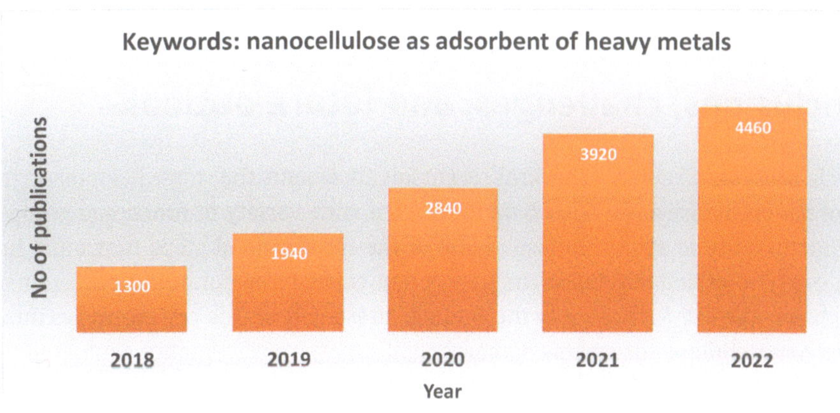

Figure 4: Trend number of publications with keywords of "nanocellulose as adsorbent of heavy metals" for the past five years.

The CNF/carboxymethyl cellulose (CMC)/branched polyethyleneimine (BPEI) aerogel was developed for the Cu(II) adsorption by Mo et al. [43]. The adsorption capacity of Cu (II) ion uptake at pH 6.0 with 452.49 mg/g after approximately 60 min was determined based on the results obtained. The porous BPEI@CNF/CMC aerogel contributed to the fast adsorption kinetics and high adsorption capacity in the static copper adsorption process [43]. This was because the aerogel possessed a high density of metal-chelating groups. A fluorescent nanocellulose-based hydrogel incorporating titanate nanofibers was developed to adsorb Cr(VI). The capacity for adsorption of the Cr(VI) ion was determined to be 648.4 mg/g [44]. The carboxyl functionalized nanocellulose was developed by Karim et al. [45] for the removal of heavy metals (Ag(I), Cu(II), Fe(II), and Fe(III). When compared to unfunctionalized nanocellulose, the adsorption capacity of Cu(II) was increased up to 374 mg/g, and the adsorption capacity of Fe(II)/Fe(III) was increased up to 456 mg/g. Compared to the increase seen in CNF and CNC that had not been functionalized, this one was 1.2–1.3 times more significant [45]. As a result, Mo et al. (2019) successfully fabricated an amino-functionalized 3D multiwall perforated CNF-based polyethyleneimine (PEI) for efficient Cu(II) adsorption. Using the Langmuir model, the developed adsorbent demonstrated a maximum adsorption capacity of 485.44 mg/g, which was calculated. It was discovered that the adsorbent that had been developed had a high level of reversibility (four cycles without significant degradation) [28]. In addition, Li et al. [46] discovered the application of nitrocellulose-based PEI. To remove Cu(II) and Pb, they fabricated a functionalized CNF with PEI (II). Research in thermodynamics and kinetics led to the discovery that the pseudo-second-order model and the Langmuir model provided the best fit for the data. At a pH of 5.5, the maximum adsorption capacity of copper(II) and lead(II) reached 175.44 and 357.44 mg/g, respectively, and it was discovered that the adsorption capacity was retained for multiple uses. EDTA could quickly regenerate the developed adsorbent and maintain its adsorption capacity even after being used multiple times.

6 Conclusions, challenges, and future outlook

This article presents a review of several recent advancements that have been made in the use of nanocellulose as an adsorbent. Utilizing a wide variety of functional groups to add functionality to nanocellulose is one of the most critical steps that must be taken in order to effectively boost the adsorption capacity of nanocellulose against multiple heavy metals. According to the findings presented in this review, most functionalized nanocellulose adsorbents demonstrated superior performance compared to other adsorbents currently on the market. Even though several studies have demonstrated that nanocellulose is a compelling new bio-based adsorbent for a variety of pollutants, some improvements can still be made to the material. The vast majority of studies on adsorption are conducted only on a batch scale, and very few works have investigated the properties of adsorption under dynamic conditions and, more specifically, for the treatment of industrial wastewater. The batch mode adsorption could be moved to adsorption columns filled with nanocellulose as a porous structure to make room for continuous filtration to treat actual water. In addition, more study is required to generate nanoscale hybrid structures on the surface of nanocellulose. These structures need to be designed to make it likely that they will interact with a variety of toxic elements. This will make it possible to develop composite adsorbents capable of absorbing several chemical species simultaneously.

The production of nanocellulose is costly, particularly at the industrial level. The amount of energy consumed during the manufacturing process of nanocellulose still needs to be solved, which prevents the production of nanocellulose from being scaled up. Despite this, there has been significant development in this field, and there are now a growing number of pilot-scale production facilities available across the globe. The biodegradability of nanocellulose is another essential factor that needs to be considered. The biodegradation of the developed adsorbents was only investigated in some of the research presented here. Because of this, the upcoming research needs to determine how long nanocellulose can function as an adsorbent. Nanocellulose's ability to degrade is impacted by various factors including the types of water it is exposed to and the specific microbes present in environments where it has been used. Therefore, these studies ought to take a life-cycle approach to analyze nanocellulose, beginning with its production and proceeding through its use and, finally, its disposal.

References

[1] Abouzeid RE, Khiari R, El-Wakil N, Dufresne A. Current state and new trends in the use of cellulose nanomaterials for wastewater treatment. Biomacromolecules 2018, 20, 573–597.

[2] Liu P, Borrell PF, Božič M, Kokol V, Oksman K, Mathew AP. Nanocelluloses and their phosphorylated derivatives for selective adsorption of Ag^+, Cu^{2+} and Fe^{3+} from industrial effluents. J Hazard Mater 2015, 294, 177–185.

[3] Tshikovhi A, Mishra SB, Mishra AK. Nanocellulose-based composites for the removal of contaminants from wastewater. Int J Biol Macromol 2020, 152, 616–632.

[4] Köse K, Mavlan M, Youngblood JP. Applications and impact of nanocellulose based adsorbents. Cellulose 2020, 27, 2967–2990.

[5] Norizan MN, Shazleen SS, Alias AH, Sabaruddin FA, Asyraf MRM, Zainudin ES, et al. Nanocellulose-based nanocomposites for sustainable applications: A review. Nanomaterials 2022, 12, 3483.

[6] Norrrahim MNF, Knight VF, Nurazzi NM, Jenol MA, Misenan MSM, Janudin N, et al. The frontiers of functionalized nanocellulose-based composites and their application as chemical sensors. Polymers 2022, 14, 4461.

[7] Mishra RK, Sabu A, Tiwari SK. Materials chemistry and the futurist eco-friendly applications of nanocellulose: Status and prospect. J Saudi Chem Soc 2018, 22, 949–978.

[8] Sehaqui H, Zhou Q, Ikkala O, Berglund LA. Strong and tough cellulose nanopaper with high specific surface area and porosity. Biomacromolecules 2011, 12, 3638–3644.

[9] Norrrahim MNF, Nurazzi NM, Jenol MA, Farid MAA, Janudin N, Ujang FA, et al. Emerging development of nanocellulose as an antimicrobial material: An overview. Mater Adv 2021, 2, 3538–3551.
[10] Si R, Pu J, Luo H, Wu C, Duan G. Nanocellulose-based adsorbents for heavy metal ion. Polymers 2022, 14, 5479.
[11] Abitbol T, Rivkin A, Cao Y, Nevo Y, Abraham E, Ben-Shalom T, et al. Nanocellulose, a tiny fiber with huge applications. Curr Opin Biotechnol 2016, 39, 76–88.
[12] Zinge C, Kandasubramanian B. Nanocellulose based biodegradable polymers. Eur Polym J 2020, 133, 109758.
[13] Pennells J, Godwin ID, Amiralian N, Martin DJ. Trends in the production of cellulose nanofibers from non-wood sources. Cellulose 2020, 27, 575–593.
[14] Sabaruddin FA, Paridah MT, Sapuan SM, Ilyas RA, Lee SH, Abdan K, et al. The effects of unbleached and bleached nanocellulose on the thermal and flammability of polypropylene-reinforced kenaf core hybrid polymer bionanocomposites. Polymers 2020, 13, 116.
[15] Shesan OJ, Stephen AC, Chioma AG, Neerish R, Rotimi SE. Fiber-matrix relationship for composites preparation. Renew Sust Compos 2019, 1–30.
[16] Norrrahim MNF, Ariffin H, Hassan MA, Ibrahim NA, Yunus WMZW, Nishida H. Utilisation of superheated steam in oil palm biomass pretreatment process for reduced chemical use and enhanced cellulose nanofibre production. Int J Nanotech 2019, 16, 668–679.
[17] Norrrahim MNF, Farid MAA, Lawal AA, Yasim-Anuar TAT, Samsudin MH, Zulkifli AA. Emerging technologies for value-added use of oil palm biomass. Environ Sci Adv 2022.
18] Rol F, Belgacem MN, Gandini A, Bras J. Recent advances in surface-modified cellulose nanofibrils. Prog Polym Sci 2019, 88, 241–264.
[19] Trache D, Tarchoun AF, Derradji M, Hamidon TS, Masruchin N, Brosse N, et al. Nanocellulose: from fundamentals to advanced applications. Front Chem, 2020, 8, 392.
[20] Zhan C, Sharma PR, He H, Sharma SK, McCauley-Pearl A, Wang R, et al. Rice husk based nanocellulose scaffolds for highly efficient removal of heavy metal ions from contaminated water. Environ Sci Water Res Technol 2020, 6, 3080–3090.
[21] Geng B, Xu Z, Liang P, Zhang J, Christie P, Liu H, et al. Three-dimensional macroscopic aminosilylated nanocellulose aerogels as sustainable bio-adsorbents for the effective removal of heavy metal ions. Int J Biol Macromol 2021, 190, 170–177.
[22] Li J, Tan S, Xu Z. Anisotropic nanocellulose aerogel loaded with modified UiO-66 as efficient adsorbent for heavy metal ions removal. Nanomaterials 2020, 10, 1114.
[23] Wei J, Yang Z, Sun Y, Wang C, Fan J, Kang G, et al. Nanocellulose-based magnetic hybrid aerogel for adsorption of heavy metal ions from water. J Mater Sci 2019, 54, 6709–6718.
[24] Rodrigues FHA, de C Magalhães CE, Medina AL, Fajardo AR. Hydrogel composites containing nanocellulose as adsorbents for aqueous removal of heavy metals: Design, optimization, and application. Cellulose 2019, 26, 9119–9133.
[25] Guo X, Xu D, Yuan H, Luo Q, Tang S, Liu L, et al. A novel fluorescent nanocellulosic hydrogel based on carbon dots for efficient adsorption and sensitive sensing in heavy metals. J Mater Chem A 2019, 7, 27081–27088.
[26] Shaheed N, Javanshir S, Esmkhani M, Dekamin MG, Naimi-Jamal MR. Synthesis of nanocellulose aerogels and Cu-BTC/nanocellulose aerogel composites for adsorption of organic dyes and heavy metal ions. Sci Rep 2021, 11, 1–11.
[27] Chai F, Wang R, Yan L, Li G, Cai Y, Xi C. Facile fabrication of pH-sensitive nanoparticles based on nanocellulose for fast and efficient As (V) removal. Carbohydr Polym 2020, 245, 116511.
[28] Mo L, Pang H, Tan Y, Zhang S, Li J. 3D multi-wall perforated nanocellulose-based polyethylenimine aerogels for ultrahigh efficient and reversible removal of Cu (II) ions from water. Chem Eng J 2019, 378, 122157.

[29] Alipour A, Zarinabadi S, Azimi A, Mirzaei M. Adsorptive removal of Pb (II) ions from aqueous solutions by thiourea-functionalized magnetic ZnO/nanocellulose composite: Optimization by response surface methodology (RSM). Int J Biol Macromol 2020, 151, 124–135.
[30] Fu F, Wang Q. Removal of heavy metal ions from wastewaters: A review. J Environ Manage 2011, 92, 407–418.
[31] Deng S, Chen S, Xue Y, Du Z, Wang P. Rapid and effective preparation of a HPEI modified biosorbent based on cellulose fiber with a microwave irradiation method for enhanced arsenic removal in water. J Mater Chem A 2016, 4, 15851–15860.
[32] Carolin CF, Kumar PS, Saravanan A, Joshiba GJ, Naushad M. Efficient techniques for the removal of toxic heavy metals from aquatic environment: A review. J Environ Chem Eng 2017, 5, 2782–2799.
[33] U. S. E. Agency. Drinking Water Contaminants, 2015. Available: http://www.epa.goc/safewater/contaminants/index.html
[34] WHO. World Health Organization International Standards for Drinking Water. ed. Geneva, 2008.
[35] Hoang AT, Nižetić S, Cheng CK, Luque R, Thomas S, Banh TL, et al. Heavy metal removal by biomass-derived carbon nanotubes as a greener environmental remediation: A comprehensive review. Chemosphere 2022, 287, 131959.
[36] M. Department of Environment. Environmental quality report 2020. Department of Environment, Minister of Environment and Water of Malaysia, 2020.
[37] Singh NB, Nagpal G, Agrawal S. Water purification by using adsorbents: A review. Environ Technol Innov 2018, 11, 187–240.
[38] Dehghani MH, Sanaei D, Ali I, Bhatnagar A. Removal of chromium (VI) from aqueous solution using treated waste newspaper as a low-cost adsorbent: Kinetic modeling and isotherm studies. J Mol Liq 2016, 215, 671–679.
[39] Worch E. Adsorption technology in water treatment: Fundamentals. Process Model 2012, 1–332.
[40] Joseph L, Jun BM, Flora JRV, Park CM, Yoon Y. Removal of heavy metals from water sources in the developing world using low-cost materials: A review. Chemosphere 2019, 229, 142–159.
[41] El-Tawil RS, El-Wakeel ST, Abdel-Ghany AE, Abuzeid HAM, Selim KA, Hashem AM. Silver/quartz nanocomposite as an adsorbent for removal of mercury (II) ions from aqueous solutions. Heliyon 2019, 5, e02415.
[42] Wang J, Guo X. Adsorption isotherm models: Classification, physical meaning, application and solving method. Chemosphere 2020, 258, 127279.
[43] Mo L, Tan Y, Shen Y, Zhang S. Highly compressible nanocellulose aerogels with a cellular structure for high-performance adsorption of Cu (II). Chemosphere 2022, 291, 132887.
[44] Peng J, Yuan H, Ren T, Liu Z, Qiao J, Ma Q, et al. Fluorescent nanocellulose-based hydrogel incorporating titanate nanofibers for sorption and detection of Cr (VI). Int J Biol Macromol 2022, 215, 625–634.
[45] Karim Z, Hakalahti M, Tammelin T, Mathew AP. In situ TEMPO surface functionalization of nanocellulose membranes for enhanced adsorption of metal ions from aqueous medium. RSC Adv 2017, 7, 5232–5241.
[46] Li J, Zuo K, Wu W, Xu Z, Yi Y, Jing Y, et al. Shape memory aerogels from nanocellulose and polyethyleneimine as a novel adsorbent for removal of Cu (II) and Pb (II). Carbohydr Polym 2018, 196, 376–384.

Mohammad Qutob, Mohd Rafatullah*, Syahidah Akmal Muhammad, Masoom Raza Siddiqui

Insight into Polycyclic Aromatic Hydrocarbons: The Occurrence, Sources, Detection, Toxicity, and Their Biodegradation

Abstract: According to the global climate change, the increase in population number, industrialization, and the increase of contaminants that are released into the environment has put pressure on our planet and increased the threat to public health. Persistent organic pollutants (POPs) are a class of harmful compounds that are resistant to natural biological, chemical, and photolytic decomposition. POPs contain many groups of pollutants, including polycyclic aromatic hydrocarbons (PAHs). In this chapter our discussion will be focused on 16 PAH compounds that were classified as a priority pollutant by USEPA, we tracked their life cycle from cradle to grave and we mentioned their toxicity and negative impacts on the environment. The PAHs could be formed through the combustion process in the presence of acetylene/vinyl acetylene or their derivatives and benzene ring. Most PAHs that are released into the environment come from incomplete combustion from anthropogenic sources. Based on the published literature, PAH compounds have been traced in various mediums, for example, soil, sediments, rivers, seas, lakes, food. Finally, we mentioned the bioremediation techniques that are used to eliminate PAHs by using different biological agents and different mediums. This chapter provides useful information for those who want to dig deeper into PAH compounds.

Keywords: polycyclic aromatic hydrocarbons, occurrence, sources, detection, toxicity, biodegradation

Acknowledgment: The authors would like to express their appreciation to the Ministry of Higher Education Malaysia for the Fundamental Research Grant Scheme with Project Code: FRGS/1/2019/STG07/USM/02/12.

***Corresponding author: Mohd Rafatullah,** Environmental Technology Division, School of Industrial Technology, Universiti Sains Malaysia, Penang 11800, Malaysia; Renewable Biomass Transformation Cluster, School of Industrial Technology, Universiti Sains Malaysia, Penang 11800, Malaysia, e-mail: mrafatullah@usm.my (MR)

Mohammad Qutob, Environmental Technology Division, School of Industrial Technology, Universiti Sains Malaysia, Penang 11800, Malaysia

Syahidah Akmal Muhammad, Environmental Technology Division, School of Industrial Technology, Universiti Sains Malaysia, Penang 11800, Malaysia; Renewable Biomass Transformation Cluster, School of Industrial Technology, Universiti Sains Malaysia, Penang 11800, Malaysia

Masoom Raza Siddiqui, Advanced Materials Research Chair, Chemistry Department, College of Science, King Saud University, Riyadh, Saudi Arabia

https://doi.org/10.1515/9783111016825-008

1 Introduction

Persistent organic pollutants (POPs) are a category of toxic substances that remain in the environment for a long time because they are resistant to biological, chemical, and photolytic decomposition [1]. POPs became a reason of concern due to their special properties, which included semivolatility, lengthy half-lives, resistive nature, negative toxicological impact, long-range transport before deposition, transformation, high lipophilicity, low water solubility, halogenated, and bioaccumulation ability [2–4]. The source points of POPs could be classified into two categories (i) natural sources such as volcanic activity, forest fires, and biological sources (plants, algae, microbes, transformation organic, etc.) and (ii) anthropogenic sources (manmade) like pesticides, agrochemicals, building destruction, the incineration of animal carcasses, equipment repair, and maintenance, recycling activities, the storage of organic chlorine pesticides, incomplete combustion of coal fuel, gas, crude oil, wood, etc. [5]. According to the transboundary property of POPs, they have been detected in many sensitive places like groundwater and polar regions [6]. In addition, POPs have been detected frequently above their recommended values in different countries around the world even in soil, sediments, food, fresh water, and in the lipid tissues also organisms' cells [7]. In this regard, several conferences and conventions have been conducted to alert the scientific community and the general masses about the adverse effects of POPs on public health along with the ecosystem and to put control around their production and release into the environment such as Aarhus protocol, Barcelona resolution, Rotterdam Convention, Stockholm Convention, Basel Convention, and Arctic environmental protection strategy [8, 9]. The researchers classified the POPs into three main groups including (i) pesticides such as dieldrin, dichlorodiphenyltrichloroethane (DDT), mirex, aldrin, chlordane, chlordecone, endrin, hexachlorobenzene, toxaphene, heptachlor, lindane; (ii) industrial chemicals such as pentachlorophenol, hexabromobiphenyl, polychlorinated biphenyls (PCB), short-chain chlorinated paraffins; and (iii) unintentional by-products like dioxins, furans, polycyclic aromatic hydrocarbons (PAHs) [10, 11]. Many fate scenarios for POPs in the environment have been proposed and concluded in Figure 1.

Among POPs, the USEPA classified many PAHs as priority pollutants according to their hazardous effects. In this chapter, we discussed the formation, sources, toxicity, detection, fates, and biodegradation of PAHs.

2 Polycyclic aromatic hydrocarbons

PAHs are compounds which have two or more benzene rings in their chemical structure. Due to their toxicity and frequent existence in ecosystems such as the atmosphere, soil, water, and sediment, researchers have been concerned about PAHs since

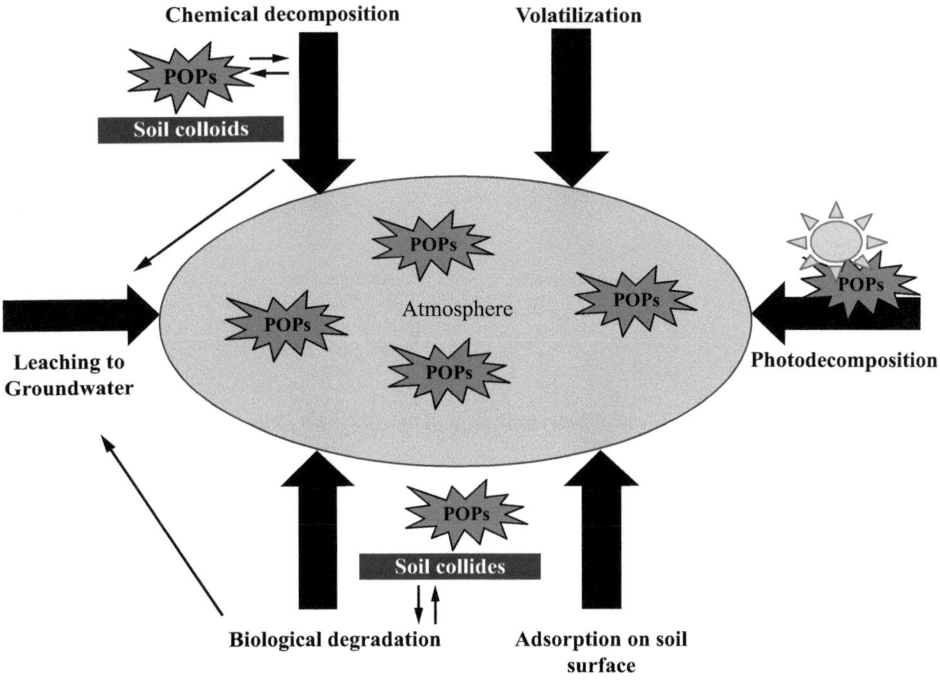

Figure 1: The fate scenarios of POPs in the environment [8].

the 1980s [12]. The environmental monitoring of PAHs began more than 40 years ago. The USEPA mentioned the following 16 PAHs as priority pollutants (Figure 2).

The emission of PAHs may occur from natural occurrences like volcanic eruptions. Forest fires where partial combustion of organic materials occurs also lead to the emission of PAHs [13, 14]. However, anthropogenic activities including burning biomass, making coke, using thermal industrial processes, driving, and burning rubbish are considered the majority of PAHs emissions [15]. Based on IARC reports, benzo[a]pyrene is one of the priority PAHs that are most likely to cause cancer. Benzo[a]pyrene is chosen as a marker for PAH pollution. On the basis of World Health Organization's recommendation, in drinking water, benzo[a]pyrene should not exceed 0.7 µg/L while the sum of selected PAHs should be in the range of 1 and 11 µg/L [16].

3 Formation mechanisms of PAHs

It has been reported that most of the PAH compounds that are released into the environment come from the incomplete combustion process either from a natural event or from an anthropogenic activity. The researchers proposed different mechanisms for the formation of PAHs and we can summarize them into three main mechanisms

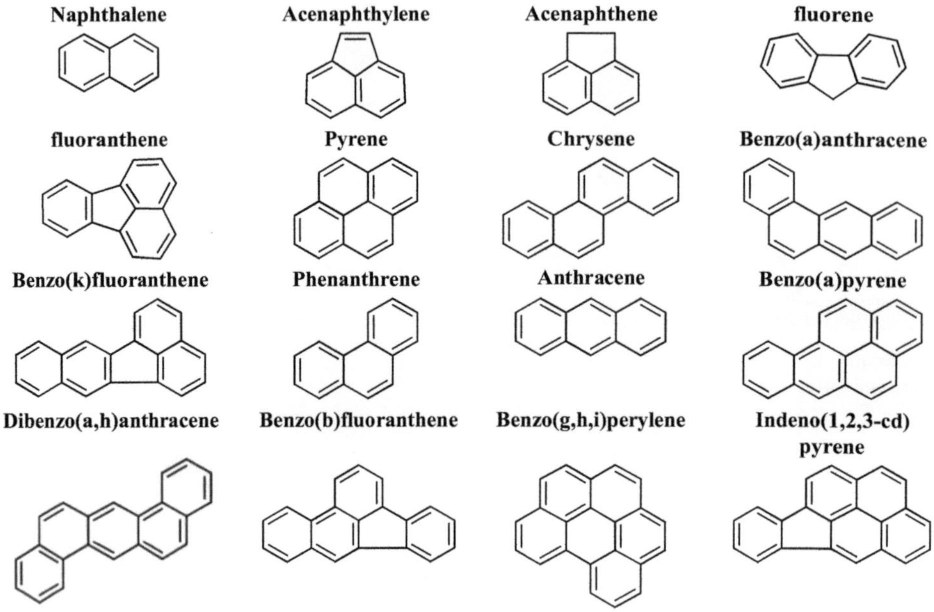

Figure 2: The PAHs that are classified as priority pollutants.

(i) the formation of PAHs via the addition of acetylene or hydrogen abstraction, (ii) the formation of PAHs via vinyl acetylene additions, and (iii) the formation of PAHs via adding radicals.

The formation of PAHs via the addition of acetylene or hydrogen abstraction and acetylene/carbon addition (HACA) could describe the formation of PAHs if the source was either anthropogenic activity or natural events (figure 3 (A)). HACA is the most common reaction mechanism for the formation of PAHs and has gained considerable attention over the last few decades as one of the key PAH formation pathways. Due to the low reaction reversibility and high affinity of hydrogen atoms, the addition of acetylene to PAHs radical promotes synthesis of PAHs. In addition, the reaction steps of the HACA also exhibit substantial exothermicity and comparatively low reaction barriers. These facts make it possible to readily overcome the energy barriers that arise during HACA reactions [17].

The presence of reactants in different forms (butadiene, butane, benzene, toluene, methane, ethane, ethylene, dimethyl ether, phenylacetylene, heptane, etc.) acts to increase the probability of the formation of PAHs.

The formation of PAHs via vinyl acetylene additions (HAVA) was proposed for the first time by Badger et al. [18]. Due to the presence of vinyl acetylene in the combustion environment, different types of flame increase the opportunity for HAVA to happen. It has been reported that the formation of a benzene ring is easy in the presence of vinyl acetylene and carbon species. Because of the double and triple bonds in vinyl

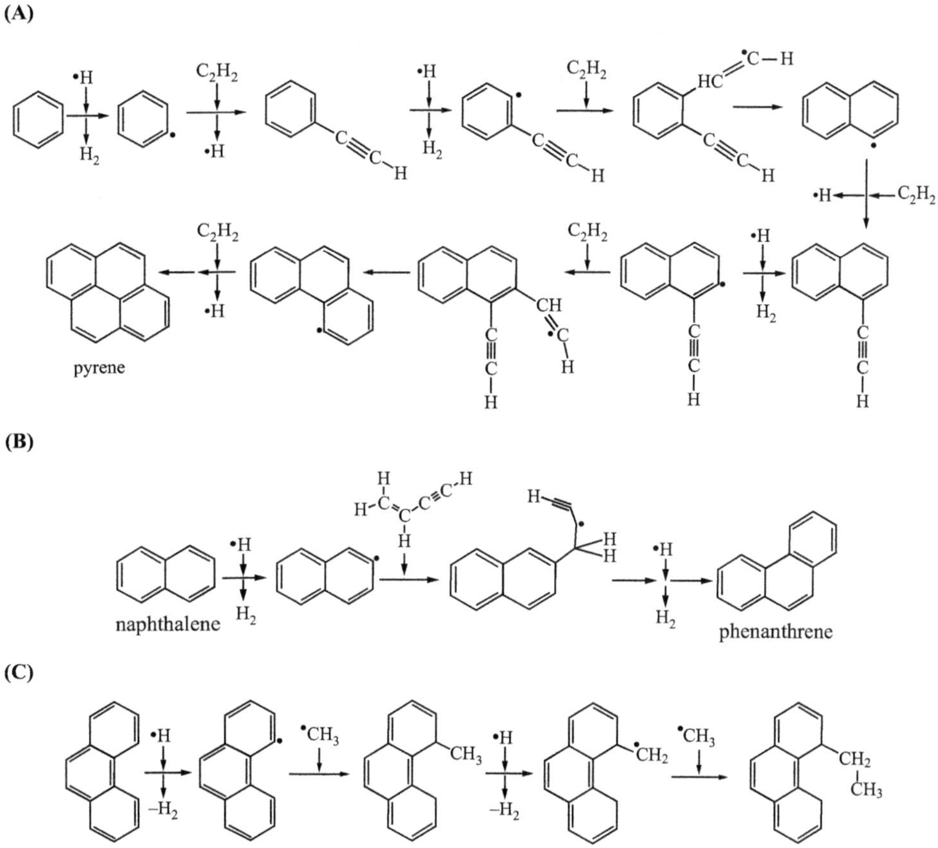

Figure 3: (A) The formation pathway of pyrene via HACA mechanism, (B) the formation of phenanthrene via HAVA mechanism, and (C) the formation of PAHs through the MAC mechanism [17].

acetylene, its carbon atoms are extremely active and are used to attack the C atoms in PAHs. The occurrence of PAHs via the HAVA mechanism can be described with two major steps illustrated in Figure 3(B); after a PAH radical is created, the HAVA mechanism allows for the formation of a new ring by (i) adding vinyl acetylene to the radical and then (ii) cyclizing it.

Radicals among the various chemical species have a special function in the development of PAHs. Different reaction processes are feasible by utilizing less or more stable carbon-centered radicals, depending on the type of radicals. Several PAH formations based on radicals have been proposed such as (i) methyl addition cyclization (MAC), (ii) ethynyl radical, (iii) phenyl addition cyclization, and (iv) resonantly stabilized radicals. To describe the mechanism, we choose methyl radical as one of the most basic, but also prevalent, alkyl radicals. As illustrated in Figure 3(C), in the MAC mechanism, two or three methyl radicals are added to the appropriate PAHs to form an ethyl or propyl chain, followed by hydrogen eliminations and cyclization to produce new aromatic rings [17].

4 The toxicity of PAHs

The toxicity of the 16 PAHs has been frequently reported. For example, Sathikumaran et al. [19] studied the toxicity symptoms of euryhaline fish named *Therapon jarbua* exposed to benzo(α)pyrene for 21 days with two distinct ambient concentrations. The authors examined the toxicity of different biomarkers such as DNA damage in the gill and liver, serotonin, acetylcholinesterase, and ethoxyresorufin-*O*-deethlylase. The results demonstrated that DNA damage and biotransformation enzyme levels increased with exposure time. Both the gill and the liver experienced dose-dependent ethoxyresorufin-*O*-deethlylase activation and DNA damage as a result of benzo(α)pyrene exposure, albeit the liver's reaction was more linear. Benzo(α)pyrene poisoning increased the neurotoxicity by reducing acetylcholinesterase activity and serotonin levels in the brain. Tani et al. [20] examined the acute toxicity test of different PAH compounds on *Hyalella azteca* with exposure 96 h, and *Daphnia magna* 48 h. They revealed that dibenz(a,h)anthracene and anthracene were nontoxic for both species, while the EC50/LC50 of phenanthrene, and benzo(a)pyrene was 330 and 0.66 µg/L, respectively for *Hyalella azteca* and *Daphnia magna*. In addition, Sun et al. [21] investigated the adverse effect of benzo(a)anthracene and fluoranthene separately and mix on the α-amylase enzyme activity of the earthworm *Eisenia fetida* in soil. The findings demonstrated that benzo(a)anthracene and fluoranthene deeply impacted the microenvironment of α-amylase's amino acid residue and its skeleton in separate and mixed pollutants systems. Furthermore, the mixed system had a greater inhibitory effect on α-amylase activity than the separated one. Nowadays, several software has been used for the toxicity prediction such as toxicity estimation software provided by USEPA. In Table 1, we predict the acute toxicity of Fathead minnow LC_{50} 96 h, toxicity development test, and mutagenicity test for the 16 PAHs.

Table 1: Different toxicity tests for 16 PAHs done via Toxicity Estimation Software.

PAH compounds	Toxicity development test (prediction result)	Fathead minnow LC_{50} 96 h (prediction value, mg/L)	Experimental mutagenicity test	Prediction mutagenicity test
Naphthalene	Developmental toxicant	8.15	–	*
Acenaphthylene	Developmental toxicant	2.45	*	*
Acenaphthene	Developmental nontoxicant	2.70	–	–
fluorene	Developmental toxicant	10.29	–	–

Table 1 (continued)

PAH compounds	Toxicity development test (prediction result)	Fathead minnow LC_{50} 96 h (prediction value, mg/L)	Experimental mutagenicity test	Prediction mutagenicity test
fluoranthene	Developmental toxicant	3.30×10^{-2}	*	+
Pyrene	Developmental NON-toxicant	7.85×10^{-2}	*	+
Chrysene	Developmental toxicant	2.27×10^{-2}	+	+
Benzo(a) anthracene	Developmental toxicant	2.39×10^{-2}	+	+
Benzo(k) fluoranthene	Developmental nontoxicant	5.54×10^{-2}	+	+
Phenanthrene	Developmental toxicant	0.81	*	+
Anthracene	Developmental toxicant	0.39	*	+
Benzo(a)pyrene	Developmental nontoxicant	5.47×10^{-2}	*	+
Dibenzo(a,h) anthracene	Developmental nontoxicant	3.56×10^{-2}	+	+
Benzo(b) fluoranthene	Developmental toxicant	4.32×10^{-2}	+	+
Benzo(g,h,i) perylene	*	*	+	+
Indeno(1,2,3 cd) pyrene	*	*	+	+

*Data unavailable.

5 The sources of PAHs

According to the literature, the PAH compounds could be released from two major sources:

(i) Natural sources such as forest fires, natural leakage, volcanic eruption, biological sources like diagenesis (organic transformation), and biogenic sources (plant, microbes, algae).

(ii) Anthropogenic sources (manmade) could be divided into two sources:
   1. Combustion sources (incomplete combustion of petroleum, gas, wood, etc.)
   2. Petrogenic sources are the products that are released from the processing and extraction of crude oil.

In general, the input of PAHs that come from natural sources is less than that from anthropogenic ones. Whatever the PAHs source, their mobility could be done through different pathways such as air deposition, river runoff, home and industrial emissions, and direct spills of petroleum [22]. In Table 2, several studies have been mentioned as a source of PAHs when they studied their concentration in different regions around the world in a different medium.

Table 2: The sources of PAHs mentioned by some studies.

Main emission sources	Medium	Sample collection	Reference
1. Fuel oil 2. Wood burning and coal 3. Oil combustion	Chinese tea	Purchased from the Chinese market	[23]
1–54% from fossil fuel combustion 2–29% from petrogenic inputs 3–16% from vehicle exhaust emissions	Sediment	Chabahar Bay, Iran	[24]
Anthropogenic sources are primarily pyrolytic nature	Surface sediments	Brunei Bay, Malaysia	[25]
The main PAH sources: 1. Motor vehicles 2. Cooking and heating in homes and power plants 3. The burning of waste 4. Tobacco smoking in the environment	Soil	Helsinki, Vilnius, Chicago, London, Tallinn	[26]
Direct petroleum combustion	Water	Lake Taihu, China	[27]
1. Historical contributions from multiple sources 2. A former coal gasification facility and petroleum product spills	Lake sediment	Owen Sound Bay Ontario, Canada	[28]

6 The detection and fates of PAHs in the environment

PAH compounds have been detected in different mediums such as soil, sediments, rivers, seas, lakes, and food

6.1 Detection of PAHs in sediment

The presence of PAHs in sediments has been detected and assessed by different researchers. The distribution and the presence of the sediment are varied from one place to another and they are extremely dependent on the physiochemical properties such as solubility, lipophilicity, and water vapor. PAHs molecules are detected in different places such as in river sediment, lake sediments, and marine sediments. For example, Tuncel and Topal [29] collected sediment samples from five sites at Oludeniz Lagoon on the Turkish Mediterranean coast. They reported that PAHs were detected in 25 samples with a total concentration of 1.85 ± 1.39 mg/kg, within the range of 0.620 and 0.515 mg/kg; acenaphthene and chrysene were the most prevalent. The authors mentioned that the majority of the points are impacted by the burning of fossil fuels, followed by the burning of biomass and human activity. The PAHs were also detected in the river sediments, Chamorro et al. [30] identified 45 organic microcontaminants in seven sediment samples in the basin of Biobío River, including monoterpenes, sesquiterpenes, diterpenes, lineal alkyl benzenes, PAHs, musk scents, sterols, and phthalate esters. The authors pointed out that the basin of Biobío River is influenced by domestic wastewater treatment, eucalyptus, or pine Kraft pulp mills, and pyrolytic and pyrogenic processes. In addition, Andersson et al. [31] identified the temporal characteristics and concentration variations of PAH during the last 5,400 years in Lille Lungegårdsvannet lake sediments. The PAH data show relatively low concentrations before the industrial era, followed by a notable increase in their concentrations. It can be observed that the sources have changed with time, moving from pure wood and coal combustion in preindustrial eras to mixed and petrogenic sources in more recent times.

6.2 Detection of PAHs in soil

The records for 120 brownfield sites contaminated with PAHs in Elmbridge Borough, a suburban area southwest of London, from 1990 to 2019 were gathered by Hellawell et al. [8]. The majority of the elevated PAH concentrations were identified in Made Ground, indicating that anthropogenic deposition was the primary source. Around 90% of the sample had a pyrogenic, unrelated-to-traffic source, according to the diagnostic ratio data. Pu et al. [32] collected 27 samples of soil from four different basins of the Yangtze River located in Yichang in the western Hubei province, China. The PAH concentrations were measured with the goal of determining their location, potential sources, and potential health concerns. The findings revealed that the surface soils contain PAHs within the range of 8.26 to 397 ng/g. They reported the average concentration of PAHs in valleys was 72.9 ng/g and 28.2 ng/g in the hilltop. They observed that PAHs with low molecular weight were dominant in both valleys and hilltops, and they suggested that the combustion of biomass and coal, followed by the burning of petroleum, is the primary source of PAHs on the soil surface. In addition, Shukla et al.

[33] gathered soil samples from the sides of the road in Lucknow, India, to examine the PAHs concentrations, origins, and potential cancer risks of 15 PAHs. The average concentration of 15 PAHs was determined to be 3748.23 ng/g, with a range between 478.94 and 8,164.07 ng/g. Four-ring PAHs made up around 32.5% of the total samples, followed by six-ring PAHs at 24.5% and five-ring PAHs at about 16.7%. The main sources of PAHs in the study region, according to the researcher, include burning of petroleum, wood, coal, and grass.

6.3 Detection of PAHs in the atmosphere

PAH concentration was detected in the ambient air of Inegol, Turkey. The samples were collected from two different locations that represent industrial and unregulated furniture manufacturing regions. The mean ambient PAH concentrations measured in the gas phase in the furniture workshops and industrial district areas were 697.82 ± 637 ng/m^3 and 772.92 ± 864.23 ng/m^3, respectively [34]. In addition, in the furniture workshops site, the average total (gas + particle phases) concentration of 16 PAH compounds was 1,111.34 ± 1,045.24 ng/m^3, whereas, in the industrial district, it was 772.92 ± 864.23 ng/m^3. These values are higher than the reported ambient levels for urban locations with significant industries around the world. Additionally, due to local combustion sources, the average particle phase percentage was 30%. The principal component analysis results revealed that the major sources of pollution in the different regions are connected (combustion, traffic, industries). However, the uncontrolled burning of furniture waste as a fuel for home heating is regarded to be the most efficient source. Both regions' health risks for the population were calculated, and it was determined that they were not very high. PAHs were quantified using gas chromatography-mass spectrometry for samples collected throughout the Delhi-National Capital Region of India [35]. The samples were obtained to examine the temporal and spatial variation of 16 PAHs. Throughout the investigations, it was shown that PAHs were strongly concentrated in the smaller size ranges of particles. In size-fractionated particles, benzo[a]pyrene (12%) was the highest mean concentration, whereas acenaphthene (2%) contributed the least. In addition, around 7,647 samples were collected in one year from 11 different urban sites in China by Li Ma et al. [36] to measure 16 PAH concentrations. They revealed that the concentration of PAHs varied from site to site and PAHs' molar weight increased along with their proportional concentration in the particle phase. The average concentrations of 16 PAHs in northern and southern Chinese cities were 83.4 ± 151 ng/m^3 and 33.4 ± 45.5 ng/m^3, respectively.

6.4 Detection of PAHs in the water

The recent levels of PAH in lake water settings are summarized by Meng et al. [37]. The findings showed that the quantities of PAHs in water ranged from 4.0 to 12,970.8 ng/L, and that naphthalene (6,525 ng/L) and indeno(1,2,3 cd) pyrene (3,452.6 ng/g) had the greatest amounts of each individual PAH. The Qinghai-Tibet Plateau Lakes District and the Mongolia-Xinjiang Lakes District were less polluted than other areas of the great lakes region in China. However, on a global scale, PAHs pollution in lakes is quite high.

El-Naggar et al. [38] provide information on the types and concentrations of PAHs found in Alexandria's coastal waters during the winter of 2015. They discovered that the mean concentration value was 991 ng/L, beyond the limit allowed by the European Union's water standard (200 ng/L), and that it ranged between 13.4 and 6,076 ng/L. According to the water samples' percentage distribution, roughly 41.7% of the tested samples contained less than 500 ng/L, while 4.2% had greater than 6,000 ng/L. The low molecular weight did not surpass 1.5%, and the four, five, and six rings made up 84%, 16.7%, and 5.5% of the total PAH, respectively. The authors mentioned that incomplete combustion of the fuel of boats and vehicle engines is the main source of PAHs, while pyrolytic origin occupied a small portion. Adeyeye et al. [39] provided insight into the accumulative of PAHs in the sediments and fish parts on a fishpond. The studied PAHs were generally more concentrated in the sediments than in the water samples. The estimated coefficient of variation % showed that the PAHs level had a low (3.51–24.7%) geographic distribution. The fish included the highest concentrations of the chemicals fluoranthene, pyrene, and anthracene.

6.5 Detection of PAHs in the food, plants, and animals

One of the PAH fates that have been mentioned is uptake by plants or detection in the food, or in the animals' tissues and organs. For example, Alves et al. [40] studied the presence of phenanthrene, anthracene, and pyrene in the tissues of *Medicago sativa L.* (alfalfa) planted in soil and artificially contaminated with 150 ppm of PAHs within 40 days. The PAHs were in trichomes, epidermal cells, damaged tissues, stomata, and other places. Guo et al. [23] calculated the concentration of PAHs concentration in the Chinese tea product; the authors revealed the 16 PAH concentrations ranged from 136.99 to 440.82 µg/kg and mainly contained between two and four rings. Furthermore, Qin et al. [41] studied the tissue distribution and bioaccumulation of PAHs in freshwater organisms from Lake Chaohu, China. For carp, the brain had the highest level of PAHs, with a concentration of 591 ng/g. For bighead fish and top mouth culter, the gills and muscles had the highest concentrations of PAHs 440 ng/g and 200 ng/g, respectively.

7 The degradation of PAHs using biological agents

The use of biological agents to treat the contaminants in different medium has attracted researchers in the environmental fields over the last decades. Bioremediation is the process that takes advantage of the ability of biological agents (plants, algae, bacteria, protozoa, etc.) to utilize the pollutants as a solo source of energy and carbon through their ability to secrete special enzymes capable of converting them into harmless substances [7]. As compared to physico-chemical methods, bioremediation techniques are affordable, cost-effective and eco-friendly [42]. Several bioremediation techniques have been applied before. These techniques could be classified into two categories, in situ and ex situ; see Figure 4. To choose the optimum bioremediation technique, many factors should be considered, such as (i) treatment cost, (ii) depth of pollution, (iii) type of pollutant, (iv) geological site, and (v) pollutant concentration [43].

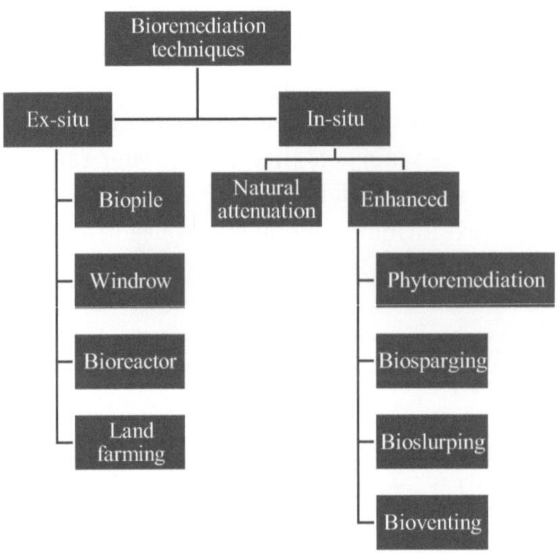

Figure 4: The classification bioremediation techniques based on in situ or ex situ treatment.

The application of bioremediation techniques has been frequently used for the elimination PAHs by using a different biological agent from a different medium. Table 3 represents the studies that implemented bioremediation methods for eliminating PAHs from different mediums.

Table 3: Studies that implemented the bioremediation methods for eliminating PAHs from different media.

PAH compounds	Biological agent	Treatment medium	Removal efficiency	PAHs	Reference
Naphthalene	*Bacillus licheniformis*	M9 agar slants	73%	100 mg/L	[44]
	Bacillus sonorensis		52%		
Acenaphthylene	*Stenotrophomonas maltophilia* CPHE1	MSM	100%	Σ[PAHs] = 10 mg/L	[45]
Acenaphthene	*Raoultella ornithinolytica*	Flask and stirred tank bioreactor scale	98.6%	50 mg/L	[46]
	Serratia marcescens		95.7%		
	Bacillus megaterium		90.0%		
	Aeromonas hydrophila		99.9%		
Fluorene	Lipopeptide biosurfactant	Sewage sludge	40.5%	484.4 mg/kg	[47]
fluoranthene	*Aspergillus aculeatus*	Batch reactor	98.4%	50 mg/L	[48]
	Mucor irregularis				
Pyrene	*Mycobacterium frederiksbergense*	MSM	100%	50 mg/L	[49]
Chrysene	*Trametes versicolor*	Soil	81%	367.94 mg/kg	[50]
Benzo(a) anthracene	*Selenastrum capricornutum*	Minibioreactor	72%	266 μg/L	[51]
	Scenedesmus acutus				
Benzo(k) fluoranthene	Microbial consortium (bacterial and fungi strains)	Soil contaminates through biostimulation	40 g of diesel spiked in 1 kg soil	93%	[52]
Phenanthrene	*Agrobacterium fabrum*	MSM	50 mg/L	85.2%	[53]
	Pseudomonas aeruginosa			85.4%	
	Bacillus subtilis			87.1%	
	Microconsortium			88.7%	
Anthracene	*Sphingomonas* sp.	Batch experimental	50 mg/L	70.5%	[54]
Benzo(a)pyrene	*Microbacterium* sp.	MSM	0.056 mM	84.2%	[55]
Dibenzo(a,h) anthracene	*Stenotrophomonas maltophilia*	BSM	13.6 mg/L	28%	[56]

Table 3 (continued)

PAH compounds	Biological agent	Treatment medium	Removal efficiency	PAHs	Reference
Benzo(b) fluoranthene	*Pseudomonas* sp. WDE11	Bushnell Haas (BH) medium	10 mg/L	97.5%	[57]
	Pseudomonas sp. WD23				
Benzo(g,h,i) perylene	*Rhodotorula* sp. NS01	Erlenmeyer flask	40 mg/L	60%	[58]
	Debaryomyces hansenii NS03				
	Hanseniaspora valbyensis NS04				
Indeno(1,2,3 cd) pyrene	*Rhodococcus aetherivorans* IcdP1	MSM	10 mg/L	70%	[59]

8 Summary and future recommendations

In this chapter we got an insight into the PAHs from cradle to grave; we mentioned their formation theories, detection places, fate, and their adverse effects on public health and organisms. Bioremediation of PAHs is one of the dominant methods to eliminate them and protect the environment from their negative impacts. However, there are some points that need further investigation, such as the following:

- There are different approaches to treat PAH. According to recent research, the actual cost of some of the approaches and monitoring of contaminated areas, however, is high.
- The processes for contaminated, risk assessment, and monitoring of PAHs must be standardized and included in legislation and laws.
- According to studies, several techniques have been tried to treat PAHs from different medium, so it is essential to develop a long-term, low-cost, low-energy remediation plan.

References

[1] Alharbi OML, Basheer AA, Khattab RA, Ali I. Health and environmental effects of persistent organic pollutants. J Mol Liq 2018, 263, 442–453. doi 10.1016/j.molliq.2018.05.029.

[2] Hussain K, et al. Monitoring and Risk Analysis of PAHs in the environment. Handb Environ Mater Manage 2019, 973–1007.

[3] El-Shahawi MS, Hamza A, Bashammakh AS, Al-Saggaf WT. An overview on the accumulation, distribution, transformations, toxicity and analytical methods for the monitoring of persistent organic pollutants. Talanta 2010, 80, 5, 1587–1597. doi 10.1016/j.talanta.2009.09.055.
[4] Adebusuyi AT, Sojinu SO, Aleshinloye AO. The prevalence of persistent organic pollutants (POPs) in West Africa – A review. Environ Challenges 2022, 7, October 2021, 100486. doi 10.1016/j.envc.2022.100486.
[5] Gutierrez-Urbano I, Villen-Guzman M, Perez-Recuerda R, Rodriguez-Maroto JM. Removal of polycyclic aromatic hydrocarbons (PAHs) in conventional drinking water treatment processes. J Contam Hydrol 2021, 243, 103888. doi 10.1016/j.jconhyd.2021.103888.
[6] Wang X, Wang C, Zhu T, Gong P, Fu J, Cong Z. Persistent organic pollutants in the polar regions and the Tibetan Plateau: A review of current knowledge and future prospects. Environ Pollut 2019, 248, 191–208. doi 10.1016/j.envpol.2019.01.093.
[7] Qutob M, Rafatullah M, Muhammad SA, Alosaimi AM, Alorfi HS, Hussein MA. A review of pyrene bioremediation using mycobacterium strains in a different matrix. Fermentation 2022, 8, 6, doi 10.3390/fermentation8060260.
[8] Gaur N, Narasimhulu K, PydiSetty Y. Recent advances in the bio-remediation of persistent organic pollutants and its effect on environment. J Clean Prod 2018, 198, 1602–1631. s 10.1016/j.jclepro.2018.07.076.
[9] Miraji H, Ripanda A, Moto E. A review on the occurrences of persistent organic pollutants in corals, sediments, fish and waters of the Western Indian Ocean. Egypt J Aquat Res 2021, xxxx, 10.1016/j.ejar.2021.08.003.
[10] Dinç B, et al. Spatial distribution and source identification of persistent organic pollutants in the sediments of the Yeşilırmak River and coastal area in the Black Sea. Mar Pollut Bull 2021, 172, July, 10.1016/j.marpolbul.2021.112884.
[11] Bull K. Protocol to the 1979 convention on long-range transboundary air pollution on persistent organic pollutants: The 1998 agreement for the UNECE region. Persistent Org Pollut 2006, 3, 1–11, 10.1007/10751132_1.
[12] Qiao M, Qi W, Liu H, Qu J. Oxygenated polycyclic aromatic hydrocarbons in the surface water environment: Occurrence, ecotoxicity, and sources. Environ Int 2022, 163, March, 107232. 10.1016/j.envint.2022.107232.
[13] Liao K, Yu JZ. Abundance and sources of benzo[a]pyrene and other PAHs in ambient air in Hong Kong: A review of 20-year measurements (1997–2016). Chemosphere 2020, 259, 127518, 10.1016/j.chemosphere.2020.127518.
[14] Okedere OB, Elehinafe FB. Occurrence of polycyclic aromatic hydrocarbons in Nigeria's environment: A review. Sci Afr 2022, 16, e01144, 10.1016/j.sciaf.2022.e01144.
[15] Howsam M, Jones KC. Sources of PAHs in the environment 3, 1998.
[16] WHO. Guidelines for drinking-water quality, 4th edition, incorporating the 1st addendum. 2017.
[17] Reizer E, Viskolcz B, Fiser B. Formation and growth mechanisms of polycyclic aromatic hydrocarbons: A mini-review. Chemosphere 2022, 291, September 2021, 10.1016/j.chemosphere.2021.132793.
[18] Formation T, et al. 2449 498, 2002, 2449–2452.
[19] Sathikumaran R, Madhuvandhi J, Priya KK, Sridevi A, Krishnamurthy R, Thilagam H. Evaluation of benzo[a]pyrene-induced toxicity in the estuarine thornfish Therapon jarbua. Toxicol Rep 2022, 9, January, 720–727. 10.1016/j.toxrep.2022.03.051.
[20] Tani K, et al. Toxicity assessment of typical polycyclic aromatic hydrocarbons to Daphnia magna and Hyalella azteca in water-only and sediment–water exposure systems. Sci Total Environ 2021, 784, p, 147156. 10.1016/j.scitotenv.2021.147156.

[21] Sun K, et al. Toxicity assessment of Fluoranthene, Benz(a)anthracene and its mixed pollution in soil: Studies at the molecular and animal levels. Ecotoxicol Environ Saf 2020, 202, January, 110864. 10.1016/j.ecoenv.2020.110864.
[22] Wolska L, Mechlińska A, Rogowska J, Namieśnik J. Sources and fate of PAHs and PCBs in the marine environment. Crit Rev Environ Sci Technol 2012, 42, 11, 1172–1189. 10.1080/10643389.2011.556546.
[23] Guo X, Chen F, Zhang W. Pollution, source and risk assessment of PAHs in Chinese tea. Lwt 2022, 167, May, 113851. 10.1016/j.lwt.2022.113851.
[24] Seyed Hashtroudi M, Aghadadashi V, Mehdinia A, Sheijooni Fumani N. Combining theoretical concepts and Geographic Information System (GIS) to highlight source, risk, and hotspots of sedimentary PAHs: A case study of Chabahar Bay. Environ Res 2023, 216, P1, 114540. 10.1016/j.envres.2022.114540.
[25] Sharifi R, et al. Source apportionment and health risk assessment of polycyclic aromatic hydrocarbons (PAHs) in the coastal ecosystem of the Brunei Bay, Brunei. Mar Pollut Bull 2022, 181, March, 113913. 10.1016/j.marpolbul.2022.113913.
[26] Saltiene Z, Brukstiene D, Ruzgyte A. Contamination of soil by polycyclic aromatic hydrocarbons in some urban areas. Polycycl Aromat Compd 2002, 22, 1, 23–35. 10.1080/10406630210371.
[27] Kong J, et al. Nitrated and parent PAHs in the surface water of Lake Taihu, China: Occurrence, distribution, source, and human health risk assessment. J Environ Sci (China) 2021, 102, 159–169. 10.1016/j.jes.2020.09.025.
[28] Buell MC, Johannessen C, Drouillard K, Metcalfe C. Concentrations and source identification of PAHs, alkyl-PAHs and other organic contaminants in sediments from a contaminated harbor in the Laurentian Great Lakes. Environ Pollut 2021, 270, 116058, 10.1016/j.envpol.2020.116058.
[29] Tuncel SG, Topal T. Polycyclic aromatic hydrocarbons (PAHs) in sea sediments of the Turkish Mediterranean coast, composition and sources. Environ Sci Pollut Res 2015, 22, 6, 4213–4221. 10.1007/s11356-014-3621-1.
[30] Chamorro S, et al. Chemical characterization of organic microcontaminant sources and biological effects in riverine sediments impacted by urban sewage and pulp mill discharges. Chemosphere 2013, 90, 2, 611–619. 10.1016/j.chemosphere.2012.08.053.
[31] Andersson M, Klug M, Eggen OA, Ottesen RT. Polycyclic aromatic hydrocarbons (PAHs) in sediments from lake Lille Lungegårdsvannet in Bergen, western Norway; appraising pollution sources from the urban history. Sci Total Environ 2014, 470–471, 1160–1172, 10.1016/j.scitotenv.2013.10.086.
[32] Pu C, Xiong J, Zhao R, Fang J, Liao Y, Song Q. "Levels, sources, and risk assessment of polycyclic aromatic hydrocarbons (PAHs) in soils of karst trough zone, Central China,". J Hydrol 2022, 614, PB, 128568. 10.1016/j.jhydrol.2022.128568.
[33] Shukla S, Khan R, Bhattacharya P, Devanesan S, AlSalhi MS. Concentration, source apportionment and potential carcinogenic risks of polycyclic aromatic hydrocarbons (PAHs) in roadside soils. Chemosphere 2022, 292, December 2021, 133413. 10.1016/j.chemosphere.2021.133413.
[34] Caliskan B, Kücük A, Tasdemir Y, Cindoruk SS. PAH levels in a furniture-manufacturing city atmosphere. Chemosphere 2020, 240, 1–10, 10.1016/j.chemosphere.2019.124757.
[35] Hazarika N, Das A, Kamal V, Anwar K, Srivastava A, Jain VK. Particle phase PAHs in the atmosphere of Delhi-NCR: With spatial distribution, source characterization and risk approximation. Atmos Environ 2019, 200, April 2018, 329–342. 10.1016/j.atmosenv.2018.11.064.
[36] Ma WL, Zhu FJ, Liu LY, Jia HL, Yang M, Li YF. PAHs in Chinese atmosphere: Gas/particle partitioning. Sci Total Environ 2019, 693, 133623, 10.1016/j.scitotenv.2019.133623.
[37] Meng Y, et al. A review on occurrence and risk of polycyclic aromatic hydrocarbons (PAHs) in lakes of China. Sci Total Environ 2019, 651, 2497–2506. 10.1016/j.scitotenv.2018.10.162.
[38] El-Naggar NA, Emara HI, Moawad MN, Soliman YA, El-Sayed AAM. Detection of polycyclic aromatic hydrocarbons along Alexandria's coastal water, Egyptian Mediterranean Sea. Egypt J Aquat Res 2018, 44, 1, 9–14. 10.1016/j.ejar.2018.02.003.

[39] Adeyeye EI, Ibigbami OA, Adesina AJ, Popoola OK, Olatoye AR, Gbolagade YA. Assessment of Polycyclic Aromatic Hydrocarbons (PAHs) distribution in water, sediments and fish parts from ponds in Ado-Ekiti, Nigeria. Polycycl Aromat Compd 2022, 0, 0, 1–12. 10.1080/10406638.2022.2064884.

[40] Alves WS, Manoel EA, Santos NS, Nunes RO, Domiciano GC, Soares MR. Detection of polycyclic aromatic hydrocarbons (PAHs) in Medicago sativa L. by fluorescence microscopy. Micron 2017, 95, 23–30, 10.1016/j.micron.2017.01.004.

[41] Qin N, He W, Liu W, Kong X, Xu F, Giesy JP. Tissue distribution, bioaccumulation, and carcinogenic risk of polycyclic aromatic hydrocarbons in aquatic organisms from Lake Chaohu, China. Sci Total Environ 2020, 749, 141577, 10.1016/j.scitotenv.2020.141577.

[42] Quintella CM, Mata AMT, Lima LCP. Overview of bioremediation with technology assessment and emphasis on fungal bioremediation of oil contaminated soils. J Environ Manage 2019, 241, March, 156–166. 10.1016/j.jenvman.2019.04.019.

[43] Azubuike CC, Chikere CB, Okpokwasili GC. Bioremediation techniques–classification based on site of application: Principles, advantages, limitations and prospects. World J Microbiol Biotechnol 2016, 32, 11, 10.1007/s11274-016-2137-x.

[44] Rabani MS, Sharma R, Singh R, Gupta MK. Characterization and identification of naphthalene degrading bacteria isolated from petroleum contaminated sites and their possible use in bioremediation. Polycycl Aromat Compd 2022, 42, 3, 978–989. 10.1080/10406638.2020.1759663.

[45] Lara-Moreno A, Morillo E, Merchán F, Villaverde J. A comprehensive feasibility study of effectiveness and environmental impact of PAH bioremediation using an indigenous microbial degrader consortium and a novel strain Stenotrophomonas maltophilia CPHE1 isolated from an industrial polluted soil. J Environ Manage 2021, 289, December 2020, 10.1016/j.jenvman.2021.112512.

[46] Alegbeleye OO, Opeolu BO, Jackson V. Bioremediation of polycyclic aromatic hydrocarbon (PAH) compounds: (acenaphthene and fluorene) in water using indigenous bacterial species isolated from the Diep and Plankenburg rivers, Western Cape, South Africa. Braz J Microbiol 2017, 48, 2, 314–325. 10.1016/j.bjm.2016.07.027.

[47] Vecino X, Cruz JM, Moldes A, Aromatic AP, Pahs H. Bioremediation of sewage sludge contaminated with fluorene using a lipopeptide biosurfactant. 2015, 9, 6, 739–742.

[48] Bankole PO, Taghoghor Omoni V, Mulla SI, Adebajo SO, Adekunle AA. Co-biomass degradation of fluoranthene by marine-derived fungi; Aspergillus aculeatus and Mucor irregularis: Comprehensive process optimization, enzyme induction and metabolic analyses. Arab J Chem 2022, 15, 9, 104036. 10.1016/j.arabjc.2022.104036.

[49] Sarma SJ, Pakshirajan K. Surfactant aided biodegradation of pyrene using immobilized cells of Mycobacterium frederiksbergense. Int Biodeterior Biodegrad 2011, 65, 1, 73–77. 10.1016/j.ibiod.2010.09.004.

[50] Vipotnik Z, Michelin M, Tavares T. Biodegradation of chrysene and benzo[a]pyrene and removal of metals from naturally contaminated soil by isolated Trametes versicolor strain and laccase produced thereof. Environ Technol Innov 2022, 28, 102737, 10.1016/j.eti.2022.102737.

[51] de Llasera MPG, Santiago ML, Flores EJL, Toris DNB, Herrera MRC. Mini-bioreactors with immobilized microalgae for the removal of benzo(a)anthracene and benzo(a)pyrene from water. Ecol Eng 2018, 121, 89–98, 10.1016/j.ecoleng.2017.06.059.

[52] Chagas-Spinelli ACO, Kato MT, de Lima ES, Gavazza S. Bioremediation of a tropical clay soil contaminated with diesel oil. J Environ Manage 2012, 113, 510–516, 10.1016/j.jenvman.2012.05.027.

[53] Sharma S, Pandey LM. Biodegradation kinetics of binary mixture of hexadecane and phenanthrene by the bacterial microconsortium. Bioresour Technol 2022, 358, June, 127408. 10.1016/j.biortech.2022.127408.

[54] Al Farraj DA, Alkufeidy RM, Alkubaisi NA, Alshammari MK. Polynuclear aromatic anthracene biodegradation by psychrophilic Sphingomonas sp., cultivated with tween-80. Chemosphere 2021, 263, 128115, 10.1016/j.chemosphere.2020.128115.

[55] Qin W, et al. Biodegradation of benzo(a)pyrene by Microbacterium sp. strain under denitrification: Degradation pathway and effects of limiting electron acceptors or carbon source. Biochem Eng J 2017, 121, 131–138. 10.1016/j.bej.2017.02.001.

[56] Juhasz A, Stanley G, Britz M. Metabolite repression inhibits degradation of benzo[a]pyrene and dibenz[a,h]anthracene by Stenotrophomonas maltophilia VUN 10,003. J Ind Microbiol Biotechnol 2002, 28, 2, 88–96. 10.1038/sj/jim/7000216.

[57] Goveas LC, Selvaraj R, Kumar PS, Vinayagam R, Sajankila SP. Biodegradation kinetics and metabolism of Benzo(a)fluorene by Pseudomonas strains isolated from refinery effluent. Chemosphere 2022, 307, P4, 136041. 10.1016/j.chemosphere.2022.136041.

[58] Notes C. Biodegradation of perylene and benzo [ghi] perylene (5-6 rings) using yeast consortium: Kinetic study, enzyme analysis and degradation pathway. 2016, 39, August, 5–15.

[59] Miao LL, Qu J, Liu ZP. Hydroxylation at multiple positions initiated the biodegradation of Indeno [1,2,3-cd]Pyrene in Rhodococcus aetherivorans IcdP1. Front Microbiol 2020, 11, September, 1–13. 10.3389/fmicb.2020.568381.

Amir Talebi, Norli Ismail*

Phytoremediation Approaches for Organic Pollutants

Abstract: Phytoremediation, environmental toxins and pollutants extraction, degradation, and sequestration using plants, is considered a green, environmentally friendly, and highly effective choice to be substituted for conventional physicochemical treatment methods. The phytoremediation process consists of several biochemical methods, such as absorption, accumulation, sequestration, transport, and degradation, in which toxic organic contaminants are metabolized into nontoxic or relatively less-toxic compounds. Each plant group can remediate a specific organic contaminant, polychlorinated biphenyls, polycyclic aromatic hydrocarbons, and halogenated hydrocarbons. In recent years, various research efforts and new findings in appropriate plant genetically engineered to improve contaminant degradation ability.

Phytoremediation shows advantages such as being cost-effective, high carbon capture capacity, assessment, and soil biological talent conservation, there are some negative, or in better words diminishing factors, that limit the phytoremediation efficiency, plants' slow growth rate, and global environmental obstacles. In addition, the laws and legislation should be modified in a way that effluent discharges are accepted by local and global authorities and of course the public mind. For this reason, there is a huge demand for broad laboratory research and investigations to utilize phytoremediation approaches for efficient and satisfactory contaminant removal.

This chapter presented a brief introduction to the phytoremediation method, followed by a review of the previous works, and the future trend for the phytoremediation method, subsequently.

Keywords: Phytoremediation, persistent organic pollutants, rhizofiltration, phytovolatilization

1 Introduction

Advances in science and technology have enabled humans to exploit natural resources to a great extent, generating unprecedented disturbances in global elemental cycles. The relatively recent introduction of anthropogenic toxic chemicals, and the massive

***Corresponding author: Norli Ismail,** Environmental Technology Division, School of Industrial Technology, Universiti Sains Malaysia, Penang 11800, USM, Malaysia, e-mail: norlii@usm.my
Amir Talebi, Schahverdy Entwicklung und Konstruktion Maschinenbau, Baden-Württemberg 79274, Germany

https://doi.org/10.1515/9783111016825-009

relocation of natural materials to different environmental compartments (soils, groundwater, and atmosphere), have caused a chaotic unbalance in the recipient ecosystems.

Organic contaminants and toxicant are the souvenirs of industrialization and the development of a globalized world. Our soil, water, air, and in one world, our environment, are exposed to millions of tons of hazardous and toxic organic compounds every day. Persistent organic pollutants (POPs) are poisonous compounds that negatively affect living species and the ecosystem. The POPs can be carried by any moving medium, such as running water or blowing wind, meaning that, for example, a POP produced in any longitude and latitude can badly influence any living creature even though a thousand miles away from where they are originally generated and released. In addition, the other problem is that the POPs are highly reluctant to degradation and decomplexation, highly prone to accumulation, and unfortunately very susceptible to entering the food chain and causing serious health issues.

International organizations such as The Stockholm Convention have counted various chemical compounds and complexes as POPs and probably the most recognized one is dichlorodiphenyltrichloroethane (DDT) [1].

Organic pollutants have already shown and brought their negative effects on our environment. Living species and biological diversities are facing severe issues exposing hazardous material by different means and from various sources, in some cases thousands of miles away from the toxic compound's original generator [2]. It is reported that due to the POPs distribution in the marine environment the number of some wild animals such as gray seals, California sea lions, Beluga whales, bottlenose dolphins, polar bears, and green sea turtles have been harshly plummeted to 86% in some cases [3].

2 Persistent organic pollutants

One of the major issues with POP loads is that they do have not only an instant effect on the exposed species but also the harm can be transferred from one generation to another generation. For example, the transportation pathway begins during breastfeeding from mother to baby in some mammals, specifically for immature placental mammals which acquire higher amounts of POPs.

The POPs harmful effect on the human being is another uprising issue. Pesticides, fungicides, herbicides, and the chemical by-products known as dioxins are widely used in agricultural industry. They are recognized for the lethal impact and exist in the food chain later accumulated in the cell bodies remains and agitated for any negative factors in a food products assortment [1, 3].

3 Phytoremediation

Phytoremediation is a treatment technique that uses plants for the removal or degradation of toxic compounds. The phytoremediation process is an environmentally friendly approach that can be substituted instead of other conventional water treatment techniques. The treatment could be either within the plant (inside the unit) or separated from the whole unit plant. An integrated process reflects the mechanisms of phytoremediation assisted either by sequestration, absorption, accumulation, degradation, and transport to metabolize harmful organic contaminants into less harmful and marginally less toxic. The uniqueness of the plant structure opens the opportunity for expansion of phytoremediation to remove recalcitrant compounds of POPs using a potential hyperaccumulator from terrestrial and aquatic plants. Figure 1 illustrates the phytoremediation subprocess of soil remediation.

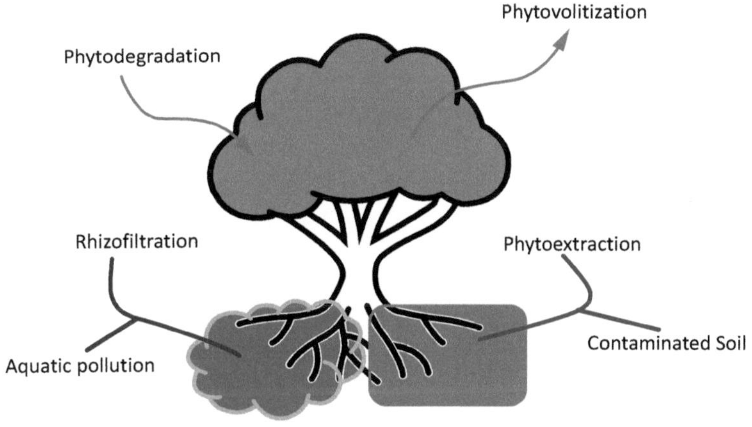

Figure 1: A schematic diagram of different phytoremediation occurrence in plant and surrounding ecosystem.

The phytoremediation method consists of a combination of physicochemical and biological systems. These methods, depending on the treatment objectives could be combined, such as absorption, accumulation, sequestration, transport, and degradation [4]. The phytoremediation objective in this case is converting toxic organic contaminants into safe and less toxic forms. Numerous organic pollutants such as polychlorinated biphenyls (PCBs), chlorinated solvents, atrazine, dieldrin, hexachlorobenzene, chlorinated solvents, petroleum hydrocarbons, polycyclic aromatic, and halogenated hydrocarbons have demonstrated an efficacious treatment outcome using diverse plant groups. Currently, many studies effectively steered to improve the performance of specific plants through manipulating the best plant DNA genes for degradability optimization [5].

Phytoremediation operates under various apparatuses or approaches such as those based on pollutant degradation (rhizodegradation, phytodegradation) or based on targeted toxic species accumulation (as phytoextraction, rhizofiltration). Phytoremediation could be designed and performed based on the dissolution of the pollutant dissipation from the mainstream of phytoremediation (phytovolatilization). The other green approach of phytoremediation is when the plantation is designed to cover the polluted media (on the surface) leading to lowering the mobility of contaminants and immobilization of the contaminants, namely as phytostabilization [6].

However, it should not be ignored that toxicity of organic pollutants such as diphenyl ether and bipyridylium can destroy (partially or completely) the plant tissues, cell biosynthesis, membrane stability, plant cell ultrastructure, plant cell DNA and consequently diminish or eventually stop the extraction process. Overall, the disadvantages and disadvantages of phytoremediation are listed:

a. Highly effective for pollution sources with low concentrations of contaminants
b. Cost-effectiveness, specifically compared with other remediation methods
c. Simplicity and ease of operation
d. Soil enrichment by organic substances and microorganisms
e. Soil protection against erosion by wind or aquatic forces
f. Demand of quite a large land and area for a rather long time
g. Specific plant choice for a specific pollutant group
h. Accumulation of toxic compounds in fruits or edible parts of the used crop

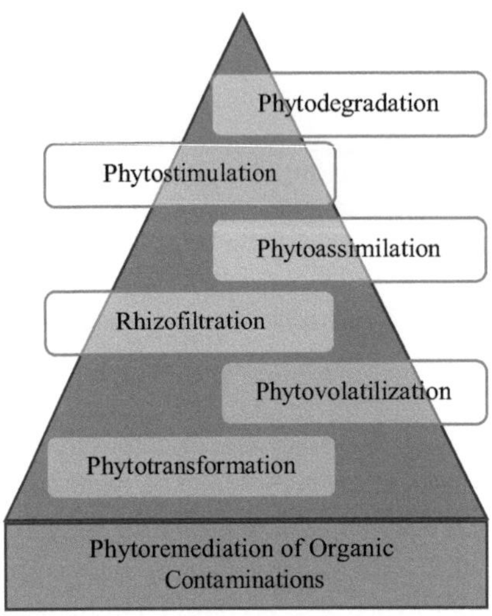

Figure 2: Outline the different types of phytoremediation mechanism.

Figure 2 describes the diverse terminology of the phytoremediation by microorganisms' inclusive fungi. The terms are referred to different treatment mechanism – degradation, stimulation, assimilation, filtration, volatilization, and transformation. Phytoremediation in plant has multiple mechanism medium that is roots, leaves, shoots, vertical, and axial stems.

4 Major factors phytoremediation of organic pollutants

The main concern in the phytoremediation of pollutants is the ability to uptake and transfer organic pollutants by plants, which is directly proportional to the organic contaminant structure and characteristics of the organic pollutants, along with the biological properties of the plants, and the medium of the environment in which the remediation take place.

It is crucial that molecular mass and hydrophobicity of the organic pollutant be in a range that the uptake can successfully occur. The molecular mass and hydrophobicity of the organic compound play a major role to this extent. Briefly described, the lower the molecular mass coefficient, the higher the organic pollutant uptake. This pattern for hydrophobicity is the other way around, in which organic compounds with high hydrophobicity are easily absorbed by plant roots. It is important to notice that organic pollutants with low molecular still can be absorbed by plants from the air [7].

Plant species are another key factor in choosing the best phytoremediation system. Not all plant species are suitable for this purpose. It can be said that commonly a commodity or crops for food supply are not appropriate for phytoremediation due to the potential risk and exposure that might enter the food chain system ended post a danger to living things [8]. Plants with high growing speed, a noticeable volume of biomass, and considerable root structure in which ease the uptake mechanism, and at the same time, the plant must be reluctant enough to resist the accumulation of the pollutant in their structure [9].

Each used plant can be modified with a certain method or methods (such as phytodegradation, phytovolatilization, rhizofiltration, rhizodegradation, and phytostabilization) to treat the pollutants from the source. For example, the following plants have been recommended by previous researchers with different objectives as organochlorines, trichloroethylene (TCE), and tetrachloroethylene, ammonium removal, and treatment of shrubs, tree, and grass. Some of the most used plants in phytoremediation are tabulated in Table 1 [10].

It is reported that the existence of carbohydrate content in plant cell walls, epicuticular waxes, and lipid content in plant roots leads to higher pollutant uptake (Gao et al., 2005). Zhang et al. [11] have reported that polychlorinated dibenzofurans (PCDD/Fs) uptake in the root was positively influenced by lipophilic adsorption, and there was a direct correlation between root accumulation of PCDD/Fs from soil media and extracted lipid in plant root.

The other significant factor in phytoremediation is the property of the existing medium in which the contaminants are introduced to the treatment system. The condition of the soil, as the key media in the phytoremediation system, has a direct effect on pollutant extraction, uptake, and overall, treatment efficiency. Media with suitable chemical characteristics such as optimal pH, and oxidation–reduction capability,

Table 1: Plants and related phytoremediation process.

Plant name	Sci-Name	Pollutant	Objective – result	Reference
Blumea	*Blumea malcolmii*	Malachite green	Phytodegradation of industrial waste	[12]
Elodea	*Elodea canadensis*	DDT	Enzyme cofactors or biomolecules	[13]
Unicellular green algae	*Chlorella pyrenoidosa*	Pentachlorophenol (PCP)	Algae activity may have been reduced by the cycling of light exposure	[14]
Coral	*Ipomoea carnea*	Textile azo dyes	Agrobacterium rhizogenesfavoured phytodegradation	[15]
Perennial reed grass	*Phragmites australis*	Ibuprofen	Wetland suitability	[16]
Water hyacinth	*Pontederia crassipes*	Sodium dodecyl sulfate (SDS)	Ascorbate peroxidase (APX) activity in the presence of *Chromolaena odorata* L. extract	[17]
Duckweed	*Spirodela polyrhiza*	Ofloxacin (OFX)	93.73–98.36% reduction efficiency	[18]

together with the existence of electrolytes have a significantly improving effect on the required condition to uptake organic compounds by the plant root system [19]. Electrical conductivity, organic matter, microbial activities, and other soil alterations can be counted as other functional parameters [20].

5 Principle of rhizofiltration

Filtering contaminated groundwater, surface water, and wastewater through a mass of roots is called rhizofiltration. This type of phytofiltration is specifically useful when the contaminant is primarily preserved in roots including its capacity to be conducted in situ or ex situ. Plant species aside from hyperaccumulators is a good phytoremediator as well. Generally, plant phytoremediator with large roots is the perfect choice for rhizoflitration due to their ability to maximize the uptake of the toxic element [21]. Thus determine the best criteria which influence the remediation efficiency in the phytoremediation techniques. Large quantities and tiny roots shall provide higher surface area and space for microbe to adhere and multiply the cell mass. Leafy vegetables and garden plants reported the ability to carry out the rhizoflitration. Other crucial elements in rhizoflitration are selective microorganisms and enzymes, which are exist and found in the plant rhizosphere.

Rhizodegradation takes place in two major phases, including detoxification and enzyme assessment. In the first stage, the organic compound chain is broken by oxygenase enzymes. Then after, the contaminants and functional groups are introduced with higher solubility and as a result, increased polarity. In this stage, the enzyme assessment for instance peptide, ether, ester and thioether linkages conjugates with polar molecules such as amino acids, carbohydrates, glutathione, and malonic acid of the crops are broken to CO_2 and H_2O [22]. Table 2 represents some previous works on the phytodegradation of some organic pollutants using the most common plants utilized in phytodegradation [10].

Table 2: Phytodegradation and common plants for organic pollutants remediation.

Plant name	Sci-name	Pollutant	Objective – remarkable result	References
Eastern Gama grass	*Tripsacum dactyloides*	Herbicities (atrazine)	Atrazine (ATR) degradation enhanced by 84–260%	[23]
Bermuda grass	*Cynodon dactylon*	Total petroleum hydrocarbons (TPHs)	Total petroleum hydrocarbons (TPHs) are degraded by 81%	[24]
Mangrove	*Kandelia candle* (L.) Druce	Phenanthrene (Ph) and pyrene (Py)	Dissipation of Ph (47.7%) and Py (37.6%) was observed in the rhizosphere.	[25]
Chinaberry tree	*Melia azedarach*	Benzo(a)pyrene	A cd-resistant plant is required for the degradation of benzo(a)pyrene	[26]
European blackberry	*Rubus fruticosus*	Polycyclic aromatic hydrocarbons (PAHs)	High molecular weight PAHs are degraded with naturally grown blueberries	[27]
Cucurbits	*Cucurbita* sp.	Polycyclic aromatic hydrocarbons (PAHs)	PAHs degradation is enhanced by plant-bacterial interaction	[28]
Maize	*Zea mays*	Petroleum inoculation	*Piriformospora indica* could enhance the degradation of petroleum hydrocarbons	[29]

Rhizofiltration is mostly suitable for heavy metal removal. The exudates of the plant's root act as natural absorbent of heavy HMs, which is capable of absorbing HMs, with a direct effect on the rhizosphere pH [30]. If the same system is used for organic pollutants treatment, the phase is called rhizodegradation. This method is used for a vast range of organic pollutants from volatile aromatic pollutants such as conventional solvents xylene, toluene, and benzene to pesticides, PAHs, PCBs, and DDT. Rhizodegradation in comparison with other phytoremediation approaches is counted as a time-consumer method and limited to the rhizosphere usage, which can affect as countereffects but advantages such as in situ detoxification, high efficiency of the targeted organic pollutant degradation, presents rhizodegradation as an interesting alternative for the researchers around the world.

According to those in the literature, most of the researchers have concluded that in rhizodegradation the crucial role played by the microbes, and plants' function is to prepare and set up a suitable atmosphere for microbial growth and improving the soil. Researchers have shown that there is a direct impact between the number of microbes and organic pollutant degradation. Various research on different organic compounds such as [^{14}C]-benzopyrene, benzene, toluene and xylene, 2-dimethylpropylbenzene, benzoic acid, pyrene, decahydronaphthalene, and phenanthrene has been carried out and agrees with the above-mentioned hypothesis [31].

6 Phytovolatilization

Among the organic pollutants, there are some which get volatilized revealing less toxicity in the atmosphere. Phytovolatilization is suitable for biodegradation of the volatile organic compounds either directly from stem and leaves or via root–soil interaction. Direct phytovolatilization consists of plant uptake, transmission, and volatilize the compound (rather hydrophobic compounds), quite opposite the indirect phytovolatilization in which plant root activities play the major role in contaminant flux rate. A plant's root can increase phytovolatilization via soil permeability, and organic pollutants transport via hydraulic redistribution or along with water flux [32].

Phytovolatilization consists of three major steps:

a. The pollutants are absorbed and uptakes by to plant body. This process is performed by the plant roots. The hazardous contaminants are decomplex into less toxic substances.
b. The broken toxic compounds are transferred toward the leaves.
c. The leaves reveal (discharge) the nontoxic compounds out to the atmosphere [33].

A few aquatic plants growth on wetland can be accommodate phytovolatilization perfectly. Direct phytovolatilization in wetlands results in a higher amount of pollutants treatment which is a great advantage for industrial wastewater treatment. However, as a disadvantage, it is observed that emissions of methane from the plants are likely significant. There have been studies on various plants such as Rabbitfoot grass, common rush, perennial reed grass, and hybrid poplar tree to examine the effect of phytovolatilization of dimethylchloroarsine, organochlorines, TCE, and tetrachloroethylene [10].

7 Phytostabilization

Among the different phytoremediation configurations, phytostabilization is particularly suitable for heavy metals remediation and treatment. In this method, metal ions are attached to various soil components and are present in different chemical forms, preventing the toxic heavy metals to enter the water sources or the food cycle. The key factors in phytostabilization are namely soil pH, the quantity of the organic mass, redox potential, and quality of the soil as well as biological variables such as the quality and quantity of the microorganisms available in the crops and plants. According to the literature, various bio-crops and plants such as giant reed and silver grass (scientific names: *Arundo donax* and *Miscanthus sinensis*, respectively) have been used for the phytostabilization of metal-contaminated media [34].

8 Conclusions

Phytoremediation, the application of plants and crops to treat, remove, extract, or detoxify polluted soil and water, is a green approach and environmentally friendly treatment method that has drawn so many researchers' attention, recently. Phytoremediation can be used for various chemical compounds and complexes including organic pollutants or minerals and heavy metals. Phytoremediation upon different configurations can be adapted to different conditions and be set up for specific environments, such as rhizoflitration, phytovolatilization, phytostabilization, and phytoextraction. As the advantages of phytoremediation can be pointed out as being cost-effective, green and pleasant appearance, and persistent and long-time usage. Additionally, phytoremediation is able to stop or at least reduce land disruption as well as significantly reduce transportation expenses and related costs compared to the other waste disposal means. As another advantage, phytoremediation can be a useful approach for recovering minerals and heavy metals as well. However, researchers have found some disadvantages of phytoremediation. It is reported that in some cases toxic compound accumulation in fruits and edible crops has been observed. Also in some cases, the treatment is highly dependent on the seasonal and climate condition, which slows down the efficiency percentage. Finally, phytoremediation faces some barriers in case of contaminant groundwater.

References

[1] Wang Z, Adu-Kumi S, Diamond ML, Guardans R, Harner T, Harte A, Kajiwara N, Klánová J, Liu J, Moreira EG, Muir DCG, Suzuki N, Pinas V, Seppälä T, Weber R, Yuan B. Enhancing scientific support for the Stockholm Convention's implementation: An analysis of policy needs for scientific evidence. Environ Sci Technol 2022, 56, 5, 2936–2949.

[2] González-Mille DJ, Ilizaliturri-Hernández CA, Espinosa-Reyes G, Cruz-Santiago O, Cuevas-Díaz MDC, Martín Del Campo CC. DNA damage in different wildlife species exposed to persistent organic pollutants (POPs) from the delta of the Coatzacoalcos River, Mexico. Ecotoxicol Environ Saf 2019, 180, 403–411.

[3] Gaur N, Dutta D, Jaiswal A, Dubey R, Kamboj DV. Role and effect of persistent organic pollutants to our environment and wildlife in Persistent Organic Pollutants (POPs). In: Rashed M (eds.). Monitoring, impact and treatment. IntechOpen, London, 2022, 10.5772/intechopen.101617.

[4] Tripathi S, Singh VK, Srivastava P, Singh R, Devi RS, Kumar A, Bhadouria R. Phytoremediation of organic pollutants: Current status and future directions. In: Singh P, Kumar A, Borthakur A (eds.). Abatement of environmental pollutants. Elsevier, 2020, 81–105.

[5] Tian K, Bilal M. Chapter 15 – Research progress of biodegradable materials in reducing environmental pollution. In: Singh P, Kumar A, Borthakur A (eds.). Abatement of environmental pollutants. Elsevier, 2020, 313–330.

[6] Bolan NS, Park JH, Robinson B, Naidu R, Huh KY. Chapter four- Phytostabilization: A green approach to contaminant containment. In: Sparks DL (eds.). Advances in agronomy. Academic Press, 2011, 145–204.

[7] Zhang C, Feng Y, Liu YW, Chang HQ, Li Z, Xue JM. Uptake and translocation of organic pollutants in plants: A review. J Integr Agric 2017, 16, 8, 1659–1668.

[8] Lajayar BA, Ghorbanpourb M, Nikabadi S. Heavy metals in contaminated environment: Destiny of secondary metabolite biosynthesis, oxidative status and phytoextraction in medicinal plants. Ecotoxicol Environ Saf 2017, 145, 377–390.

[9] Damalas CA, Eleftherohorinos IG. Pesticide exposure, safety issues, and risk assessment indicators. Inter J Environ Res Public Health 2011, 8, 5, 402–1419.

[10] Kafle A, Timilsina A, Gautam A, Adhikari K, Bhattarai A, Aryal N. Phytoremediation: Mechanisms, plant selection and enhancement by natural and synthetic agents. Environ Adv 2022, 8, 100203.

[11] Zhang M, Zhu L. Sorption of polycyclic aromatic hydrocarbons to carbohydrates and lipids of ryegrass root and implications for a sorption prediction model. Environ Sci Technol 43, 2009, 2740–2745.

[12] Kagalkar AN, Jadhav MU, Bapat VA, Govindwar SP. Phytodegradation of the triphenylmethane dye Malachite Green mediated by cell suspension cultures of *Blumea malcolmii* Hook. Bioresour Technol 2011, 102, 22, 10312–10318.

[13] Garrison AW, Nzengung VA, Avants JK, Ellington JJ, Jones WJ, Rennels D, Wolfe NL. Phytodegradation of p,p'-DDT and the Enantiomers of o,p'-DDT. Environ Sci Technol 2000, 34, 9, 1663–1670.

[14] Headley JV, Peru KM, Du JL, Gurprasad N, Mcmartin DW. Evaluation of the apparent phytodegradation of pentachlorophenol by Chlorella pyrenoidosa. J Environ Sci Health Part A 43, 361–364.

[15] Jha P, Jobby R, Desai NS. Remediation of textile azo dye acid red 114 by hairy roots of Ipomoea carnea Jacq. and assessment of degraded dye toxicity with human keratinocyte cell line. J Hazard Mater 2016, 311, 58–167.

[16] He X, Zhang J, Ren Y, Sun C, Deng X, Meng Q, Hu Z, Li R, Chen Y, Shen Z, Xia Y. Polyaspartate and liquid amino acid fertilizer are appropriate alternatives for promoting the phytoextraction of cadmium and lead in *Solanum nigrum* L. Chemosphere 2019, 237, 124483.

[17] Gong Y, Chen J, Pu. R. The enhanced removal and phytodegradation of sodium dodecyl sulfate (SDS) in wastewater using controllable water hyacinth. Int J Phytoremediation 2019, 21, 11, 1080–1089.
[18] Singh V, Pandey B, Suthar S. Phytotoxicity and degradation of antibiotic ofloxacin in duckweed (*Spirodela polyrhiza*) system. Ecotoxicol Environ Saf 2019, 179, 88–95.
[19] Saxena G, Purchase D, Mulla SI, Saratale GD, Bharagava RN. Phytoremediation of heavy metal-contaminated sites: Eco-environmental concerns, field studies, sustainability issues, and future prospects. In: de Voogt P (ed). Reviews of environmental contamination and toxicology 249. Springer, Cham., 2019, 24.
[20] Nissim WG, Palm E, Mancuso S, Azzarello E. Trace element phytoextraction from contaminated soil: A case study under Mediterranean climate. Environ Sci 2018, 25, 9, 9114–9131.
[21] Dalmiş ZLE, Andede EE. Determination of the potential of *Pistia stratiotes* L. in removing nickel from the environment by utilizing its rhizofiltration capacity. Environ Sci 2019, 62, 1–12.
[22] Surabhi LS. Sustainable and eco-friendly approach for controlling industrial wastewater quality imparting succour in water-energy nexus system. Energy Nexus 2021, 3, 100020.
[23] Lin CH, Lerch RN, Kremer RJ, Garrett. HE. Stimulated rhizodegradation of atrazine by selected plant species. J Environ Qual 2011, 40, 4, 1113–1121.
[24] Nguemté PM, Djumyom Wafo GV, Djocgoue PF, Kengne Noumsi IM, Wanko NA. Potentialities of six plant species on phytoremediation attempts of fuel oil-contaminated soils PAHs impacts on bioconcentration and translocation factors. Wat Air Soil Pollut 2018, 229, 3, Article 88.
[25] Lu H, Zhang Y, Liu B, Liu J, Ye J, Yan C. Rhizodegradation gradients of phenanthrene and pyrene in sediment of mangrove (*Kandelia candel* (L.) Druce). J Hazard Mater 2011, 196, 263–269.
[26] Kotoky R, Pandey P. Rhizosphere mediated biodegradation of benzo(A)pyrene by surfactin producing soil bacilli applied through *Melia azedarach* rhizosphere. Int J Phytoremediation 2020, 22, 4, 363–372.
[27] Alagić SC, Jovanović VPD, Mitić V, Cvetković JS, Petrović GM, Stojanović GS. Bioaccumulation of HMW PAHs in the roots of wild blackberry from the Bor region (Serbia): Phytoremediation and biomonitoring aspects. Sci Total Environ 2016, 562, 561–570.
[28] Ely CS, Smets BF. Bacteria from wheat and cucurbit plant roots metabolize PAHs and aromatic root exudates: Implications for rhizodegradation. Int J Phytoremed 2017, 19, 10, 877–883.
[29] Zamani J, Hajabbasi MA, Mosaddeghi MR, Soleimani M, Shirvani M, Schulin R. Experimentation on degradation of petroleum in contaminated soils in the root zone of maize (*Zea Mays* L.) inoculated with piriformospora indica. Soil Sediment Contamin 2018, 2, 1, 3–30.
[30] Yan A, Wang Y, Tan SN, Yusof M, Ghosh ML, Chen Z. Phytoremediation: A promising approach for revegetation of heavy metal-polluted land. Front Plant Sci 2020, 11, Article, 359.
[31] Stepanova AY, Gladkov EA, Osipova ES, Gladkova OV, Tereshonok DV. Bioremediation of soil from petroleum contamination. Processes 2022, 10, 1224.
[32] Limmer M, Burken J. Phytovolatilization of organic contaminants. Environ Sci Technol 2016, 50, 13, 6632–6643.
[33] Elehinafe FB, Agboola O, Vershima AD, Bamigboye GO. Insights on the advanced separation processes in water pollution analyses and wastewater treatment – A review. S Afr J Chem Eng 2022, 42, 188–200.
[34] Thakare M, Sarma H, Datar S, Roy A, Pawar P, Gupta K, Pandit S, Prasad R. Understanding the holistic approach to plant-microbe remediation technologies for removing heavy metals and radionuclides from soil. CRBIOT 2021, 3, 84–98.

Norli Ismail*, Harlina Ahmad

Green Technologies for Treatment of Endocrine Disruptors, Pharmaceutical Compounds, and Personal Care Products

Abstract: This chapter explores various treatment technologies deployed to treat endocrine disruptors, pharmaceutical compounds, and personal care products the so call emerging pollutant and persistent organic pollutants (POPs). Those pollutants are a large group of contaminants of emerging concern and highly persistent in the environmental ecosystem (air, soil, water). They are contributed by various anthropogenic sources or due to synergistic effect with the natural occurrence. This silent intruder is a time bomb to environment, human health, and ecosystem. It is a great concern to ensure the present treatment system is capable of treating the compounds effectively. The shortcomings, challenges, and opportunity of the treatment technologies evolution toward the green technologies approach to treat the pollutants will be discussed. The shortcomings of the treatment system design due to the complexity of the pollutants, the existing treatment system modification, and enhancement will be delineated. Finally the possible green technologies that can be adopted to minimize or remove those compounds will be suggested.

Keywords: Endocrine disruptors, pharmaceutical compounds, emerging pollutant, persistent organic pollutants, green technology treatment system

1 Introduction

Green technology is widely accepted to describe a process which finally reduces or creates friendly substances, harmless and harmonize to surrounding ecosystem in a longer period of time. A critical thought and efforts are of utmost importance, especially for emerging pollutants like endocrine-disrupting compounds (EDCs). The existence of EDCs in water body as micro pollutants exhibits the inefficiency of conventional treatment systems [1]. Thus, upgrading of conventional treatments methods including primary, secondary, and tertiary treatments is significant. Due to complexity of EDCs compounds, the combination of appropriate treatment process shall be the best option for EDCs' removal. Pharmaceuticals and personal care products (PPCPs) are one of the

***Corresponding author: Norli Ismail,** Environmental Technology Division, School of Industrial Technology, Universiti Sains Malaysia, Penang 11800, USM, Malaysia, e-mail: norlii@usm.my
Harlina Ahmad, Environmental Technology Division, School of Industrial Technology, Universiti Sains Malaysia, Penang 11800, USM, Malaysia

https://doi.org/10.1515/9783111016825-010

EDCs that have evolved tremendously and exist in different sizes up to micron size which are not easy to detect and always discharged without treatment from conventional sewage treatment plant [2].

2 Perspective of green technology and importance of endocrine disruptor to ecosystem

Green technology involves series of unit process which complement each other considering a process with less of chemical usage and produce harmless products or materials. The whole ideas and goals of green technology are for a better future of environmental ecosystem holistically and marginal waste management and damage to surroundings [3]. Green technology aims to utilize and innovate acceptable materials for contaminants elimination, or reduction in which the green chemistry contributes significant importance. Alternative green chemistry or green process seems to be significantly important and leave no options to be adapted and adopted. The EDCs in general are reported to deliver negative impacts on human and animal metabolic system, thus defecting and retarding the function of the human body system. The EDCs direct or indirectly induce the benign cell, ending up with adverse effects to human like cancer and infertile adults [4].

Most of malfunctions and unexplained metabolic system failure occur due to disrupted endocrine system. They happen in many ways in the normal endocrine system body through synthesis, storage, transport, metabolism, and release of functional hormones even at lower dosages [5–7]. Synthetic and natural hormones exist as organic pollutants indirectly flow to the conventional sewage treatment plant which is usually not completely or successfully treated and not able to achieve Sustainable Development Goal (SDG) [1].

EDCs are proven scientifically to cause negative effects as low as trace-level concentrations (<1 µg/L) and alter abnormalities of living things to aquatic organisms directly and to the human body ecosystem indirectly [8]. Due to their complexity of the chemical structure, EDCs are difficult to be treated by a single existing treatment method. Thus finding the best and doable treatment approach is significantly important to combat EDCs matters aligning with the SDG initiatives. Figure 1 illustrates the occurrences and sources of EDCs and PPCPs in daily activities.

3 Treatment technologies for EDCs and PPCPs

The conventional wastewater treatment system, neither industrial nor domestic wastewater, is capable of treating those emerging pollutants completely to comply with standard water quality parameters for drinking or final effluent discharge. A stand-alone

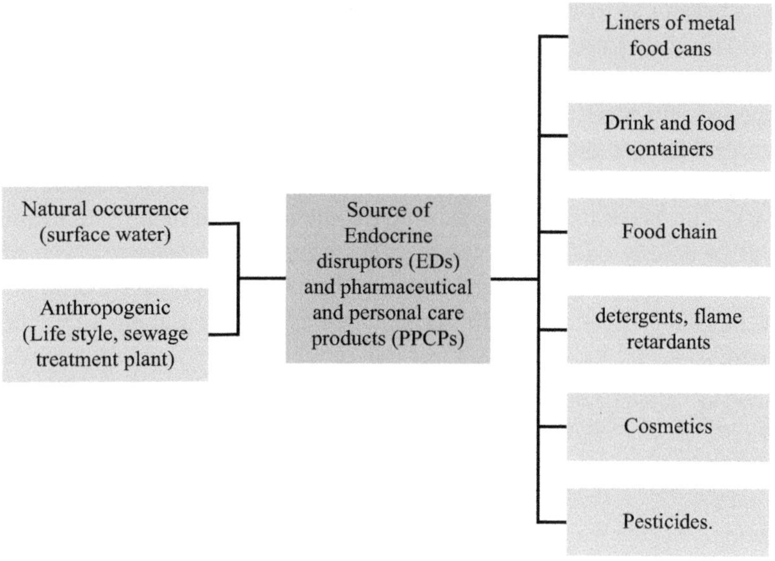

Figure 1: Distribution of endocrine disruptors and pharmaceutical and personal care products in environmental ecosystem.

physical, chemical, physicochemical and biological treatment needs upgrading to be able to receive and treat EDCs and PPCPs. It is not always the individual unit operation regardless of the type of the system, but it is more of the characteristic of the incoming influent to be treated. Profiling the pollutants' load beforehand helps to predict the treatment system performance toward EDC and PPCP removal. The wastewater treatment plant operator should be able to adjust the unit operation requirement accordingly. It is important to analyze the EDCs and PPCPs of the raw water influent for additional appropriate unit operations and configurations. Several studies elsewhere revealed excellent results for pharmaceutical wastewater through tertiary treatment oxidation-reduction, Fenton, photocatalytic, UV-adsorption, for instance.

Biological treatment depicted a possible ultimate treatment process to address the strong chemical structure of both EDCs and PPCPs. It could be a one-stage biological process with a specific bacterial strain screened and isolated from the sources of EDC- and PPCP-contaminated areas or processes. This somehow indirectly exposes the endogenous microbes to the extreme environment containing EDCs and PPCPs, creating a resistance ability to the strains. While a combination of biological and adsorption process formed an immobilized layer of the biomass rich microorganism on the surface of adsorbent. The biofiltration system commonly entails a mechanical biological system reactor, membrane biofilter, and moving bed biological reactor (MBBR).

This chapter explores various treatment technologies deployed to treat endocrine disruptors, pharmaceutical compounds, and PPCPs representing persistent organic pollutants (POPs). Those pollutants are greatly tenacious in air, soil, and water and associated with various anthropogenic sources or a result of synergistic effect with the natural compound occurrence. This silent intruder is a time bomb to environment and human health ecosystem. It is a great concern to ensure the present treatment system is capable of treating the compounds effectively. The shortcomings, challenges, and opportunity of the treatment system evolution to align with the green technology approach to treat the emerging pollutants will be discussed. The shortcomings of the treatment system design due to the complexity of the pollutants, the existing treatment system modification, and enhancement will be delineated. Finally, the possible green technologies that can be adopted to minimize or remove those compounds will be suggested. Figure 2 compiles the extension of the conventional treatment technology that has been applied for EDC and PPCP treatment.

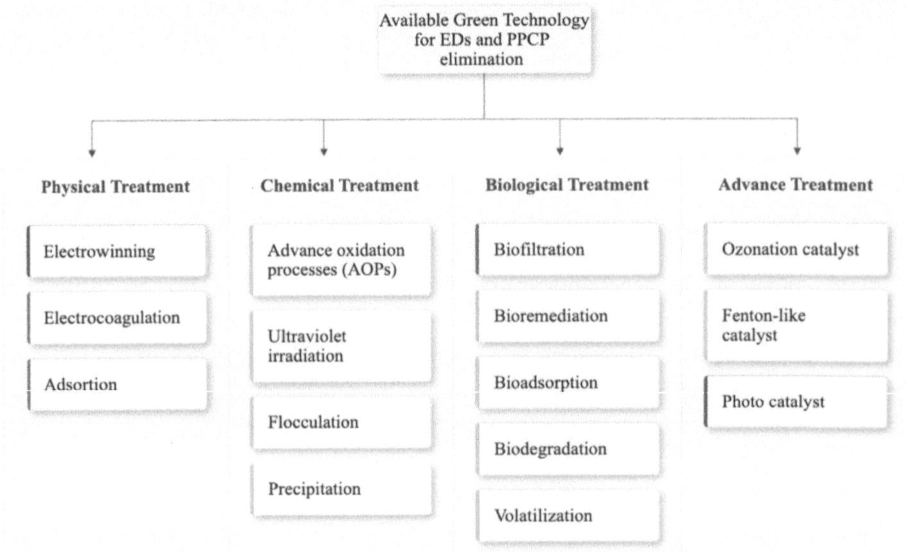

Figure 2: Compilation of different type of conventional treatment technology for EDC and PPCP treatment.

Physicochemical and biological treatment with further modification and enhancement shall remain the best option to curb the micropollutant problems to align with clean water and sanitation within SDGs' goal [1]. A continuous demand for the safe and clean water supply to ecosystem and community for daily usage summons a conventional wastewater treatment system enhancement and gradually subtle green technology approaches.

Contaminants of emerging concern, persistent organic pollutants, emerging organic and inorganic pollutants are general terminology to address EDCs. They are notorious and not easily eliminated from the water body through primary, secondary, and some tertiary treatment [9] reported by an authorized system operator treating the water, wastewater, or sewage treatment plants in particular. Hence, a combination of the existing treatment systems or advanced technologies could be the best treatment option. PPCP compound loads exhibited various removal efficiency, depending on the different types of PPCP compounds, for which ibuprofen depicted the highest level of difficulty to be eliminated compared to sulfamethoxazole. The treatment mechanism gives a significant impact on the removal efficiency associated with adsorption-biodegradation, anaerobic, anoxic-oxic system as well as the geographical area on the globe referring to the influence of the temperature in relation to a microbial consortium [10, 11]. A reliable analysis and measurement of PPCPs' and EDCs' existence in raw and treated water or wastewater is crucial especially to be able to capture the nano and micron sizes of the contaminants [12]. Low sensitivity of the measurement tools, improper sample preparation, and incorrect sampling techniques will jeopardize accurate EDC and PPCP reporting data. Table 1 summarizes the reported study by various treatment methods, including advance treatment for both compounds. The research findings of persistent organic pollutants through activated carbon-oxidation-ozonation resulted in almost satisfied outcomes [13]. The treatment method could be extended to EDCs and PPCPs.

Table 1: Treatment strategies for EDC and PPCP removal.

Physical	**Chemical**	**Physicochemical**	**Biological**	**Advance treatment**
Membrane filtration	Oxidation – e.g addition Fe(VI) work well to targeted compounds e.g phenol, olefin, amine and aniline [14]	Photocatalysis and adsorption	anoxic-anaerobic-anoxic-oxic (AAAO) process	Ozonation – addition of ozone could substantially improve their removals.
Adsorption – activated carbon, e.g., addition of PAC Could substantially improve the removals	Conventional treatment (coagulation plus chlorination) would have low removal of many EDC/PPCPs	UF + Adsorption	Moving bed biological reactor (MBBR)	Advanced oxidation processes (AOP)
Constructed wetland		Conventional water treatment with ozonation	Conventional activated sludge	UF + adsorption
		Ultrafiltration (UF)	Biological nutrient removal	UF/RO + UV

Table 1 (continued)

Physical	Chemical	Physicochemical	Biological	Advance treatment
		Filtration and adsorption	Conventional activated sludge + microfiltration	
			Membrane bioreactor (MBR)	

4 Phase-changing technology

It is an adsorption technic or process in principle. A continuous innovation of the adsorbent candidates in which the green materials are more preferable has been explored and improved timelessly. It is a continuous research working and looking into various affordable sources of plant and mineral-based adsorbent, for instance, a conventional activated carbon, mineral clay, biochar, and carbon nanotubes. The final goal is to address and align with Sustainable Goal Developments (SDGs) related to water and wastewater inland or below the sea.

The alternative green materials from various sources, biochar photocatalyst is a promising green technology and should be studied in depth to explain the degradation of organic in aqueous waste. The algae biochar has been proven to effectively remove phenolics, antibiotics and several dye solutions' constituents in both terrestrial and marine ecosystems [15–17]. Biochar of *Enteromorpha prolifera* has been improvised with potassium hydroxide for antibiotic sulfamethoxazole (SMX) adsorption [18].

Biochar possesses a larger surface area, reactive and different functional groups (C = O, COO^-, –OH, and –NH–). The reactive and functional groups serve as the foundation for surface formation, and linkages of multiple roles of BC aggregate which are flexible and adaptable preferably to allow adsorption and degradation take place on the same BC-catalyst/support platform [19]. Biochar play a role as adsorbent or catalyst, and it is mainly related to functional groups and materials of biochar sources. Biochar with high porosity will provide more active binding sites between contaminants and surface adsorbent for further degradation.

According to Shichong [20] EDC permeation through filtration or sieving mechanism is ineffective in removing EDCs. As reported elsewhere, the removal of EDCs and PhACs by nano filtration membranes is determined by many processes mechanism and factors, including screening, mesh size, pore size, charge exclusion, and adsorption. The efficient mechanisms differ from compound to compound of EDCs; thus, further study of nano filtration application is needed for a better delineation and understanding the nano filtration removal efficiency.

5 Biofiltration

Biofiltration in green technologies refers to a flow of contaminated water through a biofilm that was developed by a microbial cell mass. It is widely used as a fixed growth system as a packed bed reactor in a column reactor. The treatment process relies on endogenous microbial growth and forms a blanket of biofilm. Small voids in between the media or host media provide sufficient times for the microbes to take up and degrade the contaminants of the downflow or upflow wastewater. A trickling filter, upflow anaerobic sludge blanket (UASB), and rotating biological contactor (RBC) are examples of the conventional biofiltration. Biofiltration is commonly run in semi-aerobic conditions; thus, similar to other biological treatment processes active microbes, sludge age, surface-loading rate, organic and inorganic loads determine the system performance.

Fluctuation of incoming polluted waters to the treatment system even at lower concentration or small quantity exhibits a significant impact on the environment. Antibiotics and hormones which are known to be persistent, perpetual and hardly to remove mark a critical need to upgrade and modify the granular activated carbon (GAC) biofilter to secure water security and allow water recycling [21].

6 Bioremediation

The versatility of microorganisms in the environmental ecosystem from low, normal to extreme conditions makes them the best candidate for the ultimate process to combat the organic pollution problem. The source of potential strain of microbes available is not limited to aqueous medium but is extended to air and soils. Remediation of the contaminated area or medium by specific bacteria strain can be implemented either at the particular site or offsite. Offsite treatment involves a transportation of contaminated samples or materials to designated laboratory facilities or treatment facilities. Bioremediation can be applied both by endogenous and by exogenous microbial (bioaugmentation). The bioaugmentation is significant whenever a special or specific strain of microbes is needed to degrade a complex or recalcitrant pollutants.

However, the drawback of bioremediation is less effective in treating hazardous compounds in general. It is most probably due to the complexity and toxicity of the pollutants which can retard the microbial growth. Later it reduces the efficiency of bioremediation in relation to the decrease of activated cell microbes. Several heavy metals, trace elements, and recalcitrant compounds can retard the microbial cell growth. Microbes can be cultured or "trained" according to the degradation process needed. The integration of microbial biomass and physico-electro through winning and coagulation process will replace the conventional bioremediation which normally relies on suspended growth system. Hence, bioremediation will be a good successor to EDCs' eradication and able to achieve acceptable limit set by Environmental Protection Agency physico-electro fermentation (EPA) [22].

7 Principle of Membrane Bioreactor (MBR)

The MBR technology is considered an enhancement of the conventional activated sludge system which initially applied for sewage treatment. The MBR featured a plus point over conventional activated sludge in many ways such obtaining high effluent quality, reduced footprint, and sludge production. Membrane bioreactors (MBRs) with relevance decontamination can attain good quality and acceptable treated effluent quality for water recycling purposes. Study depicted high removal in MBRs in treating EDCs with more than 96% reduction rate compared to conventional activated sludge treatment system with only 85% reduction efficiency [23].

On the other hand, MBRs is still skeptical to most relevance stakeholders due to it high allocation for fabrication and installation of suitable membrane and additional components to ensure it is inline with green technology approaches. MBR is the integrated biological process with a physical membrane separation. The physical membrane will filtrate the particle that flows together with the incoming raw influent. All the active microbial biomass retained on the membrane surface resulted in a clarified and disinfected filtrate substantially reducing the EDC concentration [24]. Anyhow, the MBR alone is unable to remove trace recalcitrance organic pollutants that most existing in PPCPs and EDCs. The influence of precursor, electron receptor, and donors was determined to be significant factors in the MBR removal mechanism [25].

8 Electro-current assisted coagulation

Conventional coagulation-flocculation with coagulant or flocculant addition has transited to chemo-electrical assisted through positive (cathode) and negative (anode) electrodes in the ionic solution. Deposition of oxidized metals will be collected on the cathode plate. The methods are widely and practically usable for a recovery of the precious metals.

The working principle of the electro-assisted coagulation process is destabilizing the targerted compounds by electrical current. The cation charges released will bind the anionic particles that need to be removed, and settle to the bottom, leaving a clear water or less of contaminant concentration. It can be described as multiple physico-chemical process, including free-radical development [26].

9 UV irradiation and chemical assisted hydrogen peroxide

Integrated UV-H_2O_2 system can be categorized as advance treatment which is UV assisted to invigorate the hydroxyl radical's generation. Hydrogen peroxide is a powerful oxidation agent proven to lower down most pollutants that exist in wastewater. However, the study reported it being less effective for more complex pollutant compounds [27]. Additional energy sources and other associated chemical reagents depicted excellent separation for hydroxyl radical compounds. It has been demonstrated successfully for diluted basic-azodyes [28] such as in the textile industry. Photocatalytic and advance oxidation process could be a harmonize process for both environment and ecological sustainability [17].

10 Pharmaceutical compounds and personal care product – endocrine disruptor

Diverse persistent organic compounds are the major building block of pharmaceuticals and personal care products (PPCPs). They vary from regular daily products of public usage like antibiotics, drugs for allergic and asthmatic and up to complex medical hormones. The environmental concern to their existence is the ability to depreciate biotic and abiotic ecosystem at the least concentration of their accumulation [29, 30].

Endocrine disruptors (EDCs) cover a widespread array of natural and anthropogenic compounds or products in our daily life. The distribution of EDCs to many forms by the parent compounds renders a toxicity in most secondary metabolites, which sometimes pose more hazards when compared to the parent compounds [31].

Up to a certain extent under a conducive environment, the occurrence of PPCPs worsen the toxicological properties and altered the original structure to a harmful product or compound [32]. Acidic compounds of PPCPs exhibited multiples toxicological response of studied organisms [33]. The existence of elemental compounds in the water body exhibited a significant influence of toxicity elevation of fluoroquinolones and tetracyclines [34]. EDC and PPCP accumulation biologically in many platform media (air, water, soils) exposed a direct and indirect impact to human health. A significant effort by researchers to enrich the capacity and efficiency of the conventional treatment system with the ultimate goals to eradicate those emerging pollutants [35] resulted from incomplete degradation chemically or biologically [36, 37].

Figure 3 illustrates the existence of EDCs and PPCPs and the present treatment technology.

The new formation of emerging pollutants with the long list of new chemical and recalcitrant compounds seems to be no full stop and it happens all over the world [38].

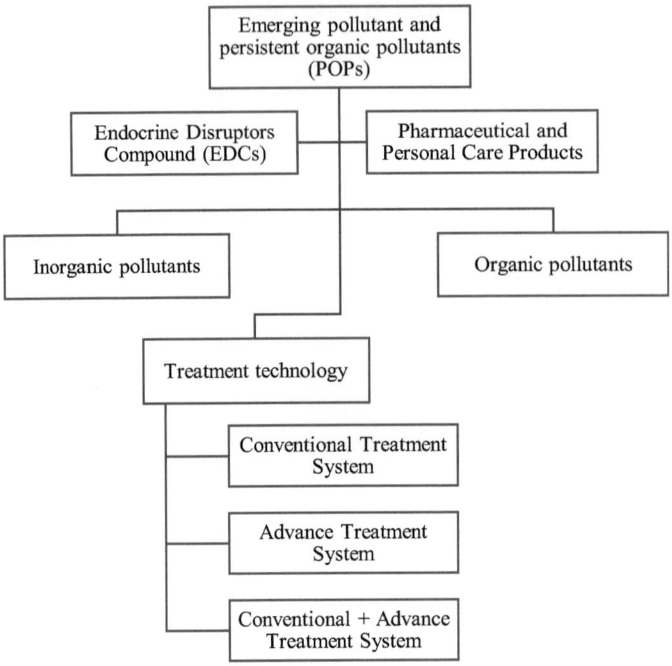

Figure 3: Overview of the EDCs and PPCPs existence and treatment approach.

More and more new registered of the new emerging pollutant has potential to create a mutation in microbial species and strain.

11 Integrated treatment: A way forward of green technology for EDCs and PPCPs removal enhancement

Referring to previous works related to EDCs and PPCPs by individual conventional and advance treatment it is best recommended to implement an integrated treatment, for example based on Fenton oxidation. A possible combination of Fenton oxidation–based treatment for various types of EDCs and PPCPs, for instance antibiotic (e.g penicillin, amoxicillin) and medicinal drug compounds (e.g., sulfamethazine, levofloxacin, ibuprofen), is best to be treated along with ozonation, Fenton oxidations, adsorption, membrane filtration, and bioremediation with endogenous or exogenous inoculum [39].

A heterogeneous Fenton system is able to minimize the drawback of the single Fenton process. The integration of the heterogeneous Fenton system with other conventional or tertiary system unit processes, as reported elsewhere, obtained and recorded high percentage of pharmaceutical compounds degradation. The Fenton heterogeneous system becomes more agile, adaptable, and caters to a broader range of persistent organic contaminants. For instance, the hybrid Fenton process through a combination of Fenton-adsorption revealed the best potential integrated system which is capable of degrading numerous classes of EDCs and PPCPs.

Full wastewater characterization conducted for EDCs and PPCPs samples will be the first screening steps for identifying the appropriate treatment system to be selected and also influence the overall treatment system performance. The wastewater characterization for EDCs and PPCPs compounds should be conducted prior to proposing any integrated or hybrid treatment method to be selected [40]. Integration of primary, secondary, and tertiary treatment in a proper and well-strategized manner as reported in previous research works shows a reliable treated effluent quality of endocrine disruptors and pharmaceutical and personal care products. Nanotechnology is reported to be the sustainable treatment method in combating EDC problems in the aquatic ecosystem. The innovation of the various types nanoparticles and nanofibers (< 100 nm) is the main structure ensuring the wider accessibility to the targeted nano size contaminant which is mostly the building block of the emerging pollutants nowadays [41].

References

[1] Nasir HM, Wee SY, Aris AZ, Abdullah LC, Ismail I. Processing of natural fibre and method improvement for removal of endocrine-disrupting compounds. Chemosphere 2022, 291, 1, 132726.

[2] Meyer MF, Powers SM, Hampton SE. An evidence synthesis of Pharmaceuticals and Personal Care Products (PPCPs) in the environment: Imbalances among compounds, sewage treatment techniques, and ecosystem types. Environ Sci Technol 2019, 53, 12961–12973.

[3] Azimi SC, Shirini F, Pendashteh AR. Advanced oxidation process as a green technology for dyes removal from wastewater: A review. Iran J Chem Chem Eng 2021, 40, 5, 1467–1489.

[4] Larsson DGJ, Adolfsson-Erici M, Parkkonen J, Pettersson M, Berg AH, Olsson P-E, Förlin L. Aquat Toxicol 1999, 45, 91.

[5] Kabir ER, Rahman MS, Rahman I. A review on endocrine disruptors and their possible impacts on human health. Environ Toxicol Pharmacol 2015, 40, 241–258.

[6] Yang O, Kim HL, Weon JI, Seo YR. Endocrine-disrupting chemicals: Review of toxicological mechanisms using molecular pathway analysis. J Cancer Prev 2015, 20, 12–24.

[7] Sarangapani C, Danaher M, Tiwari B, Lu P, Bourke P, Cullen PJ. Ecacy and mechanistic insights into endocrine disruptor degradation using atmospheric air plasma. Chem Eng J 2017, 326, 700–714.

[8] Pironti CR, Proto M, Bianco A, Montano PM, Motta, O L. Endocrine-disrupting compounds: An overview on their occurrence in the aquatic environment and human exposure. Water 2021, 13, 1347. https://doi.org/10.3390/w13101347.

[9] Khan NA, Khan SU, Ahmed S, Farooqi IH, Yousefi M, Mohammadi AA, Changani F. Recent trends in disposal and treatment technologies of emerging-pollutants – A critical review. Trends Anal Chem 2020, 122, 115744.

[10] Liu Q, Feng X, Chen N, Shen F, Zhang H, Wang S, Sheng Z, Li J. Occurrence and risk assessment of typical PPCPs and biodegradation pathway of ribavirin in wastewater treatment plants. Environ Sci Ecotechnol 2022, 11, 100184.

[11] Kallenborn R, Brorström-Lundén E, Reiersen L-O, Wilson S. Pharmaceuticals and personal care products (PPCPs) in Arctic environments: Indicator contaminants for assessing local and remote anthropogenic sources in a pristine ecosystem in change. Environ Sci Pollut Res Int 2018, 25, 33, 33001–33013.
[12] Papageorgiou M, Zioris I, Danis T, Bikiaris D, Lambropoulou D. Comprehensive investigation of a wide range of pharmaceuticals and personal care products in urban and hospital wastewaters in Greece. Sci Total Environ 2019, 694, 133565.
[13] Westerhoff P, Yoon Y, Snyder S, Wert E. Fate of endocrine-disruptor, pharmaceutical, and personal care product chemicals during simulated drinking water treatment processes. Environ Sci Technol 2005, 39, 17, 6649–6663.
[14] Yang B, Guanghao GC. Submerged membrane bioreactor in treatment of simulated restaurant wastewater. Sep Purif Technol 2012, 88, 184–190.
[15] Sharma G, Bhogal S, Gupta VK, Agarwal S, Kumar A, Pathania D, Mola GT, Stadler FJ. Algal biochar reinforced trimetallic nanocomposite as adsorptional/photocatalyst for remediation of malachite green from aqueous medium. J Mol Liq 2019, 275, 499–509.
[16] Huang W, Chen J, Zhang J. Adsorption characteristics of methylene blue by biochar prepared using sheep, rabbit and pig manure. Environ Sci Pollut Res 2018, 25, 29256–29266.
[17] Sutar S, Otari S, Jadhav J. Biochar based photocatalyst for degradation of organic aqueous waste: A review. Chemosphere 2022, 287, 132200.
[18] Wu Y, Cheng H, Pan D, Zhang L, Li W, Song Y, Bian Y, Jiang X, Han J. Potassium hydroxide-modified algae-based biochar for the removal of sulfamethoxazole: Absorption performance and mechanisms. J Environ Manag 2021, 293, 112912.
[19] Londono-Zuluaga C, Jameel H, Gonzalez RW, Lucia L. Crustacean shell-based biosorption water remediation platforms: Status and perspectives. J Environ Manag 2019, 231, 757–762.
[20] Wang S, Li L, Yu S, Dong B, Gao N, Wang X. A review of advances in EDCs and PhACs removal by nanofiltration: Mechanisms, impact factors and the influence of organic matter. Chem Eng J 2020, 406, 10, 126722.
[21] Fu GY, Rahman MA. Removal of Contaminants of Emerging Concern (CECs) from secondary biologically treated wastewater effluent using biofilters of Granular Activated Carbon (GAC). AWWA ACE19, 2019.
[22] Gao X, Kang S, Xiong R, Chen M. Environment-friendly removal methods for endocrine disrupting chemicals. Sustainability 2020, 12, 7615.
[23] Spring AJ, Bagley DM, Andrews RC, Lemanik S, Yang P. Removal of endocrine disrupting compounds using a membrane bioreactor and disinfection. J Environ Eng Sci 2007, 6, 131–137.
[24] Tang K, Xie J, Pan Y, Zou X, Sun F, Yu Y, Xu R, Jiang W, Chen C. The optimization and regulation of energy consumption for MBR process: A critical review. J Environ Chem Eng 2022, 10, 5, 108406.
[25] Tadkaew N, Hai FI, McDonald JA, Khan SJ, Nghiem LD. Removal of trace organics by MBR treatment: The role of molecular properties. Water Res 2011, 45, 8, 2439–2451.
[26] Xu Y, Liu T, Zhang Y, Ge F, Steel RM, Sun L. Advances in technologies for pharmaceuticals and personal care products removal. J Mater Chem A 2017, 5, 24, 12001–12014.
[27] Neyens E, Baeyens J. A review of classic Fenton's peroxidation as an advanced oxidation technique. J Hazard Mater B 2003, 98, 33–50.
[28] Daneshvar N, Salari D, Khataee AR. Photocatalytic degradation of azo dye acid red 14 in water: Investigation of the effect of operational parameters. J Photochem Photobiol A: Chem 2003, 157, 111–116.
[29] Ebele AJ, Abdallaf MA-E, Harrad S. Pharmaceuticals and personal care products (PPCPs) in the freshwater aquatic environment. Emerg Contamin 2017, ISSN: 2405-6650, 3, 1, 1–16.

[30] Salgado R, Noronha JP, Oehmen A, Carvalho G, Reis MAM. Analysis of 65 pharmaceuticals and personal care products in wastewater treatment plants in Portugal using a simplified analytical methodology. Water Sci Technol 2010, 62, 12, 2862–2871.

[31] García-Galán MJ, González Blanco S, López Roldán R, Díaz-Cruz S, Barceló D. Ecotoxicity evaluation and removal of sulfonamides and their acetylated metabolites during conventional wastewater treatment. Sci Total Environ 2012, 437, 403–412.

[32] Isidori M, Lavorgna M, Nardelli A, Parrella A, Previtera L, Rubino M. Ecotoxicity of naproxen and its photo transformation products. Sci Total Environ 2005, 348, 93–101.

[33] Fent K, Weston AA, Caminada D. Ecotoxicology of human pharmaceuticals. Aquat Toxicol 2006, 76, 2, 122–159.

[34] Zhang Y, Cai X, Lang X, Qiao X, Li X, Chen J. Insights into aquatic toxicities of the antibiotics oxytetracycline and ciprofloxacin in the presence of metal: Complexation versus mixture. Environ Pollut 2012, 166, 48–56.

[35] Brijesh AS, Mishra K. Chapter 2 -Treatment aspect of an emerging pollutant from Pharmaceutical industries using advanced oxidation process: Past, current, and future trends in Development in Wastewater Treatment Research and Processes: Removal of Emerging Contaminants from Wastewater Through Bio-nanotechnology. Shah M, Rodriguez-Couto S, Biswas J (eds.). Elsevier, 2022, 23–44.

[36] Mirzaei ZC, Haghighat F, Yerushalmi L. Removal of pharmaceuticals from water by homo/heterogonous Fenton-type processes – a review. Chemosphere 2017, 174, 665–688.

[37] Catalá M, Dominguez-Morueco N, Migens A, Martinez F, Valcárcel Y, Mastroianni N, López de Alda M, Barceló D, Segura Y. Elimination of drugs of abuse and their toxicity from natural waters by photo-Fenton treatment. Sci Total Environ 2015, 520, 198–205.

[38] Rodriguez-Narvaez OM, Peralta-Hernandez JM, Goonetilleke A, Bandala ER. Treatment technologies for emerging contaminants in water: A review. Chem Eng J 2017, 323, 361–380.

[39] Khan NA, Husain Khan A, Tiwari P, Zubair M, Naushad M. New insights into the integrated application of Fenton-based oxidation processes for the treatment of pharmaceutical wastewater. J Water Process Eng 2021, 44, 102440.

[40] Azizi D, Arif A, Blair D, Dionne J, Filion Y, Ouarda Y, Pazmino AG, Pulicharla R, Rilstone V, Tiwari B, Vignale L, Brar SK, Champagne P, Drogui P, Langlois VS, Blais J-F. A comprehensive review on current technologies for removal of endocrine disrupting chemicals from wastewaters. Environ Res 2022, 207, 112196.

[41] Ojha A, Tiwary D, Oraon R, Singh P. Degradations of endocrine-disrupting chemicals and pharmaceutical compounds in wastewater with carbon-based nanomaterials: A critical review. Environ Sci Pollut Res 2021, 28, 1–22.

Norli Ismail*, Nur Syamimi Selamat, Harlina Ahmad

Microbial Enzymes and Their Role in Bioremediation of Environmental Pollutants: Prospects and Challenges

Abstract: A vast number of papers have presented the significance and application of microbial enzymes in biological process degradation. Bioremediation is a term used to explain the biological process and it can be expended to many other associate processes (biodegradation, bio detoxification, bio mineralization, and biotransformation). Application of microbial enzyme in bioremediation seems a promising treatment technology and an ultimate strategy to combat the wide range of environmental pollutants. In this chapter bioremediation is defined as a biological process which actively involves varies consortium of microorganism mostly bacteria. Bioaugmentation and biotransformation are part of bioremediation which transforms the contaminants to lesser concentration. The challenges and prospects of microbial enzyme applicability in environmental pollutants' bioremediation of various microorganisms will be dissected. The important factors affecting and determining the bioremediation efficiency and possible mechanisms of microbial enzymes responsible for bioremediation will also be discussed in this chapter, thus providing extensive and comprehensive information for further research alignment and way forward dealing with a wider range of emerging environmental pollutants.

Keywords: Microbial enzyme, bioremediation, emerging environmental pollutant, biological reduction mechanism

1 Introduction

It is known for a fact that the Earth is endowed with a variety of natural resources. The exploration and exploitation of these resources, both renewable and non-renewable, has contributed to the advancement of human civilization through rapid urbanization and industrialization, which eventually results in environmental pollution. The surrounding air, water, and land are natural elements of the environment that are essential for living for both biotic and abiotic ecosystems. Environmental pollution is a phenomenon in which harmful and noxious substances are, intentionally or unintentionally,

***Corresponding author: Norli Ismail,** Environmental Technology Division, School of Industrial Technology, Universiti Sains Malaysia, Penang 11800 USM, Malaysia, e-mail: norlii@usm.my
Nur Syamimi Selamat, Harlina Ahmad, Environmental Technology Division, School of Industrial Technology, Universiti Sains Malaysia, Penang 11800 USM, Malaysia

https://doi.org/10.1515/9783111016825-011

released into the environment. Two major sources of environmental pollution are natural events, such as volcanic activity and forest fire, and human activities, such as industrial wastewater discharge and greenhouse gaseous emission. However, the term "pollution" often refers to the negative impact on the ecosystem due to anthropogenic sources.

Organic and inorganic substances, isotopes, gases, nanoparticles, and organometallic compounds are only a few examples of the various forms of environmental pollutants [1]. The organic compounds commonly will undergo degradation or minerilization thus produced water, carbon dioxide and energy in form of heats as by-products [2]. The limited natural degrading capability of the organic waste, for instance, can result in environmental issues including air and water pollution as well as soil contamination [3]. These issues, if left unresolved, could pose a threat to natural ecosystems and eventually to human via pollutant transfer pathway.

Conventional treatment methods (physical, physicochemical, biological) as primary and secondary treatment process can be operated individually or incorporated to each other for the best treatment efficiency. Always the biological treatment exhibited the ultimate treatment approach for a sustainable treated effluent quality or zero effluent discharge [1]. Physical, chemical, and physicochemical treatment is reported to be less time-consuming, easy to operate, providing fast results compared to bioremediation. Sometimes it is more effective than biological process, even as a stand-alone process. However, it has created a secondary pollution problem such as high operating, maintenance, and construction costs and energy utilization [4]. The by-product of secondary pollutants subsequently summons another operational cost for disposal and management. A biological process enhancement through various types of sustainable bioremediation is significantly importance to align with exponential growth of inhabitants in rural and urban areas due to rapid development and health level quality.

It has long been understood how soil microbial communities affect ecosystem processes like nutrient cycling [5] and the biodegradation of contaminants is generally seen as a technique with high societal acceptance [6]. In order to eliminate pollutants of the affected areas through bioremediation and bioaugmentation of biotransformation, various types of microorganisms with potential enzyme secretion and abilities are believed to be the ultimate process, effective and ecologically friendly [7]. In addition, reports have claimed endogenous bioremediation to be able to remove incalcitrant pollutants, thus restoring and cleaning the contaminated areas [8]. Microbial degradation frequently involves consortia of different species of microorganisms, some of which are capable of breaking down complexes compounds by a consortium of microbes of their individual roles [9]. The integrated process and reaction by associated microbes induce the degradation or remediation process through enzymatic activities. The interaction process among the microbes remains dynamic and is controlled by many factors [10]. Microbial enzymes are widely used in a variety of industries and other applications since they are known to be essential as metabolic catalysts. Besides that, these enzymes are also necessary for the bioremediation of pesticides and herbicides, as well as for cosmetic industry ingredients [3].

This chapter intends to highlight challenges factors and impending microbial enzyme solicitation by various microbial applications of so-called bioremediation. This chapter will also cover key elements influencing and determining the effectiveness of bioremediation as well as potential mechanisms of microbial enzymes involved in bioremediation, consequently providing extensive and comprehensive information for future research alignment and way forward dealing with a wider range of emerging environmental pollutants.

2 Bioremediation, microbial enzymes, and their applications in treatment of environmental pollutants

Degradation is described by the American Society for Testing and Materials (ASTM) as an irreversible process for a material that went through a major alteration in its structure and is often characterized by a loss of properties such as integrity, molar mass, structural, mechanical, or chemical properties [11]. Bioremediation generates carbon dioxide and water as by-product throughout the degradation process by microbial enzymes [12]. Bioremediation can be characterized as in situ and ex situ bioremediation. The in situ bioremediation comprises engineered and intrinsic bioremediation, whereby ex situ bioremediation is encompassed of solid and slurry phase bioremediation [13]. Furthermore, the engineered bioremediation includes biosparging, biostimulation, bioaugmentation, bioventing, natural attenuation, and bioslurping; the solid phase bioremediation includes biopiling, composting, land farming, and biofilter; and slurry phase bioremediation includes bioreactor. The hierarchal flow diagram of the categorization of bioremediation is as presented in Figure 1.

Conventionally, a variety of remediation technologies have been utilized to remove pollutants from the environment. However, the remedy of harmful substances, which costs the loss and malfunctions of many environmental systems, caused dissatisfaction and apprehensions in the global citizens. A push factor in providing a more promising treatment approach which is able to diminish pollution problems thus uplifts the bioremediation as assuring process for neutralizing and reinstating the living environment [14]. Currently, special attention is given to so-called green technology in manufacturing which generates less harmful materials until the end of their life cycles. This is possibly achievable through enzymatic degradation [15], and this technology has gained more attention since it is more cost-effective, efficient, and ecologically benign as compared to the conventional physical and chemical pollutant removal approaches [16]. Since

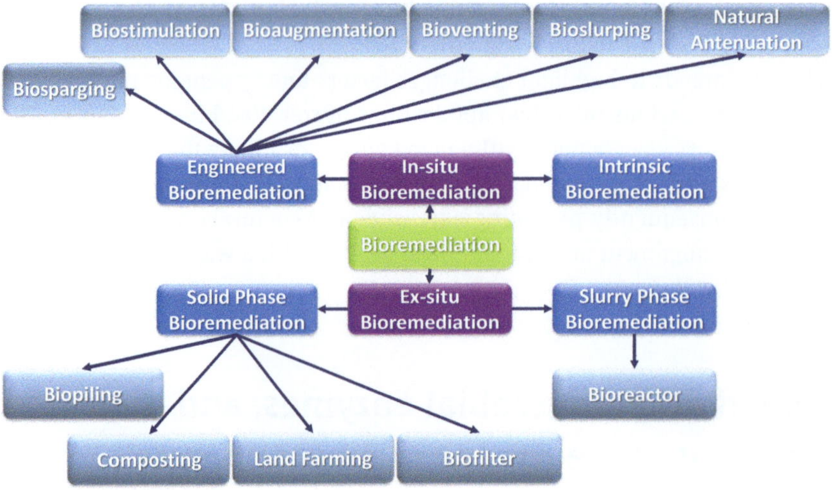

Figure 1: Hierarchal flow diagram showing categorization of bioremediation.

microbial enzymes are known to serve as essential metabolic catalysts, in addition to other applications, they are frequently used in a variety of sectors [3].

Microorganisms in soil have a role in the breakdown of both organic and inorganic materials which occur naturally or anthropogenically, giving the soil microbes a potent explorer to undertake, adapt, and rehabilitate the changes of the terrestrial and aquatic ecosystem [17]. Numerous bacteria have the capability to increase the metals' precipitation in various metabolic pathways with and without the presence of oxygen. The tolerance ability of individual microbes determines the removal efficiency of the treatment process [18]. Previous research depicted homogenous or heterogeneous inoculums exist as aerobes, anaerobes, or facultative. *Bacillus*, *Pseudomonas*, *Ruminococcus*, *Bacteroides*, and *Lactobacillus*, for instance, accelerate in solubilizing or liquefying wastes. Protease, amylase, lipase, and cellulase are examples of microbial enzymes that stimulate microbial growth and are frequently found in many of biological products [19].

The term "biological products" used to describe microorganisms containing microbial enzymes, however, did not only apply to bacteria but also to other microbes such as fungi and yeast. Previous research has demonstrated that many fungal species were also capable of transforming diverse environmental contaminants, including xenobiotic extraterrestrial vehicles and nitrogenous explosives, employing cellular degrading enzymes. Additionally, numerous studies have already documented how bacteria and fungi may effectively alter or degrade nitroaromatic contaminants while maintaining their viability [8]. Thus, such a naturally occurring phenomenon has numerous benefits over physical and chemical methods, including being environmentally friendly, economical, and producing less incompletely degraded wastes or sludge, which contributes to the burden on the environment [20].

Enzymes are greatly efficient as biocatalysts or bio-stimulants that are practically important at full-scale operation in various manufacturing process. They have become in high demand due to green process to the environment [21]. Enzymes are present in all living things and important compounds for living things' survival through metabolic reactions. Pharmaceutical, healthcare, and food industries benefit a lot from enzyme applications [22].

Over the years, many bacterial strains and species of fungi that produce various microbial enzymes have been investigated and examined for their effectiveness in removing various environmental contaminants. For instance, a variety of nitroaromatic and nitramine-type explosives have been reported to support the growth of *Escherichia, Enterobacter, Citrobacter, Clostridium, Desulfovibrio, Klebsiella,* and *Pseudomonas*. Notwithstanding, a small number of enzymes produced by microbes are found to be involved in the catalysis of nitroaromatics' breakdown including TNT [23]. Trichoderma species is the mitosporic fungus that transforms TNT components and possesses higher tolerance than other species. The species exhibited a greater capacity to defecate transient intermediates (amines), TNT, and DNT in homogenous inoculums compared to other fungal species [8]. These findings imply that many more fungi species can be used in bioremediation.

Several Bacillus species, including *Bacillus licheniformis, Bacillus megaterium, Bacillus subtilis, Bacillus cereus,* and *Bacillus pumilus*, are classified as keratinase producers. Keratinases are serine proteases that are part of the subtilisin group (S8 family) [24]. Keratinase is an example of protease enzyme that is widely applied in many industrial applications. These extracellular inducible enzymes are typically produced by different microorganisms in selected culture media and mass produced nowadays for biotechnology industry, for example, in feed meals, organic fertilizer, leather tanning processing, health and personal care product industry [25], and waste biodegradation.

Carboxylesterases, lipases, and cutinases are enzymes primarily linked to the hydrolysis of plastics [9], while numerous hydrolyzing enzymes from various classes, including esterase from Comamonas acidovorans TB-35, and TfCut2, Tcur0390, are the enzymes which hydrolyze polyester. Tcur178 has been isolated from *Thermobifida fusca* KW3, *Thermomonaspora curvata* DSM43183, and LC cutinase is reported to be capable of hydrolyzing polyurethane (PU) [26]. The bacterium strains, fungus species, and enzymes described are some examples of different microorganisms and microbial enzymes used to bioremediate environmental pollutants. More examples are as presented in Table 1.

Table 1: Studies done on biodegradation of various environmental pollutants by different types of microorganisms through microbial enzymes.

Microorganism	Pollutant	Enzyme	Reference
Bacillus cereus	Feather	Keratinase	[24]
Trichoderma viride	2,4,6-trinitrotoluene (TNT)	Dioxygenase	[8]
Humicola insolens	Polyurethane-polyester (PU-PE)	Cultinase	[26]
Pseudomonas putida	Trinitrotoluene (TNT)	Nitroreductase Xenobiotic reductase Azoreductase	[23]
	Organophosphorus compounds (OP)	Phosphotriesterase	[10]
Xylaria feejeensis *Trametes versicolor*	Essential oil extraction residues	Laccase Cellulase Xylanase	[3]
Rhodoccocus sp. *Streptomyces sp.*	Plastic		[9]
Acinetobacter *Escherichia-Shigella* *Bacteroides* *Microbacterium* *Parabacteroids*	Petroleum hydrocarbons (PH)	Catalases Lipases Dehydrogenases Phosphatase	[27]
Bacillus sp. 8E1A	Lignocellulosic waste	Carboxymethyl cellulase (CMCase) Filter paper hydrolase (Fpase) Microcrystalline cellulose Avicel hydrolase (Avicelase)	[28]
Bacillus spp.	Feather	Keratinase	[25]
Proteobacteria *Actinobacteriota* *Bacteroidota* *Verrucomicrobiota* *Myxococcota* *Plantomycetota* *Firmicutes*	Oat Straw Pine Sawdust Birch Leaf Litter	Cellulolytic enzymes CAZymes (Carbohydrate-Active Enzymes)	[29]
Aspergillus sp. *Penicillium sp.* *Emericellopsis sp.*	Organic Matter	Proteases Lipases Amylases Cellulases Xylanases Pectinases	[2]

Table 1 (continued)

Microorganism	Pollutant	Enzyme	Reference
Faoltella planticola *Ochrobactrum* *Thiophenivorans* *Bacillus flexus* *Staphylococcus* *xylosus*	Methyl Orange Dye	Azoreductase DCIP reductase NADH reductase Laccase LiP MnP Nitrate reductase Tyrosianase	[20]
Rhodotorula sp. *Exophiala sp.*	Aromatic Hydrocarbons (AH) in oil polluted site		[30]
Pseudomonas pseudoalcaligenes	Furfural		[6]
Escherichia coli	Azo Dye Basic Orange 2	Azoreductase Deaminase	[31]
Bacterial Consortium	Maize Straw	Hydrolases Lyases Oxidoreductases Auxiliary enzymes	[32]
Serratia marcescens	Mycotoxin fumonisin	Hydrolase Transferase	[33]
Terribacillus goriensis	Polybutylene succinate (PBS)	Esterase	[34]
Proteobacteria	Polychlorinated biphenyls (PCBs)	Dehydrogenase	[35]

3 Challenges and prospects of microbial enzymes

The numerous applications of microbial enzymes to treat various environmental pollutants are not without significant drawbacks. One of the major drawbacks in the application of microbial enzymes is the endurance and deflection of contaminants to biodegradation, chemical oxidation, and hydrolysis [8]. For example, the nitroaromatic compound structures strengthen preventing any transformation even by oxygenase enzymes [23]. Previous studies showed that similar challenges are encountered while degrading pollutants comprising polymer chains as the chains' proximity to one another causes the creation of crystalline areas, which restricts the accessibility to the chains by microorganisms [11].

Despite of the fact that abundant strains of isolated bacteria from many sources or substrates which demonstrates excellent secretion of cellulolytic Carbohydrate-Active

Enzymes, again the effectiveness of the treatment is determine by a variety of bacteria which existed in the dedicated environments [29]. This would imply that multiple bacterial strains are required to isolate sufficient enzymes to degrade a specific contaminant. Additionally, investigations have shown that the oxidation of a number of xenobiotic substances can occasionally result in the production of hazardous chemicals such as quinones [30]. Apart from that, the usage of enzymes on an industrial scale is also constrained by some of their major limitations, including their high solubility, poor stability at high temperatures, severe pH values, or need for organic co-solvents, and their inability to be utilized for a successive reaction cycle [36].

However, despite these significant drawbacks, the use of microbial enzymes in the treatment of environmental contaminants also displayed promising future potential, thanks to the ongoing research and exploration of new and innovative ideas to improve the characteristics and efficiency of the enzymes. For instance, the newly discovered keratinases, which were isolated in a study by [24], demonstrated at par quality and effectiveness to chemicals used in keratin hydrolysis and produced less impact on the environment. A great degree of ordinary disparity in trinitrotoluene (TNT) acceptance, degradation, and transformation biologically has been observed in several fungus species, according to studies, and these fungi utilize the TNT explosives as a nitrogenous source for normal growth [8].

In a different study, Humicola insolens cutinase (HiC) was found to have a broad enough substrate specificity to enable it to hydrolyze esther and urethane bond, demonstrating the adaptability in breaking down of polyurethanes [26]. It is important to continue researching the novel non-pathogenic Pseudomonas species and evaluate their potential for in situ bioremediation of polluted areas [23]. The properties of bio enzymes can also be modified by immobilizing them on macro and micro support materials for immobilization carriers [36].

Studies have also shown that microbes exhibit steadiness when overwhelming with inorganic pollutants. This is due to the metabolic load of the degradation of the pollutants distributed among several microbial consortia with a unique mechanism and they together accelerate the biodegradation rate [20]. Using microbial consortia can therefore be a way to decompose pollutants quickly and effectively [29]. Additionally, the combination of plants and microbes to remediate soil has also gained popularity since this technique combines both phyto and microbial remediation for accelerating organic contaminants degradation. Consequently, the combination of phyto-microbial remediation of soil polluted with organic contaminants seemed promising [14]. The findings of previous investigations have outlined both the significant setbacks and promising future uses of microbial enzymes in the remediation of environmental pollutants.

The knowledge of wastewater treatment using microbial or so-called bioremediation is not easy to understand by lay people. The selection of a wastewater treatment process will depend on the type of pollutants that existed in the water body. It can be through chemical, physical, and biological process of treatment or their combinations. Microbiological studies will give information on what the bacteria are doing through the wastewater treatment process. They also allow us to make perturbations in the environment and examine how these affect microbial activities. Finally, the conclusion can be drawn not only about what chemical reactions are occurring in the microbial new habitat, but also about which organisms are responsible for the improvement of the water quality (BOD, COD, heavy metals, etc.).

It is good to be able to visualize and profile the bioremediation, bioaugmentation, or biotransformation throughout the treatment process. Microbes–wastewater interaction can be further understood with the application of machine learning tools using microbial wastewater treatment data, new algorithms, and/or approaches for classification of microbial wastewater treatment data. It is a way forward in the microbial enzyme activities and approaches. Knowledge of processes at the microbial level allows better utilization of wastewater. A visualization and data mining approach based on research data from microbes-wastewater interaction using machine learning can expedite the damage control action in advance.

4 Important factors affecting microbial enzymes' efficiency in environmental pollutants bioremediation

An effective microbial bioremediation of environmental pollutants is affected by several different factors. These elements are crucial for ensuring that the microorganisms and microbial enzymes are functioning at their peak levels. Bioremediation through microbial enzymatic activity attainment relies on many factors and parameters such as temperature, pH, dissolved oxygen, nutrition availability, toxicity, and microbial pressures, to name a few [20]. Specific elements influencing the efficacy of bioremediation vary depending on the microorganisms and microbial enzymes used. For instance, the bioremediation of petroleum hydrocarbons (PHs) was greatly determined by the soil conditions physically and chemically; thus, the removal efficiency in bioremediation may be monitored and evaluated using enzymes and physicochemical soil parameters [27].

Temperature, time, pH, and source of the cellulose are the primary parameters impacting Bacillus and related genera in degrading or breaking down cellulase, including production and enzymatic activity. Cellulase of 8E1A were reported to be active in the temperature range of 30–90 °C; the highest CMCase, Avilase, and FPase activity was recorded at 60 °C [28]. On the contrary, the chemical properties of the polymer, structure, crosslink density, crystallinity, and the fact that most plastics are xenobiotics are variables contribute to the biodegradation resistance of polyurethanes (PU). Despite this, research indicates that, in contrast to the polyester hydrolysis of ester bonds observed during microbial degradation, the kind of polyol utilized as the starting material affects the degradation of Pus caused by microorganisms [11].

A summary of main factors affecting microbial bioremediation of environmental pollutants is as shown in Table 2. The bioremediation or bioaugmentation is reported having a capability to lessen pathogen, β-lactam antibiotics, and inorganic elements in one stage treatment process. For instance, at dedicated temperature and addition of inoculum B. subtilis resulting a treated sewage effluent that is suitable for safe disposal and for agriculture irrigation purposes [37].

The findings reveal under the known treatment conditions and specific inoculum can effectively inactivate the pathogens. Hence open research pathway for other potential pathogens with enzymatic activities exploration.

5 Mechanisms of microbial enzymes

Depending on the enzyme utilized and the type of pollutant, different mechanisms are involved in the biodegradation of environmental contaminants by microbial enzymes. Organic material is used by bacteria as a source of sustenance for their metabolic processes and later will be oxidized, resulting a simple sub-metabolite which will be discharged into environment. The existing of the sub-metabolite in the aquatic and terrestrial environment can induced the secretion of polysaccharides that is the protein compounds, exhibit a glue-like material [38].

Fungi generally play a role as decomposer, will attach to the substrates through hyphae. The fungi will take up all the appropriate nutrients for their growth and created enzymes such as cellulase, amylases, pectinases, proteases, and lipases. Those enzymes capable of hydrolyzing plant polysacharides (cellulose, starch, pectin, xylan), proteins and lipids [2].

In the case of polymers, the process entails two basic processes, the first of which is known as primary degradation and is referred to as abiotic. The second step, known as secondary degradation, is facilitated by the enzymatic attack of microorganisms, and is referred to as biotic. PU, for instance, can go through hydrolysis in the first stage of biodegradation, which is catalyzed by specific enzymes like esterases, proteases, and ureases. Esterases, for example, catalyze the hydrolysis of ester bonds, proteases, the hydrolysis of amine peptide bonds, and ureases, and the hydrolysis of urea functional groups. Although some bacteria are attributed for the microbiological attack on the PU surface, the predominant microorganisms that oversee changing the surface of the PU are filamentous fungi, such as Aspergillus and Pestalotiopsis [11].

Decolorization and deprivation of azo dyes by microbes occur in two phases. Azo's linkages are broken down in the first stage or phase by reductive cleavage, which creates mostly colorless, but potentially dangerous, aromatic amines, while the degradation of aromatic amines takes place in the second stage. The azo bond (-N = N-) is split by the enzyme during azoreductase, enzymatic degradation, involving transferring of four electrons to reduce equivalent in which two electrons play a role as electron acceptor to azo dye in each stage. The decolorization takes place when the colorless solution is formed. The resultant intermediate is a harmful aromatic amin, which is subsequently broken down aerobically or occasionally microaerophilically [39].

An O-glycoside hydrolases-class enzyme complex is necessary to completely degrade cellulose. Cellulolytic enzymes comprise three different enzymes including endoglucanases such as carboxymethyl cellulase, which randomly separate the -1,4 glycosidic bonds located in the amorphous regions of the cellulose. In general, exoglycanases releases glucose and cellobiose units from reducing ends of the cellulose chain. Cellobioses then transform cellobiose to glucose [28]. The mechanism of the microorganism's community that able to metabolizes organophosphorus (OP) associated with contaminated soils by pesticides needs extensive research works. Anyhow both gram-positive and gram-negative bacteria under aerobic conditions revealed a positive result for OP degradation. The reaction pathways illustrated a good degradation model by both bacteria with hydrolytic enzymes lead a role in the OP degradation pathway [10]. In another study bioremediation or biotransformation using organic solvent tolerant bacteria (OSTB) used for solvent treatment reported elsewhere was explained to take place in two possible routes of mechanism. The secreted enzyme will be used to further react with solvent through biotransformation, resulting in a harmless final product. The bioremediation and biotransformation complement each other for the used solvent treatment.

Table 2: Factors affecting microbial biodegradation.

Factors	Remarks
Biological Factors	- Interaction (competition, predation, succession) among organisms, population size and composition - Types and toxicity of contaminants - Catalysts levels - Enzymes
Oxygen	- For microorganism respiration - Better biodegradation rates can be achieved by using anaerobes
Moisture Content	- For microbial growth
Nutrients	- Affects microbial growth and reproduction - Affects biodegradation rate and effectiveness
Temperature	- Affects the metabolic turnover of degradative enzymes - Affects the microbial physiological characteristics, which influences how quickly bioremediation occurs
pH	- Acidity, alkalinity, and basicity of a substance have an impact on microbial metabolism and the removal process
Site Characterization and Selection	- Procedures for site selection involve figuring out the extent of contamination, both horizontally and vertically, setting parameters and sample sites, and specifying sampling and analysis techniques
Metal Ions	- Essential for bacteria and fungi, but excessive amounts inhibit cell metabolism

6 Conclusions

Based upon the findings of previous studies, this review could conclude that microbial enzymes have great potential and prospect for the future way forward in both industrial applications and treatment of environmental pollutants. This is due to their great many potential applications for various processes in the industrial sector and their ability to degrade and utilize pollutants as source for microbial growth. Though there have been some significant drawbacks in the applications of microbial enzyme due to the characteristics of the pollutants, such as the presence of crystalline regions, and some of the microbial enzyme characteristics such as being instable at higher solubility, being instable at high temperatures, extreme pH values, organic co-solvents, and being incapable for successive reaction cycles, there are still opportunities for the adoption of bioremediation by microbes for an ultimate solution to eradicate the environmental pollutants owing to the continuous research and investigation of fresh and creative ideas for enhancing the properties and effectiveness of the enzymes.

References

[1] Alotaibi BS, Khan M, Shamim S. Unraveling the underlying heavy metal detoxification mechanisms of bacillus species. Microorganisms 2021, *9*, 8, 1628. 1–31.

[2] Gomoiu I, Cojoc R, Ruginescu R, Neagu S, Enache M, Maria G, Dumbrăvician M, Olteanu I, Rădvan R, Ratoiu LC, Atanassova V, Ghervase L. Brackish and hypersaline lakes as potential reservoir for enzymes involved in decomposition of organic materials on frescoes. Fermentation 2022, *8*, 9, 462. 1–24.

[3] Carrión-Paladines V, Fries A, Caballero RE, Daniëls PP, García-Ruiz R. Biodegradation of residues from the palo santo (Bursera graveolens) essential oil extraction and their potential for enzyme production using native xylaria fungi from Southern Ecuador. Fermentation 2019, 5, 3, 1–20.

[4] Filote C, Roşca M, Hlihor RM, Cozma P, Simion IM, Apostol M, Gavrilescu M. Sustainable application of biosorption and bioaccumulation of persistent pollutants in wastewater treatment: Current practice. Processes 2021, 9, 10, 1–38.

[5] Carrasco-Espinosa K, Avitia M, Barrón-Sandoval A, Abbruzzini TF, Salazar Cabrera UI, Arroyo-Lambaer D, Uscanga A, Campo J, Benítez M, Wegier A, Rosell JA, Reverchon F, Hernández G, Boege K, Escalante AE. Land-use change and management intensification is associated with shifts in composition of soil microbial communities and their functional diversity in coffee agroecosystems. Microorganisms 2022, 10, 9, 1763. 1–15.

[6] Igeño MI, Macias D, Blasco R. A case of adaptive laboratory evolution (ALE): Biodegradation of furfural by pseudomonas pseudoalcaligenes CECT 5344. Genes 2019, 10, 7, 499. 1–14.

[7] Miloloža M, Uki Š. Optimization of polystyrene biodegradation by bacillus cereus and pseudomonas alcaligenes using full factorial design. 2022, 14, 4299. 1–18.

[8] Alothman ZA, Bahkali AH, Elgorban AM, Al-Otaibi MS, Ghfar AA, Gabr SA, Wabaidur SM, Habila MA, Hadj Ahmed AYB. Bioremediation of explosive TNT by Trichoderma viride. Molecules 2020, 25, 6, 1–13.

[9] Charnock C. Norwegian soils and waters contain mesophilic, plastic-degrading bacteria. Microorganisms 2021, *9*, 1, 1–18.

[10] Cardozo M, De Almeida JSFD, De A Cavalcante SF, Salgado JRS, Gonçalves AS, França TCC, Kuca K, Bizzo HR. Biodegradation of organophosphorus compounds predicted by enzymatic process using molecular modelling and observed in soil samples through analytical techniques and microbiological analysis: A comparison. Molecules 2020, 25, 1, 8. 1–21.
[11] Lopes RVV, Loureiro NPD, Quirino RL, Gomes ACM, Pezzin APT, Manzur LP, Dos Santos ML, Sales MJA. Biodegradation study of polyurethanes from linseed and passion fruit oils. Coatings 2022, 12, 5, 617. 1–15.
[12] Ren L, Men L, Zhang Z, Guan F, Tian J, Wang B, Wang J, Zhang Y, Zhang W. Biodegradation of polyethylene by enterobacter sp. D1 from the guts ofwax moth galleria mellonella. Int J Environ Res Public Health 2019, 16, 11, 1941. 1–11.
[13] Bala S, Garg D, Thirumalesh BV, Sharma M, Sridhar K, Inbaraj BS, Tripathi M. Recent strategies for bioremediation of emerging pollutants: A review for a green and sustainable environment. Toxics 2022, 10, 8, 1–24.
[14] Xie H, Liu R, Xu Y, Liu X, Sun F, Ma Y, Wang Y. Effect of in situ bioremediation of soil contaminated with DDT and DDE by stenotrophomonas sp. Strain DXZ9 and ryegrass on soil microorganism. Microbiol Res 2022, 13, 1, 64–86.
[15] Matula M, Kucera T, Soukup O, Pejchal J. Enzymatic degradation of organophosphorus pesticides and nerve agents by EC: 3.1.8.2. Catalysts 2020, 10, 12, 1–18.
[16] Zheng X, Wang L, Qi L, Dong Z. A novel organophosphorus acid anhydrolase from deep sea sediment with high degradation efficiency for organophosphorus pesticides and nerve agent. Microorganisms 2022, 10, 6, 1112. 1–16.
[17] Maclean J, Mayanna S, Benning LG, Horn F, Bartholomäus A, Wiesner Y, Wagner D, Liebner S. The terrestrial plastisphere: Diversity and polymer-colonizing potential of plastic-associated microbial communities in soil. Microorganisms 2021, 9, 9, 1–19.
[18] Raveh-Amit H, Tsesarsky M. Biostimulation in desert soils for microbial-induced calcite precipitation. Appl Sci (Switzerland) 2020, 10, 8, 2905. 1–9.
[19] Loiko N, Kanunnikov O, Serdyukov D, Axelrod V, Tereshkin E, Vishnyakova A, Litti Y. Didecyldimethylammonium chloride- and polyhexamethylene guanidine-resistant bacteria isolated from fecal sludge and their potential use in biological products for the detoxification of biocide-contaminated wastewater prior to conventional biological treat. Biology 2022, 11, 9, 1332. 1–16.
[20] Ibrahim A, El-Fakharany EM, Abu-Serie MM, Elkady MF, Eltarahony M. Methyl orange biodegradation by immobilized consortium microspheres: Experimental design approach, toxicity study and bioaugmentation potential. Biology 2022, 11, 1, 76. 1–27.
[21] Josu López-Fernández MDB, V F. Rhizopus oryzae lipase, a promising industrial enzyme: Biochemical characteristics, production and biocatalytic applications. Catalysts 2020, 10, 11, 1277. 1–50.
[22] Ruginescu R, Enache M, Popescu O, Gomoiu I, Cojoc R, Batrinescu-Moteau C, Maria G, Dumbravician M, Neagu S. Characterization of some salt-tolerant bacterial hydrolases with potential utility in cultural heritage bio-cleaning. Microorganisms 2022, 10, 3, 1–15.
[23] Cabrera MÁ, Márquez SL, Pérez-Donoso JM. Comparative genomic Analysis of Antarctic Pseudomonas Isolates with 2,4,6-Trinitrotoluene transformation capabilities reveals their unique features for xenobiotics degradation. Genes 2022, 13, 8, 1354. 1–26.
[24] Almahasheer AA, Mahmoud A, El-Komy H, Alqosaibi AI, Aktar S, Abdulazeez S, Borgio JF. Novel feather degrading keratinases from bacillus cereus group: Biochemical, genetic and bioinformatics analysis. Microorganisms 2022, 10, 1, 1–27.
[25] Gafar AA, Khayat ME, Ahmad SA, Yasid NA, Shukor MY. Response surface methodology for the optimization of keratinase production in culture medium containing feathers by bacillus sp. Upm-aag1. Catalysts 2020, 10, 8, 848. 1–18.

[26] Di Bisceglie F, Quartinello F, Vielnascher R, Guebitz GM, Pellis A. Cutinase-catalyzed polyester-polyurethane degradation: Elucidation of the hydrolysis mechanism. Polymers 2022, 14, 3, 411. 1–12.

[27] Cui JQ, He QS, Liu MH, Chen H, Sun MB, Wen JP. Comparative study on different remediation strategies applied in petroleum-contaminated soils. Int J Environ Res Public Health 2020, 17, 5, 1–17.

[28] Dobrzyński J, Wróbel B, Górska EB. Cellulolytic properties of a potentially lignocellulose-degrading Bacillus sp. 8E1A strain isolated from bulk soil. Agronomy 2022, 12, 3, 1–11.

[29] Gladkov GV, Kimeklis AK, Afonin AM, Lisina TO, Orlova OV, Aksenova TS, Kichko AA, Pinaev AG, Andronov EE. The structure of stable cellulolytic consortia isolated from natural lignocellulosic substrates. Int J Mol Sci 2022, *23*, 18, 10779. 1–19.

[30] Ide-Pérez MR, Fernández-López MG, Sánchez-Reyes A, Leija A, Batista-García RA, Folch-Mallol JL, Sánchez-Carbente MDR. Aromatic hydrocarbon removal by novel extremotolerant exophiala and rhodotorula spp. From an oil polluted site in Mexico. J Fungi 2020, 6, 3, 1–17.

[31] Ikram M, Naeem M, Zahoor M, Hanafiah MM, Oyekanmi AA, Ullah R, Farraj DAA, Elshikh MS, Zekker I, Gulfam N. Biological degradation of the azo dye basic orange 2 by Escherichia coli: A sustainable and ecofriendly approach for the treatment of textile wastewater. Water (Switzerland) 2022, 14, 13, 2063. 1–23.

[32] Jin M, Guan H, Zhang W, Tian D, Wei J, Kalkhajeh YK, Gao H. High level of iron inhibited maize straw decomposition by suppressing microbial communities and enzyme activities. Agronomy 2022, 12, 6, 1–13.

[33] Keawmanee P, Rattanakreetakul C, Pongpisutta R. Microbial reduction of fumonisin b1 by the new isolate serratia marcescens 329-2. Toxins 2021, 13, 9, 1–16.

[34] Kim SH, Cho JY, Cho DH, Jung HJ, Kim BC, Bhatia SK, Park S-H, Park K, Yang YH. Acceleration of polybutylene succinate biodegradation by Terribacillus sp. JY49 isolated from a marine environment. Polymers 2022, 14, 19, 3978. 1–16.

[35] Lenola MD, Caracciolo AB, Ancona V, Laudicina VA, Garbini GL, Mascolo G, Grenni P. Combined effects of compost and medicago sativa in recovery a PCB contaminated soil. Water (Switzerland) 2020, 12, 3, 860. 1–17.

[36] Patila M, Athanasiou PE, Kortessis L, Potsi G, Kouloumpis A, Gournis D, Stamatis H. Immobilization of laccase on hybrid super-structured nanomaterials for the decolorization of phenolic dyes. Processes 2022, 10, 2, 233. 1–14.

[37] Al-Gheethi AA, Mohamed RMSR, Efaq AN, Norli I, Abdullah AH, Amir HK, Ab Kadir MO. Bioaugmentation process of secondary effluents for reduction of pathogens, heavy metals and antibiotics. J Water Health 2016, 14, 5, 780–795.

[38] Abe MM, Branciforti MC, Brienzo M. Biodegradation of hemicellulose- cellulose-starch-based bioplastics and microbial polyesters. Recycling 2021, 6, 1, 22. 1–21.

[39] Muliadi FNA, Halmi MIE, Wahid SBA, Gani SSA, Zaidan UH, Mahmud K, Shukor MYA. Biostimulation of microbial communities from Malaysian agricultural soil for detoxification of metanil yellow dye; a response surface methodological approach. Sustainability (Switzerland) 2021, 13, 1, 1–15.

Debbie Dominic, Ooi Wen Ching, Lim Wai See, Mohd Asyraf Kassim, Siti Baidurah*

Overview of Biological Treatment Technologies for Palm Oil Wastes and the Resultant Product Application as Biomass Fuel toward Sustainable Environment

Abstract: Ponding systems are commonly implemented at an industrial scale as a treatment method for palm oil wastes before discharge to the environment. Wastewater from palm oil mills is one of the largest waste crises in the palm oil industry. The primary goal of this chapter is to explore different aspects and different ranges of promising current POME treatments, including the exploitation of biological techniques in the presence of: (1) bacteria, fungi, yeasts, and (2) microalgae. The present chapter elaborates the practicability of vast non-conventional and highly efficient treatments for palm oil wastes pertaining to the optimal treatment output with low COD, BOD, total suspended solids, and decolorization with reduce treatment time. The final part discusses the potential application of the resultant products toward sustainable environment.

Keywords: Palm oil mill effluent treatment, biodegradation, bioprocess, biomass fuel

1 Introduction

Oil palm is an estate crop that provides important roles in the development of Malaysia's economy. The introduction of oil palm commodity to Malaysia improved the country's agricultural sector and economy standard. The oil palm agriculture has developed from a modest crop since 1917, and nowadays is among the most important contributors to Malaysia's Gross Domestic Product (GDP), with acreage allocated to oil

Acknowledgements: This research is funded by the Ministry of Higher Education Malaysia, through the Fundamental Research Grant Scheme (FRGS), with the project code of FRGS/1/2021/STG01/USM/02/12.

***Corresponding author: Siti Baidurah**, Bioprocess Technology Division, School of Industrial Technology, Universiti Sains Malaysia, Penang 11800, Malaysia; Green Biopolymer, Coatings & Packaging Cluster, School of Industrial Technology, Universiti Sains Malaysia, Penang 11800, Malaysia, e-mail: sitibaidurah@usm.my

Debbie Dominic, Ooi Wen Ching, Lim Wai See, Mohd Asyraf Kassim, Bioprocess Technology Division, School of Industrial Technology, Universiti Sains Malaysia, Penang 11800, Malaysia

https://doi.org/10.1515/9783111016825-012

palm cultivation reaching 5.7 million hectares in 2016 [1]. Despite being among the major palm oil producers, this industry generated tons of wastes every year which in turn created waste disposal issues and contributed to environmental degradation.

The oil palm wastes originated from mills and plantations that categorized into two groups: (1) liquid form and (2) solid biomass. Palm oil mill effluent (POME) is group as liquid waste while empty fruit bunch (EFB), mesocarp fruit fibers (MF), oil palm frond (OPF), and palm kernel shells (PKS) are grouped as solid waste [2]. POME is accumulated voluminously at the final stage of palm oil milling activities with approximately 2.5 tons production for every 1 ton of crude palm oil produced [2, 3]. Although POME is non-toxic waste, the release of raw POME into the surrounding ecosystem without any extensive treatment will cause undesirable environmental pollutions due to its high chemical oxygen demand (COD), biochemical oxygen demand (BOD), oil and grease content as well as its acidic and viscous characteristics. EFB, the second most voluminous oil palm waste produced, is a highly fibrous biomass disposed into the environment either by burning or by being left to decompose due to its bulky nature and low commercial value. Additionally, the burning and decomposition of EFB in an open environment cause air pollution and insect manifestation.

Until the present time, open ponding is the most common treatment approach used to treat POME prior to discharge into the environment at industrial scale due to low technology requirements and convenience. Researchers invent various treatment technologies to biovalorize POME and EFB as an alternative energy source such as biomass fuel. However, there are few challenges that need to be addressed, namely, hydraulic retention time (HRT), land space, and final discharge parameter. Therefore, various alternative approaches with shorter HRT and small space requirement which produces sustainable by-products are of great interest.

This chapter elaborates various aspects of promising current POME treatments, including the exploitation of biological techniques with the aid of: (1) bacteria, fungi, yeasts, and (2) microalgae. The exploitation of biological techniques to biovalorize POME and EFB into sustainable biomass fuel in the presence of various indigenous microorganisms will significantly enhance the effectiveness of utilization of palm oil waste due to its high carbon composition. These initiatives are parallel with various Malaysian Government agendas, namely: (1) Sustainable Development Goals (SDG) adopted by United Nation of Educational Scientific and Cultural Organization (UNESCO) Number 12 of Responsible Consumption and Production, (2) 10–10 Science, Technology and Innovation (STI) Malaysia framework, and (3) The National Policy on Science, Technology and Innovation (DSTIN) Malaysia 2021–2030.

2 Malaysian oil palm industry

Oil palm industry is among the driving forces in the Malaysia agricultural sector. It plays the role as the fourth largest contributor which drives the economic growth of the country [4, 5]. The oil palm industry developed almost a century ago when the oil palm tree, *Elaeis guineensis*, was brought to Malaysia by the British in 1870. The oil palm tree adapted well in Malaysia due to its tropical climate. In year 2007, Malaysia became the second largest oil palm producer and exporter in worldwide after Indonesia. Malaysia accounted for 24.77% and 31.94% of palm oil production and exports, respectively, worldwide from the marketing year of October 2020 to September 2021 [6].

Until now, almost 71% of agricultural land bank has been used for oil palm tree plantation [5]. In 2009, this plantation field in Malaysia recorded 4.7 million hectares [5] while it has increased enormously throughout the years and increased to 5.9 million hectares in the year 2020 [7]. In Malaysia, the oil palm industry is regulated by the Ministry of Plantation Industries and Commodities (MPIC). The main responsibilities of MPIC are to create plans and strategies for the development of plantation and its product. In addition, agencies such as Malaysian Palm Oil Board (MPOB), Malaysian Palm Oil Council (MPOC), and Malaysian Palm Oil Association (MPOA) help in growing and development of the oil palm industry in Malaysia.

2.1 Palm oil production and sources of POME

Palm oil processing begins with the transportation of fresh fruit bunch (FFB) to the milling factories. To produce high quality of palm oil, the spoilt FFB is excluded. The selected FFB then undergoes sterilization at high temperature of 140 °C, pressurized steam at 40 psi for a duration of 75–90 min [8]. Sterilization helps in deactivation of enzyme activities, separation of kernels from shells, elimination of external impurities, and softening of the fruitlets [9]. The sterilization process generates an enormous amount of condensate which needs to be discarded occasionally. Condensate from sterilization accounts for 36% of total POME [10]. To detach the fruitlets from bunches, sterilized FFB are transferred into the rotating drum. The mesocarps are pressed to extract the crude palm oil (CPO). The crude palm oil produced at this stage comprises 35–45% of palm oil, 45–55% of water and impurities [9, 10]. The CPO produced is fed to the clarification tank to settle down and separate impurities and waste oil. The CPO formed on top of the tank is removed and stored at crude oil tanks. The sludge at the bottom of the clarification tank is transferred to the sludge separator, the decanter heavy phase tank, and de-oiling tank for further extraction of the oil. The recovered oil is restored to the clarification tank. The remaining sludge and waste oil are transferred into the trapping pond which then undergoes ponding treatment of POME. The wastewater from the purification process accounts for the largest amount (60%) of total POME. The third source of POME was from hydro-cyclone wastewater, which contributes to 4% of total POME [10].

2.2 Biomass wastes in oil palm industry

Due to the high demand for palm oil, the oil palm industry is growing wider and bigger, and the wastes generated also increased gradually. During the production of palm oil, approximately 90% is contributed by biomass wastes while only 10% is contributed by crude palm oil (CPO) [11]. To produce one ton of palm oil, various types of solid biomass and liquid wastes will be generated. The solid biomass wastes include mesocarp fiber (MF), empty fruit bunch (EFB), palm kernel shell (PKS), oil palm fronds (OPF), oil palm trunks (OPT), and liquid waste, palm oil mill effluent (POME) [12].

2.2.1 Palm oil mill effluent

Palm oil mill effluent (POME) is the only liquid waste produced from the oil palm industry. POME is usually produced from oil extraction, washing, and cleaning processes [13]. POME is made up of wastewater from sterilization process, clarification tank, and hydrocyclone process [10]. In producing 1 ton of crude palm oil (CPO), 5–7 tons of water are used in the extraction process. However, over 50% of the water used ends up as POME [14] whereby 2–3.5 tons of POME produced per ton of CPO produced are usually left underutilized [15].

POME is a brownish and high-viscosity liquid made up of 95–96% of water, 0.6–0.7% of oils, and 4–5% of total solids whereby 2–4% are suspended solids [2]. It has an acidic pH of 3.4–5.2, with an unpleasant odor, temperature of 80–90 °C, and high values of COD (15,000–100,000 mg/L), BOD (10,250–43,750 mg/L), and TSS (5,000–-54,000 mg/L) [9]. POME is considered a non-hazardous effluent as there are no chemicals being added during the processes. However, due to the high COD, BOD, and unpleasant odor, disposal of POME into the waterway will threaten aquatic life.

Concurrently, POME contains high organic content with vast essential minerals, such as potassium, calcium, magnesium, aluminum, and other nutrients which are vital for the growth of microorganisms [17]. It contains a high amount of degradable organic matter [18] as the main element present in POME is carbon (C), which made up of around 51.0% of POME's composition. Toxic heavy metals such as lead, cadmium, mercury, manganese, or chromium does not exist in POME [19]. Generally, the characteristic of POME is varied across different batches, duration of time, milling factories, and processing line. It also depends on the quality of oil palm fruits used, the discharge limit of the factory, climate (rainy or drought season), and conditions during oil palm processing [13]. The general characteristic of raw POME is temperature from 80–90 °C, pH of 3.4–5.2, COD concentration of 15,000–100,000 mg/L, BOD concentration of 10,250–43,750 mg/L, oil and grease content of 130–18,000 mg/L, and TSS concentration of 5,000–54,000 mg/L [9].

The elemental analysis indicated that the highest percentage of elements present in POME are carbon (C) (51.0%) and oxygen (O) (35.3%). The minerals and trace elements are present in a small amount in POME such as potassium (K) (6.77%), chlorine (Cl) (2.75%), calcium (Ca) (1.09), magnesium (Mg) (1.09%), sulfur (S) (0.553%), silicon (Si) (0.552%), phosphorus (P) (0.429%), aluminum (Al) (0.215%), ferrum (Fe) (0.141%), sodium (Na) (0.0632%), rhodium (Rb) (0.0286%), and manganese (Mn) (0.0243%) [9]. High carbon content in POME increases the combustible carbon, which in turn increases the CEV and combustion efficiency [20]. Upon drying, the moisture content will be reduced, and the output will be in powder form, containing high carbon content that is favorable to be applied as biomass fuel [17].

3 Obsolete and current POME treatment method

Throughout the years, various efforts have been made by researchers to develop effective treatment method of POME prior to discharge to the surrounding ecosystem. Treatment of POME is needed to ensure it does not exceed the standard discharge limit and reduce the negative impacts toward the environment. The Department of Environment (DOE), Malaysia, has set the standard discharge limits for POME and legislated the Environmental Quality Act (EQA) 1974 to mitigate and reduce the pollution toward the surrounding ecosystem [9].

3.1 Obsolete ponding system

The conventional method of ponding system or land application is extensively implemented as POME treatment method in Malaysia. More than 85% of oil mills adopted this method owing to its simple design, economical aspect, simplicity of handling, and low maintenance costs [9, 10]. This treatment method employs both aerobic and anaerobic systems. POME needs to go through several ponding systems before it can be discharged. The obsolete ponding system consists of various pond stages such as acidification pond, cooling pond, anaerobic pond, aeration (aerobic pond), facultative pond, polishing pond, and, finally, the treated POME becomes the final discharge [9].

In the anaerobic ponding, the hydrolytic bacteria secrete various hydrolytic enzymes such as cellulase, lipase, and xylanase with the function to break down the complex carbohydrates, proteins, and lipids into their monomers [21]. This simpler monomer is further broken down by acidogenic bacteria, acetogenic bacteria, and methanogens into the final products such as carbon dioxide, hydrogen, and methane via the processes of acidogenesis, acetogenesis, and methanogenesis. The organic content in POME is further biodegraded in the facultative ponds to ensure it meets the discharging limit which complies with the Environmental Quality Act (EQA) 1974. The microbe from the aerobic pond is being clarified at the polishing pond prior to discharge of effluent.

However, this method has several downsides such as lengthy hydraulic retention time (HRT) of 100–120 days, large pond area needed (30–45 acres) [9], and production of sludge in a vast amount that needs to be discarded [22]. Furthermore, POME cannot be completely decolorized through the ponding system [23] and the end product is unable to fit with the standard discharge limit set by the Department of Environment (DOE), Malaysia [24]. Most importantly a huge amount of uncaptured greenhouse gases, methane and carbon dioxide (CH_4 and CO_2), is generated during the anaerobic degradation processes and released to the atmosphere, which aggravates the global warming [22, 25].

3.2 Biological treatment method

Biological treatment of POME (aerobic and anaerobic biodegradation) is performed by inoculation of microbes such as bacteria, fungi, yeast, and microalgae [26]. The microbes being inoculated are of mixed culture or pure culture (after POME sterilization) which function to break down organic matter in POME [27]. It is an eco-friendly and cost-effective method that could be applied in an industrial scale [9]. Aerobic biodegradation of POME can be conducted in laboratory scale using shake flasks. In the study conducted by Mohammad et al., the aerobic fermentation of POME is carried out using the shake flask and the bacteria *Lysinibacillus* sp. LC556247 was inoculated [17]. The fermentation condition is set at agitation speed of 180 rpm and temperature of 35 ± 2 °C for 5 days. Based on the study, the COD, BOD, and TSS removal efficiency is 50.83%, 71.73%, and 42.99% respectively. Upon the designated period of fermentation, the samples are then dried to obtain powdered output with moisture content less than 5%, which is applicable as a biomass fuel.

The efficiency of the biological treatment can be determined through the characterization of POME through analysis of BOD, COD, TSS and oil and grease after fermentation. BOD is the measure of amount of organic matter while COD measures the total amount of organic and inorganic matter in POME. TSS refers to the solids which do not dissolve and remain on the filter after filtration process. Oil and grease refer to the oil contents that are present in POME. During the fermentation, complex compounds, namely carbohydrates, lipids, and protein, are being degraded by the microorganisms inoculated in POME. The degradation of these complex compounds will lead to decrease in percentage of these parameters. POME can be discarded if these parameters do not exceed the standard discharge limit set by the Department of Environment (DOE), Malaysia. The higher reduction efficiency of these parameters indicates higher efficiency of the biological treatment.

In the anaerobic treatment of POME, the biogas generated could be captured and utilized as an energy source. The heat generated from the methane gas is converted into electricity and delivered via the combined heat and power (CHP) systems. This approach can mitigate the greenhouse gases emissions from the oil palm industry. In lieu of the conventional anaerobic treatment, researchers have developed several alternative anaerobic treatments for POME, namely anaerobic suspended growth processes, membrane separation anaerobic treatment processes, anaerobic sludge blanket processes (in which up-flow anaerobic sludge blanket reactors and anaerobic baffled reactors are being utilized), and hybrid anaerobic treatment processes [22].

3.2.1 Current POME treatment with the aid of various bacteria and fungus

Table 1 delineates various co- and mixed-culture fermentation of biological treatment utilizing POME and its degradation efficiency. Bala et al. described the COD and BOD removal efficiencies of 90.65% and 93.11%, respectively, using a mixed culture of *Bacillus cereus* 103 PB and *Bacillus subtilis* 106 PB [28]. Another research by Said et al. which utilized the mixed culture of *Bacillus toyonensis* strain BCT-71120 and *Stenotrophomonas rhizophila* strain e-p 10 achieved COD, BOD, and TSS removal efficiency of 86.00%, 94.00%, and 80.00%, respectively, which is reported to be more efficient than single pure culture [29].

According to both researchers, high removal efficiency achieved in mixed culture is due to the synergism between the mixed culture bacteria strains [28, 29]. Synergism between microorganisms is achieved when the growth of mixed culture is higher than the single pure culture [30]. Furthermore, the isolated bacteria strains have been pre-adapted to the POME environment as they are indigenous to POME. The mixed or consortium culture which has been pre-adapted to the fermentation environment will become more competitive among each other which leads to higher degradation efficiency than single pure culture [31].

Suseela and Muralidhar utilized mixed culture of fungal strains (*Emericella nidulans, Aspergillus niger* and *Aspergillus fumigatus*) isolated from POME dump site and showed high removal efficiency of COD and BOD at 91.43% and 93.43% respectively [32]. The same study also showed that the removal efficiency of mixed culture fungal strains was higher than single pure culture. The synergistic effect between *Aspergillus niger, Aspergillus fumigatus*, and *Emericella nidulans* degraded the complex polysaccharide faster than single pure culture and led to higher degradation efficiency. In the presence of complex substrates such as lignocellulosic materials, it will enhance the synergism of consortium microbes while the limit in substrates will cause competition between the consortium of microbes [30]. Research conducted by Suseela and Muralidhar are also in agreement with the previous study as POME is rich in complex polysaccharides [32]. The higher removal efficiency of COD and BOD in the study

Table 1: Various co- and mixed-culture fermentation of bacteria and fungi for POME treatment and its removal efficiency.

Microbes	Working volume (L)	Fermentation conditions	Removal efficiency (%)	References
Mixed culture of *Bacillus cereus* 103 PB and *Bacillus subtilis* 106 PB Mixed culture of *Micrococcus luteus* 101 PB and *Stenotrophomonas maltophilia* 102 PB	0.25 0.25	Aerobic fermentation using shake flask at 150 rpm, 37 °C for 5 days	COD (90.65), BOD (93.11) BOD (67.33), COD (71.84)	[28]
Mixed culture of *Emericella nidulans, Aspergillus niger* and *Aspergillus fumigatus*	0.25	Aerobic fermentation using shake flask at 150 rpm, 30 °C for 5 days	COD (91.43), BOD (94.34)	[32]
Mixed culture of *Bacillus toyonensis* strain BCT-71120 and *Stenotrophomonas rhizophila* strain e-p 10	3	Aerobic fermentation using bioreactor for 18 days	COD (86.00), BOD (94.00), TSS (80.00)	[29]
Co-culture of *Lipomyces starkeyi* and *Bacillus cereus*	0.2	Aerobic fermentation using shake flask at 150 rpm, 30 °C for 6 days	COD (83.66), BOD (77.34), TSS (71.43), Oil and grease (79.23), TOC (81.34), TN (65.30)	[33]
Microbial consortium from POME compost	0.1	Aerobic fermentation using shake flask at 100 rpm, 30 °C for 5 days	COD (90.30)	[34]

evinced the synergism effect between the fungal strains due to the degradation efficiency during fermentation.

3.2.2 Current POME treatment with the aid of various microalgae

Microalgae are unicellular photosynthetic microorganisms; for example blue-green algae, prokaryotic cyanobacteria, and eukaryotic protists can thrive in the aquatic environment as well as on various types of soil. Due to their simple structure, microalgae have a high growth rate and high photosynthetic efficiency. In biotechnology processes, these microorganisms were identified as beneficial microalgae and evinced good characteristics as it can utilize carbon dioxide, light, water, and various nutrients to convert into energy for cell growth [35]. Microalgae are highly adaptable and flexible to proliferate in different types of environmental conditions, including open ponds and closed photobioreactors [36]. Microalgae mainly consist of lipid/fatty acid, proteins, and carbohydrates as the main chemical component [37]. The ratio of lipid, proteins, and carbohydrates constituents may vary depending on the types of microalgae. For example, *B. braunii* contains 33–86% of lipid, 4–40% of protein, and 20% carbohydrate. *C. pyrenoidosa* can cultivate high amount of carbohydrate within the range of 20–57%, 31–47% protein, and 8–35% lipid. *S. platensis* can accumulate a high amount of protein within 42–63%, 8–30% carbohydrate, and 4–13% lipid.

In the treatment of POME using microalgae, POME was used as a source of lipid/fatty acid in the substrate media and microalgae become prominent as one of the promising sources for biodiesel production. Moreover, the demand for palm oil is increasing rapidly among tropical countries, especially in Malaysia [38]. Along with the increase of palm oil production, the POME waste will correlatedly increase. Compared to the conventional wastewater treatment processes that introduce activated sludge and biofloc to biodegrade organic carbonaceous materials to CO2, algae are able to assimilate organic pollutants into cellular components such as lipids and carbohydrates, therefore enabling the attaining pollutant reduction in more sustainable and environmentally friendly approaches. As early as the 1950s, microalgae studies had been explored in great detail for both wastewater treatment and biomass production [39].

Photoautotrophic [40], heterotrophic [41], mixotrophic [42], and photo-heterotrophic [40] are the microalgae cultivation methods used by researchers. Photoautotrophic is defined as utilization of light energy and inorganic carbon for growth. Heterotrophic relies on organic carbon source for growth, and mixotrophic is the cultivation mode that combines heterotrophic and autotrophic modes for simultaneous working, where inorganic and organic source was used in the presence of light. Photo-heterotrophic is defined as growing mode that uses light energy to generate ATP and utilizes organic carbon source from other organisms. *Coccomyxa dispar* and *Scenedesmus parvus* are able to cultivate under autotrophic, heterotrophic, and mixotrophic modes [43].

Many studies related to wastewater treatment reported that microalgae are able to eliminate nitrogen and phosphorus. Microalgae can proliferate rapidly, namely, they are able to divide once every 3 to 4 h, but mostly divide every 1 to 2 days under favorable cultivation conditions [44]. The average lipid content of algal cells varies between 1 and 70%; however it can increase to 90% of dry weight under certain favorable cultivation conditions [45]. Various parameters can affect both lipid composition and lipid content in many microalgae, namely light, temperature, pH, salinity, irradiance, and, most markedly, nutrient availability such as carbon [46] nitrogen [47], phosphorus [48], and metal elements [49, 50]. For example, microalgae species *C. pyrenoidosa* achieved 71.16% in nutrient reduction and 68% lipid production in POME. Table 2 delineates the experiments conducted by researchers with various types of microalgae in POME treatment with nutrient reduction and lipid production.

Additionally, many studies described the enhancement of microalgae for biofuel production. Lü et al. reported the application of bioengineered microalgae of *Synechocystis*

Table 2: Microalgae cultivation performance in POME.

Microalgae species	Nutrient reduction (%)	Lipid (%)	Growth rate (day)	Biomass productivity (g/L/d)	Duration (d)	Ref.
Chlamydomonas	68.0	60	1.2	0.08	7	[51]
Chlorella sp.	–	–	0.066	0.06	15	[52]
S. plantesis	–	–	–	9.80	13	[53]
C. pyrenoidosa	71.16	68	1.8	0.13	10	[54]
T. suecica	95.0	27	–	0.21	7	[55]
N. oculata	93.6	39	–	0.52	7	[56]

sp. PCC 6803 for the bioethanol production, with a theoretical yield of 0.696 g ethanol/g CO_2 [57]. Different type of microalgae will accumulate various lipid contents. Both *C. vulgaris* and *C. pyrenoidosa* were used to examine the effect of different proportion of raw POME to algae ratio on the lipid production [54]. In this context, the medium containing POME was prepared with the ratio of 10:90, 50:50, and 95:5 (raw POME:algae ratio) [54]. As a result, both microalgae achieved their optimal growth under POME, and algae ratio of 95:5 and concentration ratio of 10:90 were observed to be inapt for microalgae cultivation. This observation evinced POME is a carbon-enriched medium that able to improve microalgae proliferation and simultaneously mitigate wastewater pollution. Conversely, oleaginous microalgae can efficiently produce oil in the presence of light due to their high lipid content (30–70%) [58].

3.2.3 Bioremediation of POME using microalgae for the potential production of biomass fuel as renewable energy

Due to depletion of fossil fuel reserves and pollution from the greenhouse gases caused by fossil diesel, lipid-rich biological materials have gained much attraction nowadays as they have the potential to produce biodiesel. Because of their unique characteristics and intricate role in the wastewater treatment, microalgae bring various advantages and can be manipulated for POME treatment. Various algae portray unique characteristics that are able to be manipulated for wastewater treatment application, for example, POME. This is due to the fact that microalgae can accumulate high biomass yield and utilize small-scale land area compared to the ponding system. Various microalgae are able to reduce undesired products in the wastewater, namely COD, BOD, phosphorous, ammonia, nitrogen, nutrients and so on. This method provides a cost-efficient and sustainable alternative especially for the POME treatment. Furthermore, carbon-neutral biofuels can be produced by taking advantage of microalgae's ability to utilize emitted carbon dioxide for photosynthesis process [40]. In this context, POME contains unrefined palm oil and therefore comprises a vast degradable organic matter. Hence, POME is deemed suitable to sustain the microbial growth including microalgae.

Microalgae has the ability to undergo photosynthesis with the nutrients and phosphorous provided in POME that contribute to renewable and sustainable fuel production in the form of gas, liquid, or solid. Through photosynthesis process by the microalgae, the carbon dioxide level in the system will decrease. This approach solves the problem of high levels of carbon dioxide in POME. Evidently, POME treatment with microalgae can successfully produce biomethane, biohydrogen, and bioethanol [12]. Due to the global energy security concern, POME treatment with the aid of microalgae to produce biofuel is evinced to be an extremely viable approach. Moreover, the high-quality biofuel and bioactive compounds produced can be useful for pharmaceutical and energy industries. Various parameters can affect the quality of biodiesel produced, namely, the raw biomassic material itself, lipid productivity, and fatty acid compositions of the microalgae. *Chlorella* is extensively studied for the POME treatment, which is of main interest for biodiesel production due to its ability to accumulate significant amounts of lipids under stress conditions [59].

Moreover, microalgae can serve as a biosorbent to remove heavy metals within the wastewater via the biosorption process. Microalgae have high metal binding capabilities due to their characteristics of having properties such as lipids, polysaccharides, and protein on the surface of microalgae cell wall [60]. These properties have functional groups, namely amino, carboxyl, hydroxyl, and sulfate, that act as active binding sites for heavy metals [60].

4 Biomass fuel

The demand for fossil fuels increasing over the years correlates with the increase of populations in the world. The depletion of fossil fuels and emission of greenhouse gases has become the key for the researchers to seek for renewable energy sources. Biomass fuel is an alternative renewable fuel source which is produced from biomass such as wood products, crop residues, or dead vegetation. It is extensively utilized as renewable energy source as it is relatively cheap and indigenous nature. Unlike fossil fuels, this renewable energy is clean, non-depleting, sustainable, and renewable. Moreover, it does not emit greenhouse gases into the atmosphere unlike fossil fuels. The energy inside the biomass wastes is stored in a chemical form. This chemical energy is released and converted to heat energy during combustion process. Biomass fuel can be used in residential boilers and stoves, and internal combustion engines, and it combines heat and power plant for delivery of electricity [20].

Biomass fuels have a relatively low calorific energy value (usually 15–20 MJ/kg) compared to coal fuel (30 MJ/kg) [17]. Another research shows CEV value of pellet obtained from the mixture of POME and oil palm frond (OPF) at different ratios ranging from 18.763 MJ/kg to 30.850 MJ/kg [16]. High CEVs can be achieved by utilizing a high ratio of POME and monitoring by a high amount of carbon element. The highest CEV value was obtained from ratio of 90:10 POME and OPF [16].

4.1 Characterization of biomass fuel

Biomass fuel is characterized by the physical and chemical properties which determine the pellet quality. Physical properties of biomass fuel include moisture content, CEV, ash content, fixed carbon, volatile matter, crack analysis, particle size, and bulk density. Moisture content is among the dominating parameters in determining the quality of biomass fuel and pellet as it can influence the CEV of a fuel. Biomass fuel with high moisture content needs extra energy to evaporate the moisture before it can ignite. This will lower the energy released during the combustion process, which results in a lower CEV [20].

4.1.1 Moisture content

Moisture content is among the significant parameters contributing to the compactness of biomass fuel pellet. At the optimal moisture content, it helps in self-bonding of individual particles, which enhances the binding of each particle. At too low moisture content, the pellet formed will crack easily, while moisture content beyond the optimal level will decrease the inter-particle force of the pellet. A relatively higher moisture content will cause the formation of biphasic mixture (solid and liquid mixture), which leads to disintegration of the pellet. The pellet is stable in the range of 5–15% of moisture content. The standard limit of moisture content for a biomass fuel pellet is ≤10% [17]. The oil, lignin, protein, and starch also act as natural binders in pellet [17]. The research from Hassan et al. [16] shows that the briquettes produced from POME and OPF in the ratio of 90:10 exhibit the least cracks. This is because POME has a high oil content which could hold the particles more tightly [16].

4.1.2 Volatile matter

Volatile matter are the components in solid biomass fuel, which are evaporated at a high temperature (except moisture). Volatile matter consist of short- and long-chain hydrocarbon and aromatic compounds in biomass fuel. Pellet with high volatile matter will ignite easily, oxidized and gasified at a subsequently lower temperature [20]. Oil palm biomass has a higher volatile matter compared to the coal fuel and other biomass. Hence, oil palm biomass is burnt more easily under typical combustion conditions.

4.1.3 Calorific energy value

Another dominating parameter in determining the biomass fuel and pellet quality is calorific energy value (CEV). The CEV is the energy that is being released during the combustion of a fuel. It also represents the maximum amount of energy recovered from the biomass. High CEV released from the biomass fuel indicates high quality of the fuel. Ash content can be defined as a non-combustible substance which remains as powdery residues upon combustion. In biomass combustion, ash mainly contains of inorganic elemental oxides. High ash content of biomass fuel acts as a catalyst and increases the emission of NOx, which is toxic to the environment [20]. Biomass fuel with high ash content has a lower CEV due to the low formation of C-H bond. Furthermore, ash content causes ash slagging, inhibition of combustion process, overheat, and blockage and leads to corrosion of boiler. Hence, a high-quality biomass fuel and pellet should have a lower ash content which produces less residues after combustion. The standard limit of pellet ash content is less than 5%.

Fixed carbon content is the crucial parameter in determining the CEV of biomass fuel pellet. Biomass with higher fixed carbon content will have a higher CEV [17]. Pellet compactness is also affected by the moisture content and natural binders present in the pellet. A pellet with high compactness indicates a strong pellet. A strong pellet with optimal moisture content and natural binders will exhibit less cracks. The strong pellet is more convenient for storage, handling, and transportation with less dust formation. The formation of dust and fine particles will harm the workers' health and affect the combustion efficiency.

4.1.4 Chemical properties

The chemical properties of biomass fuel and pellet are determined through ultimate analysis. The ultimate analysis determines the percentage of elements such as carbon (C), oxygen (O), hydrogen (H), nitrogen (N), and sulfur (S). Biomass fuel with high percentage of C and H will have a higher CEV as energy is released through the breaking of C-H bonds. High N and S content is undesirable in biomass fuel and pellet owing to the emission of toxic nitrogen oxides (NO_x) and sulfur oxides (SO_x) during combustion. Hence, most of the biomass fuel contains less than 0.2% of nitrogen and sulfur. This indicates that the biomass fuel is more environmentally friendly than coal fuel.

4.1.5 Pelletization

Pelletizing is a densification process which compresses the loose biomass into a cylindrical pellet dimension of 5–15 mm diameter and 10–50 mm height. The biomass fuel may also briquette into a compacted brick shaped. This biomass will have higher density, lower moisture content, and higher CEV. Pelletization of the biomass is a favorable approach for industries with boilers as the pellets are convenient to be handled and enhances the flowability of the pellet when feed into the boilers either manually or automatically. Furthermore, system blockages, risk of explosions, and dust emission of biomass fuel pellets could be reduced during logistics such as handling, storage, and transport. This process is necessary to ensure the pellets are economically viable, do not degrade easily, have high efficiency energy conversion during burning, and are able to retrofit fossil-based industrial power plant.

5 Conclusion

Utilization of agricultural waste such as palm oil waste into biomass fuel using naturally occurring microbes is sustainable rather than leaving the waste in open spaces such as open ponding systems without extensive treatment. This also means the use of new advance biotechnology treatments improves and provides a more sustainable opportunity for palm oil industry to achieve a zero-discharge concept and sustainable production. The absence of these approaches causes the uncontrollable value of greenhouse gas (GHG) emissions into the atmosphere. Whilst bacteria are usually exploited in POME treatment, the data indicate a lower COD removal efficiency comparatively to fungi. This observation is due to the robustness of the fungi itself and the ability of the fungi to perform anaerobic digestion of POME. Microalgae showed a remarkable performance in the POME treatment; however, the HRTs for microalgae are extended as compared to the fungus and bacteria.

References

[1] Kushairi A, Singh R, Ong-Abdullah M. The oil palm industry in Malaysia: Thriving with transformative technologies. J Oil Palm Res 2017, 29, 4, 431–439.

[2] Abdullah N, Sulaiman F. The oil palm wastes in Malaysia In: Matovic MD (ed.). Biomass now sustainable growth and use. Intech Open, London, UK, 2013, 75–100.

[3] Madaki YS, Seng L. Palm oil mill effluent (POME) from Malaysia palm oil mills: Waste or resource. Int J Environ Sci Technol 2013, 2, 6, 1138–1155.

[4] Begum H, Choy EA, Alam ASAF, Siwar C, Ishak S. Sustainability of Malaysian oil palm: A critical review. Int J Environ Sustain Dev 2019, 18, 4, 387–408.

[5] Ferdous Alam ASA, Er AC, Begum H. Malaysian oil palm industry: Prospect and problem. J Food Agric Environ 2015, 13, 2, 143–148.

[6] Oilseeds: World markets and trade market signals to sell soybean stocks. United States Department of Agriculture, 2022. (Accessed March, 2022, at https://apps.fas.usda.gov/psdonline/circulars/oilseeds.pdf).

[7] Kate AT, Kuepper B, Piotrowski M. Oil palm expansion in Peninsular Malaysia is guided by non- transparency. In: Chain reaction research. 2021, 1–14.

[8] Akhbari A, Kutty PK, Chuen OC, Ibrahim S. A study of palm oil mill processing and environmental assessment of palm oil mill effluent treatment. Environ Eng Res 2020, 25, 2, 212–221.
[9] Mohammad S, Baidurah S, Kobayashi T, Ismail N, Leh CP. Palm oil mill effluent treatment processes – A review. Processes 2021, 9, 5, 1–22.
[10] Liew WL, Kassim MA, Muda K, Loh SK, Affam AC. Conventional methods and emerging wastewater polishing technologies for palm oil mill effluent treatment: A review. J Environ Manage 2015, 149, 222–235.
[11] Kaniapan S, Hassan S, Ya H, Nesan KP, Azeem M. The utilisation of palm oil and oil palm residues and the related challenges as a sustainable alternative in biofuel, bioenergy, and transportation sector: A review. Sustainability 2021, 13, 6, 3110.
[12] Ahmad A, Buang A, Bhat AH. Renewable and sustainable bioenergy production from microalgal co-cultivation with palm oil mill effluent (POME): A review. Renew Sust Energ Rev 2016, 65, 214–234.
[13] Kamyab H, Chelliapan S, Din MFM, Rezania S, Khademi T, Kumar A. Palm oil mill effluent as an environmental pollutant. In: Waisundara V (ed.). Palm oil. Intech Open, London, UK, 2018, 13–28.
[14] Kospa HS, Lulofs KR, Asdak C, Rahim SE. Strategies to reduce water footprint in palm oil production: A case of PTP Mitra Ogan, Baturaja, South Sumatra. InMATEC Web of Conferences, EDP Sciences, 2017, 97, 01122.
[15] Hosseini SE, Wahid MA. Pollutant in palm oil production process. J Air Waste Manag Assoc 2015, 65, 7, 773–781.
[16] Hassan S, Kee LS, Al-Kayiem HH. Experimental study of palm oil mill effluent and oil palm frond waste mxture as an alternative biomass fuel. J Eng Sci Technol 2013, 8, 6, 703–712.
[17] Mohammad S, Baidurah S, Kamimura N, Matsuda S, Abu Bakar NAS, Muhamad NNI, Ahmad AH, Dominic D, Kobayashi T. Fermentation of palm oil mill effluent in the presence of *Lysinibacillus* sp. LC 556247 to produce alternative biomass fuel. Sustainability 2021, 13, 21, 11915.
[18] Kamyab H, Din MFM, Hosseini SE, Ghoshal SK, Ashokkumar V, Keyvanfar A, Shafaghat A, Lee CT, Bavafa AA, Majid MZA. Optimum lipid production using agro-industrial wastewater treated microalgae as biofuel substrate. Clean Technol Environ Pol 2016, 18, 8, 2513–2523.
[19] Saad MS, Wirzal MDH, Putra ZA. Review on current approach for treatment of palm oil mill effluent: Integrated system. J Environ Manag 2021, 286, 112209.
[20] Loh SK. The potential of the Malaysian oil palm biomass as a renewable energy source. Energy Convers Manag 2017, 141, 285–298.
[21] Tan KM, Liew WL, Muda K, Kassim MA. Microbiological Characteristics of Palm Oil Mill Effluent. In International Congress on Chemical, Biological, and Environmental Sciences (ICCBES) 2016, 186–200.
[22] Lee ZS, Chin SY, Lim JW, Witoon T, Cheng CK. Treatment technologies of palm oil mill effluent (POME) and olive mill wastewater (OMW): A brief review. Environ Technol Innov 2019, 15, 100377.
[23] Tamrin KF, Zahrim AY. Determination of optimum polymeric coagulant in palm oil mill effluent coagulation using multiple-objective optimisation on the basis of ratio analysis (MOORA). Environ Sci Pollut Res 2017, 24, 19, 15863–15869.
[24] Parthasarathy S, Mohammed RR, Fong CM, Gomes RL, Manickam S. A novel hybrid approach of activated carbon and ultrasound cavitation for the intensification of palm oil mill effluent (POME) polishing. J Clean Prod 2016, 112, 1218–1226.
[25] Loh SK, Lai ME, Ngatiman M, Lim WS, Choo M, Zhang Z, Salimon J. Zero discharge treatment technology of palm oil mill effluent. J Oil Palm Res 2013, 25, 3, 273–281.
[26] Dominic D, Baidurah S. Recent Developments in Biological Processing Technology for Palm Oil Mill Effluent Treatment – A Review. Biology 2022, *11*, 4, 525.
[27] Cheng YW, Chong CC, Lam MK, Ayoub M, Cheng CK, Lim JW, Yusup S, Tang Y, Bai J. Holistic process evaluation of non-conventional palm oil mill effluent (POME) treatment technologies: A conceptual and comparative review. J Hazard Mater 2021, 409, 124964.

[28] Bala JD, Ismail N. Studies on the reduction of organic load from palm oil mill effluent (POME) by bacterial strains. Int J Recycl Org Waste Agric 2015, 4, 1, 1–10.

[29] Said M, Sitanggang AS, Julianda R, Estuningsih SP, Fudholi A. Production of methane as bio-fuel from palm oil mill effluent using anaerobic consortium bacteria. J Clean Prod 2021, 282, 124424.

[30] Deng YJ, Wang SY. Synergistic growth in bacteria depends on substrate complexity. J Microbiol 2016, 54, 1, 23–30.

[31] Bala JD, Lalung J, Ismail N. Biodegradation of palm oil mill effluent (POME) by bacterial. Int J Sci Res 2014, 4, 1, 2250–3153.

[32] Suseela L, Muralidhar P. Reduction of organic load from palm oil mill effluent (POME) using selected fungal strains isolated from POME dump sites. Afr J Biotechnol 2018, 17, 36, 1138–1145.

[33] Karim A, Islam MA, Khalid Z, Yousuf A, Khan MMR, Faizal CKM. Microbial lipid accumulation through bioremediation of palm oil mill effluent using a yeast-bacteria co-culture. Renew Energy 2021, 176, 106–114. https://doi.org/10.1016/j.renene.2021.05.055.

[34] Nwuche CO, Aoyagi H, Ogbonna JC. Treatment of palm oil mill effluent by a microbial consortium developed from compost soils. Int Sch Res Notices 2014, 2014, 1–8.

[35] Cheah WY, Show PL, Chang JS, Ling TC, Juan JC. Biosequestration of atmospheric CO_2 and flue gas-containing CO_2 by microalgae. Bioresour Technol 2015, 184, 190–201.

[36] Francisco ÉC, Neves DB, Jacob-Lopes E, Franco TT. Microalgae as feedstock for biodiesel production: Carbon dioxide sequestration, lipid production and biofuel quality. J Chem Technol Biotechnol 2012, 85, 395–403.

[37] Cheng SY, Show PL, Lau BF, Chang JS, Ling TC. New prospects for modified algae in heavy metal adsorption. Trends Biotechnol 2019, 37, 1255–1268.

[38] Onoja E, Chandren S, Abdul Razak FI, Mahat NA, Abdul Wahab RA. Oil palm (Elaeis guineensis) biomass in Malaysia: The present and future prospects. In: Nzihou A (ed.). Waste and biomass valorization. Springer, Netherlands, 2019, 2099–2117.

[39] Show PL, Tang MSY, Nagarajan D, Ling TC, Ooi CW, Chang JS. A holistic approach to managing microalgae for biofuel applications. Int J Mol Sci 2017, 18, 215.

[40] Chen CY, Liu CH, Lo YC, Chang JS. Perspectives on cultivation strategies and photobioreactor designs for photofermentative hydrogen production. Bioresour Technol 2011, 102, 8484–8492.

[41] Liang Y, Sarkany N, Cui Y. Biomass and lipid productivities of Chlorella vulgaris under autotrophic, heterotrophic and mixotrophic growth conditions. Biotechnol Lett 2009, 31, 7, 1043–1049.

[42] Mata TM, Martins AA, Caetano NS. Microalgae for biodiesel production and other applications: A review. Renew Sust Energ Rev 2010, 14, 1, 217–232.

[43] El-Sheekh M, Abomohra AE, Hanelt D. Optimization of biomass and fatty acid productivity of *Scenedesmus obliquus* as a promising microalga for biodiesel production. World J Microbiol Biotechnol 2013, 29, 915–922.

[44] Huang G, Chen F, Wei D, Zhang X, Chen G. Biodiesel production by microalgal biotechnology. Appl Energy 2010, 87, 1, 38–46.

[45] Li X, Hu H, Gan K, Sun Y. Effects of different nitrogen and phosphorus concentrations on the growth, nutrient uptake, and lipid accumulation of a freshwater microalga *Scenedesmus* sp. Bioresour Technol 2010, 101, 14, 5494–5500.

[46] Wu TY, Mohammad AW, Jahim JM, Anuar N. A holistic approach to managing palm oil mill effluent (POME): Biotechnological advances in the sustainable reuse of POME. Biotechnol Adv 2014, 27, 1, 40–52.

[47] Perez-Garcia O, Escalante FME, de-Bashan LE, Bashan Y. Heterotrophic cultures of microalgae: Metabolism and potential products. Water Res 2011, 45, 1, 11–36.

[48] Niklas KJ. Carbon/nitrogen/phosphorus allometric relations across species. In: White PJ, Hammond JP (eds.). The ecophysiology of plant-phosphorus interactions. New York, USA, 2008, 9–30.

[49] Zhang L, Happe T, Melis A. Biochemical and morphological characterization of sulfur-deprived and H_2-producing *Chlamydomonas reinhardtii* (green alga). Planta 2014, 214, 4, 552–561.

[50] Rodolfi L, Chini Zittelli G, Bassi N, Padovani G, Biondi N, Bonini G, Tredici MR. Microalgae for oil: Strain selection, induction of lipid synthesis and outdoor mass cultivation in a low-cost photobioreactor. Biotechnol Bioeng 2016, 102, 1, 100–112.

[51] Kamyab H, Din MFM, Keyvanfar A, Majid MZA, Talaiekhozani A, Shafaghat A, Lee CT, Shiun LJ, Ismail HH. Efficiency of microalgae chlamydomonas on the removal of pollutants from Palm Oil Mill Effluent (POME). Energy Proc 2015, 75, 2400–2408.

[52] Hadiyanto H, Azimatun Nur MM, Hartanto GD. Cultivation of *Chlorella* sp. As biofuel sources in palm oil mill effluent (POME). Int J Renew Energy Dev 2012, 1, 45–49.

[53] Maulana M, Nur A, Ganang DH. Enhancement of biomass production from spirulina sp cultivated in POME medium. In Proceedings of the International Conference on Chemical and Material Engineering 2012.

[54] Kamyab H, Md Din MF, Lee CT, Keyvanfar A, Shafaghat A, Majid MZ, Ponraj M, Yun TX. Lipid production by microalgae *Chlorella pyrenoidosa* cultivated in palm oil mill effluent (POME) using hybrid photo bioreactor (HPBR). Desalin Water Treat 2015, 55, 13, 3737–3749.

[55] Nur MM, Garcia GM, Boelen P, Buma AG. Influence of photodegradation on the removal of color and phenolic compounds from palm oil mill effluent by *Arthrospira platensis*. J Appl Phycol 2021, 33, 2, 901–915.

[56] Shah SM, Ahmad A, Othman MF, Abdullah MA. Effects of palm oil mill effluent media on cell growth and lipid content of *Nannochloropsis oculata* and *Tetraselmis suecica*. Int J Green Energy 2016, 3, 2, 200–207.

[57] Lü J, Sheahan C, Fu P. Metabolic engineering of algae for fourth generation biofuels production. Energy Environ Sci 2011, 4, 7, 2451–2466.

[58] Pal P, Chew KW, Yen HW, Lim JW, Lam MK, Show PL. Cultivation of oily microalgae for the production of third-generation biofuels. Sustainability 2019, 11, 19, 5424.

[59] Idris NA, Loh SK, Lau HL, Mustafa EM, Vello V, Tan CY, Phang SM. Cultivation of microalgae in medium containing palm oil mill effluent and its conversion into biofuel. J Oil Palm Res 2017, 29, 291–299.

[60] Schiewer S, Volesky B. Biosorption by marine algae. In: Valdes JJ (ed.). Bioremediation. Springer, Netherlands, 2000, 139–169.

Joydeep Das, Snehannita Sarkar, Kousish Saha, Sandeep Roy,
Soma Nag*

Emerging Environmental Contaminants: Sources, Consequences, and Future Challenges

Abstract: Emerging contaminants, also known as pollutants of emerging concern, comprise a new category of compounds found in both surface water and groundwater. Emerging environmental contaminants reach the environment from various sources such as industrial and pharmaceuticals wastes, municipal wastewater, agricultural practices, and various human activities. These contaminants are scattered throughout the environment and pose serious threats to the ecosystem, although great progresses have been made in the discernment and analysis during the past few years. These emerging contaminants include traces of a number of chemicals, nanomaterials pharmaceuticals, illicit drugs, antibiotics, hormones, various disinfection, personal care products, etc. Low concentrations of these toxins may nevertheless be mobile or persistent in air, water, soil, sediments, and ecological receptors. Thus, the entire aquatic species and populations are exposed directly or indirectly to these contaminants. As a result, environmental alteration can lead to a powerful indirect interaction that can control the entire ecosystem. The consequences of emerging contaminants can be exacerbated by climate change, which can send them floating down rivers during dry spells and bring them back to the surface during wetter times. The goal of this chapter is to give readers a bird's-eye view of the new and emerging sources of pollution, and how they get into the environment, as well as the threats they pose to humans and aquatic life.

Keywords: Contaminants, ecosystem, environment, wastewater, pharmaceutical

1 Introduction

Environmental contaminants or pollutants, such as pesticides, metals, and other toxic organic compounds, can cause serious damage to the living beings as well as the nonliving objects. A fair share of research is being conducted to understand the various harmful effects of toxic substances on the physiology of living beings [1]. Water quality has deteriorated due to anthropogenic activities such as the increase in manufacturing and the subsequent discharge of industrial effluents, as well as house and hospital discharges [2].

***Corresponding author: Soma Nag,** Chemical Engineering Department, National Institute of Technology Agartala, Tripura, India, e-mail: somanag.nita@gmail.com
Joydeep Das, Snehannita Sarkar, Kousish Saha, Sandeep Roy, Chemical Engineering Department, National Institute of Technology Agartala, Tripura, India

https://doi.org/10.1515/9783111016825-013

Heavy metals are one of the most hazardous components which are mostly found in industrial effluents. The most carcinogenic heavy metals like Cr, Cd, As, and Pb induce significant contamination and health concerns [3]. Water pollution and soil pollution caused by heavy metals from contaminated water and industrial effluents endanger both living and nonliving things in the natural world. Heavy metal pollution has a direct impact on agricultural production that is caused in part by the interaction of food and raw materials used in labelling and packaging, including text ink patterning and graphics [4]. Emerging pollutants such as heavy metals, pesticides and other pollutants from various industries continue to be a major concern especially for water pollution as is depicted in the Figure 1.

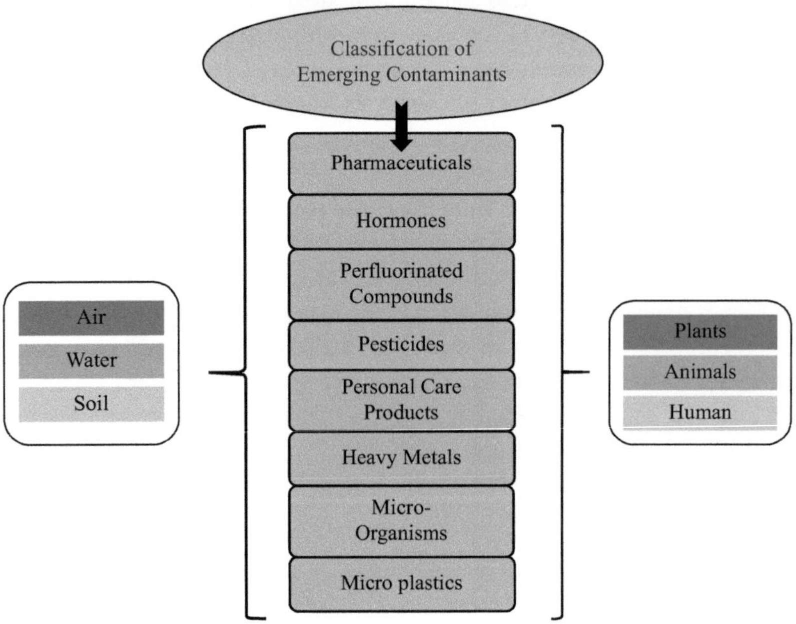

Figure 1: Classification of emerging pollutants.

So, the availability of safe drinking water is a major problem in developing nations. Lack of proper infrastructure and economical constraints cause dearth of potable water. There were nearly 785 million people in the world who lacked access to safe drinking water in 2019, with 144 million of those individuals relying solely on surface water sources [5].

Researchers are working in different new directions for exclusion of toxic pollutants, heavy metals, organic and other emerging contaminants from water and wastewater, and adsorption of the pollutants by some natural agents present in the environment itself is believed as one of the most sustainable and eco-friendly processes for solving these problems [6–9]. Cellulose is a cheap natural polymer, and its main sources are fibers from most common plants available worldwide. It has commendable metal- and other pollutant-binding capacity. The microbes, viruses, and plant materials also act as biodegrading agents for the emerging contaminants [10].

The human interference releases the biological contaminants like germs and viruses into water supplies. These agents, known as emerging or re-emerging pathogens due to their potential toxicity, have garnered considerable attention. Many water-related diseases continue to be a major global killer because of the prevalence of biological micro-pollutants like mycoplasmas, enteric bacteria, viruses, and protozoa. Many of these illnesses originate from microorganisms found in water, and the range of disorders is expanding. Algae-based treatment is found to be effective and inexpensive in extracting nutrients and heavy metals from water and waste water. The microalgae-based pre-treatment systems include primary decantation, active sludge, oxidation ditches, etc. Aerobic bacteria with oxygen are delivered by algae to break the organic matters, and carbon dioxide is absorbed by bacterial respiration. Biological fixation by microalgae is a capable technique to handle the carbon dioxide. The method is highly flexible to poisonous organic pollutants [11]. Moreover, pollutants may have effects on a wide variety of tolerant species via a spectrum of ecological mechanisms. The term "secondary contaminant impacts" describes this type of consequence. Secondary contamination impacts cannot be detected by single-species toxicity tests in the lab; instead, studies at the population, community, or ecosystem level are necessary. The direct effects of pollutants on predators and grazers may have indirect consequences for resistant species. The direct effects of pollutants on vulnerable species can also change competitive dynamics among resistant producer and consumer populations [12].

Behavior plays as the bridge between ecological and physiological processes; it can be ideal for understanding emerging environmental contaminants effects. Fish are an outstanding model for this case, as quite several ecologically persistent fish behaviors are effortlessly observed and evaluated in a controlled setting. Furthermore, fish physiology, a crucial test for integrative studies, is acknowledged to be a big concern. The use of behavioral indicators in fish for ecologically appropriate monitoring of emerging environmental pollution has been proposed by a sizable number of researchers [13]. Fish exhibits natural behavior according to predetermined physiological sequences that are set off by outside stimuli working through neural networks. Any disruption of these sequences before they are finished is likely to cause negative behavioral changes. The beginning of these sequences is also influenced by a variety of physiological and environmental factors. Because of the harmful effects of aquatic contaminants, inappropriate behavioral responses to environmental and physiological stimuli can have severe consequences for survival [14].

The motive of this review is to assess the effects of emerging pollutants exposure in the environment. The mechanism of their activities to the communities of each tropic, living or nonliving domains, and their everlasting affects on the earth are also evaluated.

2 Consequences of the emerging contaminants

Over hundreds of thousands of various substances are manufactured and used by industries every year; all these chemical substances are a major part of everyday life. The global economy as it relates to these chemical markets has a significant impact on employment and, consequently, on economic growth. Nevertheless, a variety of compounds are discharged into the environment, and chemical contamination on both land and water is now widespread. These new environmental toxins have the potential to have a wide range of terrible and long-term consequences on human health as well as the health of the entire ecosystem (cancer, neurological problems, etc.). To counteract these severe harms, a number of nations have created risk-based pollutant management systems and legislation employing a series of systematized assays designed specifically to separate the toxicological and eco-toxicological risks of contaminants [15].

The effects of micro pollutants in aquatic habitats are of major concern because living species present are exposed, with potential ramifications for future generations. When micro pollutants are present in freshwaters (surface and ground waters) at low (trace) quantities (nano or microgram/L), depending on their source, the situation becomes more complex [16]. Their detection and removal have become rather difficult, but are of the utmost importance, as they pose a significant threat to the reuse of treated wastewater and the continuity of water cycle management. In addition, they pose severe hazards to human health, such as the development of infection resistance and disruption of the endocrine system [14]. Recent estimates indicate that roughly 1000,000 chemicals are currently in use. Over four hundred thousand of these chemicals have been recognized as persistent organic pollutants. When one considers technological advancements and the rampant release of harmful contaminants into the environment, this number becomes even more alarming. The use of pesticides is a further significant factor. Almost all pesticides are typically sprayed on crops in a combination. They are thus directly released into the environment since they are applied directly to the soil or occasionally sprayed over crops [17]. Pesticides usually have high solubility, and because of that organisms can easily absorb them by several means. After its absorption all these toxic chemicals are spread and distributed throughout their whole body and then undergo several biotransformation reactions [18]. These toxic poisons directly change and alter the endocrine systems [19]. Due to the severe impacts on human health, the use of nanoparticles by humans is a major cause for concern. The severe effects of nanoparticles on human health depend on a number of variables, including exposure, genetics, preexisting conditions, and the chemistry, shape, size, electromagnetic characteristics, and aggregation state of the particles. A recent study demonstrates a strong association between levels of particle and air pollution. They are responsible for a number of health issues affecting the human race, including mortality, cardiovascular and respiratory disorders, and a number of malignancies [20].

3 Sources of emerging pollutants

3.1 Chemical contaminants in drinking water

Food consumption has progressed from a simple trading channel between producer and consumer to a complicated chain involving numerous participants. Our drinking water, like food, is at risk of pollution, with major health consequences for humans and marine life, as well as other organisms that ingest the dirty water. Industrial and municipal discharges, natural geological formations, urban and rural runoff, water delivery materials, and the drinking water treatment process are all potential sources [21]. Human operations such as drilling and hydraulic fracturing have enhanced energy output, but they have also raised the likelihood of drinking water contamination. Heavy metals (e.g., nickel, mercury, copper, and chromium) can contaminate drinking water, leading to a rise in incidence of carcinogenic and non-carcinogenic health abnormalities [22]. Pharmaceutical waste materials are another known chemical polluter of water supplies [23]. Multiple pollutants, including aluminum, arsenic, lead, fluoride, radon, and pesticides, are known to contaminate drinking water. Mothers' consumption of chemical water has been linked to significant effects on infant gestation and birth weight, among other negative health outcomes, including an increased risk of cancer and cardiovascular diseases [24]. The potential routes of the chemical contamination are shown in Figure 2.

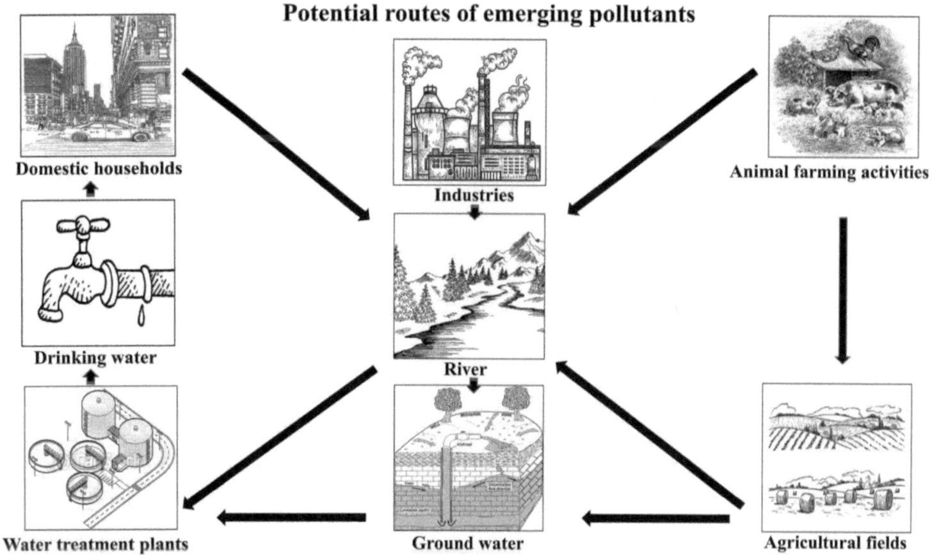

Figure 2: Potential routes of emerging pollutants.

3.2 Pharmaceuticals

Pharmaceuticals have been drawing more and more interest as possible bioactive compounds in the environment during the past 15 years [25]. They are considered emergent contaminants in waterways because they are either currently unregulated or through a regularization process for which the directives and regulatory frameworks are not yet in place. Water quality, as well as ecosystem and human health, can be negatively impacted by the continuous release and widespread presence of pharmaceuticals in the environment at low concentrations [26–28]. The therapeutic classes that are most frequently found in water include (i) anti-inflammatory and analgesics drugs such as paracetamol, ibuprofen, diclofenac, (ii) antidepressants, (iii) antiepileptic (iv) lipid-lowering drugs (v) β-blockers (atenolol, propranolol, and metoprolol) and other substances [29]. Many nations now recognize pharmaceutical product contamination of surface and ground water as an environmental issue, and as a result, the study of pharmaceuticals in the environment has emerged as a distinct area of study.

3.3 Hormones

Hormones are commonly utilized for treatment of human, domestic animal and livestock, and they are likely to enter the environment. The risk of presence of hormones and pharmaceuticals in drinking water is concerning because even at low doses, the non-deliberate contact of these bioactive compounds has harmful, long-lasting consequences on human health, such as endocrine disruption and antibiotic resistance. Most of these pollutants are not significantly eliminated by standard wastewater treatment techniques, and septic systems are prospective causes of hormones to groundwater. Several investigations have documented these sources and their probable roles in adding chemicals to groundwater. Other potential sources are seeping out of sewer lines, landfills, cropland treated with bio solids, etc. [30].

3.4 Pesticides

In the last 50 years, use of agrochemicals and insecticides has a sizeable contribution to the food security. Regardless of increased food availability, these agrochemicals have produced pollution and have a harsh effect on the ecosystem and human health. Similarly, pesticides conceal into soils and groundwater and end up in drinking water. Pesticides are equally harmful for humans and animals, and their consequences depend on the toxicity levels and the length of exposure. Pesticide hazard studies are usually site-specific, with only a few focusing on regional and global scales to assess the risks associated with a single pesticide class.

3.5 Personal care products

Health, beauty, and cleaning items all fall under the umbrella termed as "personal care products" (PCPs). Among these are antibacterials, perfumes, insecticides, preservatives, and UV filters. Because of their existence and deleterious influence on aquatic ecosystems, particularly in regard to endocrine disruption and reproductive diseases, several of them are regarded chemicals of growing concern. Sewage effluents from wastewater treatment plants are a major entry point for these chemicals into aquatic environments due to inadequate or ineffective treatment.

3.6 Heavy metals

Pollution from heavy metals is a major problem because of the harm it does to people and the planet. Lead, chromium, cadmium, mercury, and arsenic are only a few examples of the most dangerous heavy metals to humans. These toxic heavy metals are non-biodegradable, and their accumulation in human bodies may occur either by direct intake or through food chains. As a result, different health problems such as lung cancer, kidney, liver and gastric damages may happen [31].

3.7 Micro-Organisms

A number of foodborne diseases are caused by microbial contamination. Foodborne infections and outbreaks are recorded in which fresh produce and animal products are contaminated with pathogenic bacteria, viruses, and protozoa from polluted sources. Specific sources of contamination that occurred at the early stages of the food chain are investigated in meat manufacturing and fresh produces. Wild and domestic animals, flies, and rats can infect the food-growing fields and cause human sickness by direct contact. The majority of fresh produce-related outbreak has been caused by animal encroachment into growing fields or fecal contamination from nearby animal production facilities, which has likely resulted in crop contamination, dirty irrigation water, and incorrect manure.

3.8 Micro plastics

In many cases, plastics will make their way into freshwater ecosystems via unintended pathways. Land littering is a major public and environmental health concern, especially in protected regions where garbage levels are mostly determined by the number of visitors. Consequently, it is crucial to take steps to lessen its impact on the environment. It's possible that this is a more significant source in some parts of the world than others due to differences in waste management procedures. The same is true for micro plastics; there is more than one way for them to get into the environment. Examples include those found in consumer cosmetics, which are likely to have a higher profile in more affluent areas [32].

3.9 Other pollutants

Surfactants are a wide variety of organic chemicals added to soaps, dish washing liquids, shampoos also used in industries as lubricants, hydraulic fluids, and fluids used in petroleum industry for cleaning oil spills. These surfactants enter the aquatic ecosystems when they are washed through the drains and not entirely cleaned by the wastewater treatment systems. Alkylphenols are widely used in surfactants, and they are highly toxic to aquatic organisms and have been shown to possess estrogenic properties [33]. Sugar substitutes made from artificial sweeteners are used extensively in foods, drinks, medicines, and hygiene items like mouthwash all around the world. They are used in food because they have no or very little effect on energy levels [34]. Most common artificial sweetening agents are acesulfame, aspartame, cyclamate, neotame, neohesperidine dihydrochalcone, saccharin, and sucralose [35]. As food additives, artificial sweeteners undergo thorough testing for potential detrimental health consequences in humans and their maximum limit is fixed; even then some have been found at high quantities in surface water, groundwater, and drinking water, causing health consequences [36].

3.10 Impact of climate change

Climate changes have resulted in prolonged and severe droughts, which concentrate contaminants in rivers and cause adverse effects on the aquatic and human life [37]. Recently, drinking water system in the City of Teledo had to be shut down for some time due to large algal bloom in western lake Erie in the United States [38]. Algal blooms release a high percent of algal toxins that are harmful to humans and aquatic life.

4 Future challenges and directions

While there has been some success in identifying and quantifying new environmental pollutants as well as gaining a better knowledge of their environmental fate and developing strategies for cleanup, much work remains to be done. To begin, there are likely many environmental pollutants that have not yet been discovered. Particularly illustrative of this are the newly identified per-and polyfluoroalkyl substances (PFAS) which are a group of chemicals used to make special coatings to the products to resist heat, oil, stains, grease, and water. This category includes pollutants that have been lurking in the environment for years without being named. There are still many obstacles to overcome in the areas of new environmental pollutants detection, quantification, understanding of environmental destiny, and development of remediation solutions. First, there are a lot of pollutants in the environment that are still undiscovered. This is especially clear in the new PFAS classes that keep being found in the environment. These include environmental pollutants that have been around for a long time but have not yet been properly identified [39]. Last but not least, there is still a need to develop enhanced AOPs and other treatment methods to more effectively remove these contaminants. The development of anti-fouling and self-healing membranes is a recent area of research that shows great potential for membranes. However, it is to make sure that the development of harmful by-products from these procedures should not have any significant influence on human or aquatic environment.

5 Possible solutions

Nowadays reuse of wastewater is more acceptable as it is less expensive than seawater desalination and is getting popular in many countries due to scarcity of freshwater resources. People try various biodegradation techniques such as micro-algae based biodegradation, photo-degradation, biosorption, volatization for wastewater treatment. Some use physical methods, like microfiltration (MF), ultrafiltration (UF), reverse osmosis (RO), and UV disinfection/ UV-H_2O_2 treatment, prior to industrial use of wastewater. A small portion of the treated water is added to a drinking water reservoir or into a groundwater aquifer [40].

All the emerging contaminants that are found in the environment widely differ in several aspects such as the chemical, biological, and physical properties, and some of the properties such as biodegradability, chemical reactivity, and hydrophobicity. Many engineering processes are deployed to study all these properties, for development of better ways for their eradication from the environment. But a single process cannot remove all of these contaminants and it is nearly impossible, so we use a multistep process so that a clean removal can be achieved. There are several treatments developed for contaminant removal which include oxidation and advanced oxidation processes, adsorption on GAC or PAC, and rejection by membrane technologies. The degradation of the contaminants mainly depends on the type of oxidant used for the process, but hydroxyl radicals are non-selective [41].

To adequately address the issue of emerging pollutants, we require a very interdisciplinary approach. There are now significant gaps in our understanding of new toxins that must be filled by chemical, biological, and environmental experts. Toxicologists will be crucial in determining how these substances harm human health and the health of the ecosystem. More research is needed to determine how to prevent the creation and exposure to these compounds. Finally, it is critical that the public become aware of the growing problem of emerging toxins so that regulatory agencies such as the EPA may be pressured to address it.

6 Conclusion

Emerging pollutants are introduced into the environment and human populations through a wide range of recognized and unknown pathways. The review reveals that these toxins continuously cause emerging and serious concerns to the environment, particularly human health. Furthermore, the emergence of novel chemical outputs often outpaces the capabilities of safety remediation, risk assessment methodologies, monitoring approaches, and existing preventative measures. To estimate the human health impacts of pollutant exposure, the following various interacting factors should be considered: contaminant concentration, category and attributes of pollutants, scale and level of pollutants, and the danger intensity generated for health and accessible resources. Future methodologies and investigations are needed to build general remediation theories and methods, which are essential for the eradication of new toxins in the environment and the human body. The emerging pollutants are suspected of being mutagenic, teratogenic, and carcinogenic to humans and other animals based on ongoing research and studies. On the one hand, new pollutants and harmful effects on the human body have not been linked through significant and extensive research. On the other hand, even though long-term chronic exposure is rarely used in animal models and the exposure levels in animal trials may not accurately reflect human exposure levels in reality, we cannot disregard the negative impacts they have shown. More extensive research with better designs is required to get results that are more convincing and understandable.

References

[1] Nag S, Mondal A, Bar N, Das SK. Biosorption of chromium (VI) from aqueous solutions and ANN modelling. Environ Sci Pollut Res 2017, 24, 23, 18817–18835.

[2] Das R, Saha A, Nath H, Das J. Water quality assessment of different samples collected from various locations of Tripura university campus and selection of appropriate point of use water purification technology. SARASWAT: Tripura Univ Res J 2022, 2, 1, 52–61.

[3] Nag S, Das J. Viability of membrane separation techniques in the detoxification of recalcitrant pollutants. In: Removal of refractory pollutants from wastewater treatment plants. CRC Press, 2021, 289–298.

[4] Nath PC, Nandi NB, Tiwari A, Das J, Roy B. Applications of nanotechnology in food sensing and food packaging. In: Nanotechnology applications for food safety and quality monitoring. Academic Press, 2023, 321–340.
[5] Das J, Debnath C, Nath H, Saxena R. Antibacterial effect of activated carbons prepared from some biomasses available in North East India. Energy Source A: Recovery Util Environ Eff 2019, 1–11.
[6] Nag S, Biswas S. Accumulation and detoxification of metals by plants and microbes. In: Removal of emerging contaminants through microbial processes. Springer, Singapore, 2021, 359–372.
[7] Nag S, Biswas S. Removal of emerging contaminants through bionanotechnology-Elsevier. In: Development in wastewater treatment research and processes. Elsevier, 2022, 291–310.
[8] Nag S, Mondal A, Mishra U, Das SK. Removal of Chromium (VI) from aqueous solutions using mango leaf powder: Process optimization with Response Surface Methodology. J Ind Chem Soc 2017, 93, 11, 1279–1298.
[9] Das J, Mondal A, Biswas S, Nag S. The eco-friendly treatment of rubber industry effluent by using adsorbent derived from Moringa oleifera bark and Pseudomonas sp, cultured from effluent. Water Sci Technol 2022, 86, 11, 2808–2819.
[10] Nag S, Biswas S. Cellulose-based adsorbents for heavy metal removal. In: Green adsorbents to remove metals, dyes and boron from polluted water. Springer, Cham, 2021, 113–142.
[11] Deb S, Nag S. Use of microalgae for the removal of emerging contaminants from wastewater. In: Biodegradation and detoxification of micropollutants in industrial wastewater. Elsevier, 2022, 193–210.
[12] Fleeger JW, Carman KR, Nisbet RM. Indirect effects of contaminants in aquatic ecosystems. Sci Tot Environ 2003, 317, 1–3, 207–233.
[13] Atchison GJ, Henry MG, Sandheinrich MB. Effects of metals on fish behavior: A review. Environ Biol Fishes 1987.
[14] Eisler R. Aquatic toxicology: Molecular, biochemical, and cellular perspectives. Trans Am Fisheries Soc 1995.
[15] Sanchez W, Egea E. Health and environmental risks associated with emerging pollutants and novel green processes. Environ Sci Pollut Res 2018, 25, 7, 6085–6086.
[16] Aamand J, Gavrilescu M, Agathos S, Fava F. Emerging pollutants in the environment: Present and future challenges in biomonitoring, ecological risks and bioremediation, 2014, 1–10.
[17] Biziuk M, Przyjazny A, Czerwinski J, Wiergowski M. Occurrence and determination of pesticides in natural and treated waters. J Chromatogr A 1996.
[18] Hashimi MH, Hashimi R, Ryan Q. Toxic effects of pesticides on humans, plants, animals, pollinators and beneficial organisms. Asian Plant Res J 2020.
[19] Colborn T, Vom Saal FS, Soto AM. Developmental effects of endocrine-disrupting chemicals in wildlife and humans. Environ Health Perspect 1993.
[20] Buzea C, Pacheco II, Robbie K. Nanomaterials and nanoparticles: Sources and toxicity. Biointerphases 2007.
[21] Calderon RL. The epidemiology of chemical contaminants of drinking water. Food Chem Toxicol 2000.
[22] Nag S, Mondal A, Mishra U, Bar N, Das SK. Removal of chromium (VI) from aqueous solutions using rubber leaf powder: Batch and column studies. Desalin Water Treat 2016, 57, 36, 16927–16942.
[23] Shen R, Andrews SA. Demonstration of 20 pharmaceuticals and personal care products (PPCPs) as nitrosamine precursors during chloramine disinfection. Water Res 2011.
[24] Currie J, Graff Zivin J, Meckel K, Neidell M, Schlenker W. Something in the water: Contaminated drinking water and infant health. Can J Econ 2013.
[25] Kümmerer K. Drugs in the environment: Emission of drugs, diagnostic aids and disinfectants into wastewater by hospitals in relation to other sources – A review. Chemosphere 2001.

[26] Kolpin DW, et al. Pharmaceuticals, hormones, and other organic wastewater contaminants in U.S. streams, 1999–2000: A national reconnaissance. Environ Sci Technol 2002.
[27] Yuan F, Hu C, Hu X, Qu J, Yang M. Degradation of selected pharmaceuticals in aqueous solution with UV and UV/H2O2. Water Res 2009.
[28] Sirés I, Brillas E. Remediation of water pollution caused by pharmaceutical residues based on electrochemical separation and degradation technologies: A review. Environ Int 2012.
[29] Bush K. Antimicrobial agents targeting bacterial cell walls and cell membranes. OIE Rev Sci Tech 2012.
[30] Bexfield LM, Toccalino PL, Belitz K, Foreman WT, Furlong ET. Hormones and pharmaceuticals in groundwater used as a source of drinking water across the United States. Environ Sci Technol 2019.
[31] Nag S, Mondal A, Mishra U, Bar N, Das SK. Removal of chromium (VI) from aqueous solutions using rubber leaf powder: Batch and column studies. Desalin Water Treat 2016, 57, 36, 16927–16942.
[32] Hutzinger O, Springer P, Barceló ED. The handbook of environmental chemistry. Umweltwiss Schadstoff-Forsch 2010, 22, 4, 511–512.
[33] Tremblay L, Stewart M, Northcott G. Emerging contaminants: Should we be worried? Revolve 2016, August, 2008, 19–21.
[34] Kroger M, Meister K, Kava R. Low-calorie sweeteners and other sugar substitutes: A review of the safety issues. Compr Rev Food Sci Food Saf 2006.
[35] Lange FT, Scheurer M, Brauch HJ. Artificial sweeteners-A recently recognized class of emerging environmental contaminants: A review. Anal Bioanal Chem 2012.
[36] Harpaz D, et al. Measuring artificial sweeteners toxicity using a bioluminescent bacterial panel. Molecules 2018.
[37] La Kind JS, et al. Exposure science in an age of rapidly changing climate: Challenges and opportunities. J Expo Sci Environ Epidemiol 2016.
[38] Lee JJ. Driven by climate change, algae blooms behind Ohio water scare are new normal. Nat Geo 2014.
[39] Houtz EF, Sedlak DL. Oxidative conversion as a means of detecting precursors to perfluoroalkyl acids in urban runoff. Environ Sci Technol 2012.
[40] Richardson SD, Plewa MJ, Wagner ED, Schoeny R, DeMarini DM. Occurrence, genotoxicity, and carcinogenicity of regulated and emerging disinfection by-products in drinking water: A review and roadmap for research. Mutat Res Rev Mutat Res 2007.
[41] Lee Y, von Gunten U. Oxidative transformation of micropollutants during municipal wastewater treatment: Comparison of kinetic aspects of selective (chlorine, chlorine dioxide, ferrateVI, and ozone) and non-selective oxidants (hydroxyl radical). Water Res 2010.

Nikhila Mathew, Abha Tirpude, Anupama M Pillai,
Pabitra Mondal, Tanvir Arfin*

Emerging Contaminants in Air Pollution and Their Sources, Consequences, and Future Challenges

Abstract: Emerging contaminants are dispersed during environmental matrices and enter the environment through various anthropogenic causes. Even though there have been significant improvements in the detection and examination of suggested pollutants in recent years, various undiscovered contaminants that pose a threat to the environment must be found and measured in diverse environmental components. These contaminants may be dispersed and persistent in the air and ecological receptors even at low concentrations. There still needs to be more reliable information on their destiny and environmental behavior, as well as the dangers they pose to the ecosystem and human well-being. The environmental impact of several developing micropollutants still needs to be clarified since there is frequently insufficient data to assess their danger.

Keywords: Emerging pollutants, anthropogenic sources, environmental matrices, VOC, air pollution

1 Introduction

Due to its harmful impact on ecosystems, electronegative substance environmental contamination is today one of the most significant issues [1]. Some of the conventional contaminants of the environment include pesticides, heavy metals, polychlorinated biphenyls (PCBs), and other industrial pollutants. The accidental release of mine waste, approximately 1 million gallons, into the Animas River, which contains zinc, iron, and copper, turns the river orange and causes downstream contamination and

Acknowledgment: Authors acknowledge the Knowledge Resource Centre (CSIR-NEERI/2022/DEC/APC/1) for their support.

***Corresponding author: Tanvir Arfin,** Air Pollution Control Division, CSIR-National Environmental Engineering Research Institute (CSIR-NEERI), Nehru Marg, Nagpur 440020, India, e-mail: tanvirarfin@gmail.com
Nikhila Mathew, Abha Tirpude, Anupama M Pillai, Pabitra Mondal, Air Pollution Control Division, CSIR-National Environmental Engineering Research Institute (CSIR-NEERI), Nehru Marg, Nagpur 440020, India

https://doi.org/10.1515/9783111016825-014

further into other rivers. Consequently, those conventional contaminants, including heavy metals, are a significant problem that must be considered.

Although many previously unknown contaminants have been identified in recent environmental studies [2] many of these do not cause severe toxicity, different conventional, legal impurities, outcomes in various exquisite methods, for example, or endocrine disruption. Also, numerous biomagnifications are indifferent to traditional pollutants. Such emerging contaminants (EC) were not considered in earlier times but existed globally in the environment. Subsequently, most of these rising impurities exhibit distinct possessions and outcomes. Much more particularly, we have reviewed its health aspect, but interesting significant criteria related to marine existence are not mentioned. Such impurities are frequently known as "emerging" but were not determined earlier. Many have existed in the environment for decades because of the unavailability of careful and significant measures. A remarkable increase in air pollutant emission in the atmosphere was observed [3]. Persistent organic pollutants (POPs) are categorized as adverse composites since they resist numerous biochemical and photolytic degradation factors. POPs have persisted in air, soils, and sediments [4]. Since POPs are highly toxic and show top endurance in nature, they compile into the adipose tissues of humans and wilds and generate plenty of observable, developmental, or reproductive mutations [4]. POPs attained global importance because of their feasibility in movement over an extended range of sources. Several researchers show beyond doubt that POPs are a menace because they are toxic, bioaccumulative, and persistent. These adulterants include many contaminants comprising various classes of organic compounds, comparatively polyfluoroalkyl substance (PFAS) and flame retardant (FR). Figure 1 shows the chemical structure and formula of non-identical emerging contaminants in theatmosphere.

These compounds are being released into the environment in high quantities as other dangerous POPs due to urbanization and their expanding home, industrial, and agricultural usage; as a result, they can acquire fundamental constituents, as shown in Figure 2.

PFAS has been utilised in products since 1950 but only became recognised as a contaminant in the 2000s. Also called PFAS. The modification and progress of electrospray ionization (ESI)-mass spectrometry (MS) in the 1980s and the upgradation in mass spectrometer susceptibility enabled the identification and measurement of such contaminants, many among them essential liquid chromatography (LC)-MS alternatively gas chromatography (GC)-MS since they are polar and have high molecular weight.

2 Volatile organic compounds

Volatile organic compounds (VOCs) occur in many diversified and all-pervasive forms; they are produced both clearly and mannerly in the environment. Although learning about VOCs on one's own rather than in groups is essential, their impacts on wellness

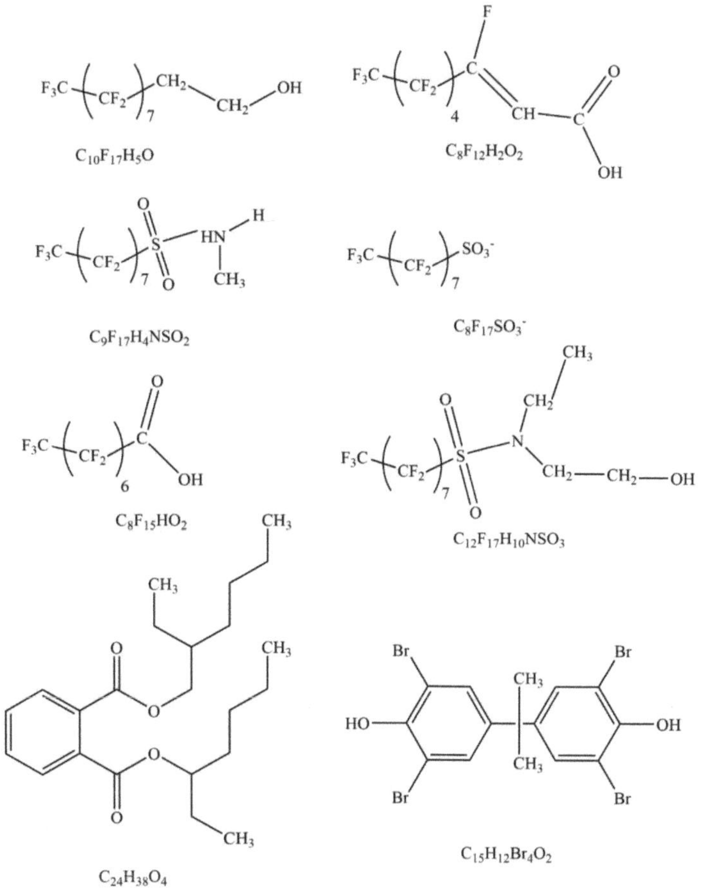

Figure 1: Chemical structure and formula of non-identical emerging contaminants establish in the atmosphere.

and the habitat may differ. Various VOCs are presented in Figure 3. The Environmental Protection Agency (EPA) and the World Health Organization (WHO) recognize these.

Table 1 introduces the arrangement of volatile organic compounds according to their essential aspect: volatility and their holdings on ecological community and human well-being.

The increasing VOCs used in this book chapter are acrylonitrile, ethylene oxides, 1,4-dioxane, toluene, chloroform, naphthalene, 1-3-butadiene, trichloroethylene, formaldehyde, 1,1,2,2-tetrachloroethane, and dichloromethane. Table 2 outlines VOC components and its sources in the atmosphere.

Table 3 details the VOC concentration investigations conducted in three countries: Spain, Cyprus, and Rome.

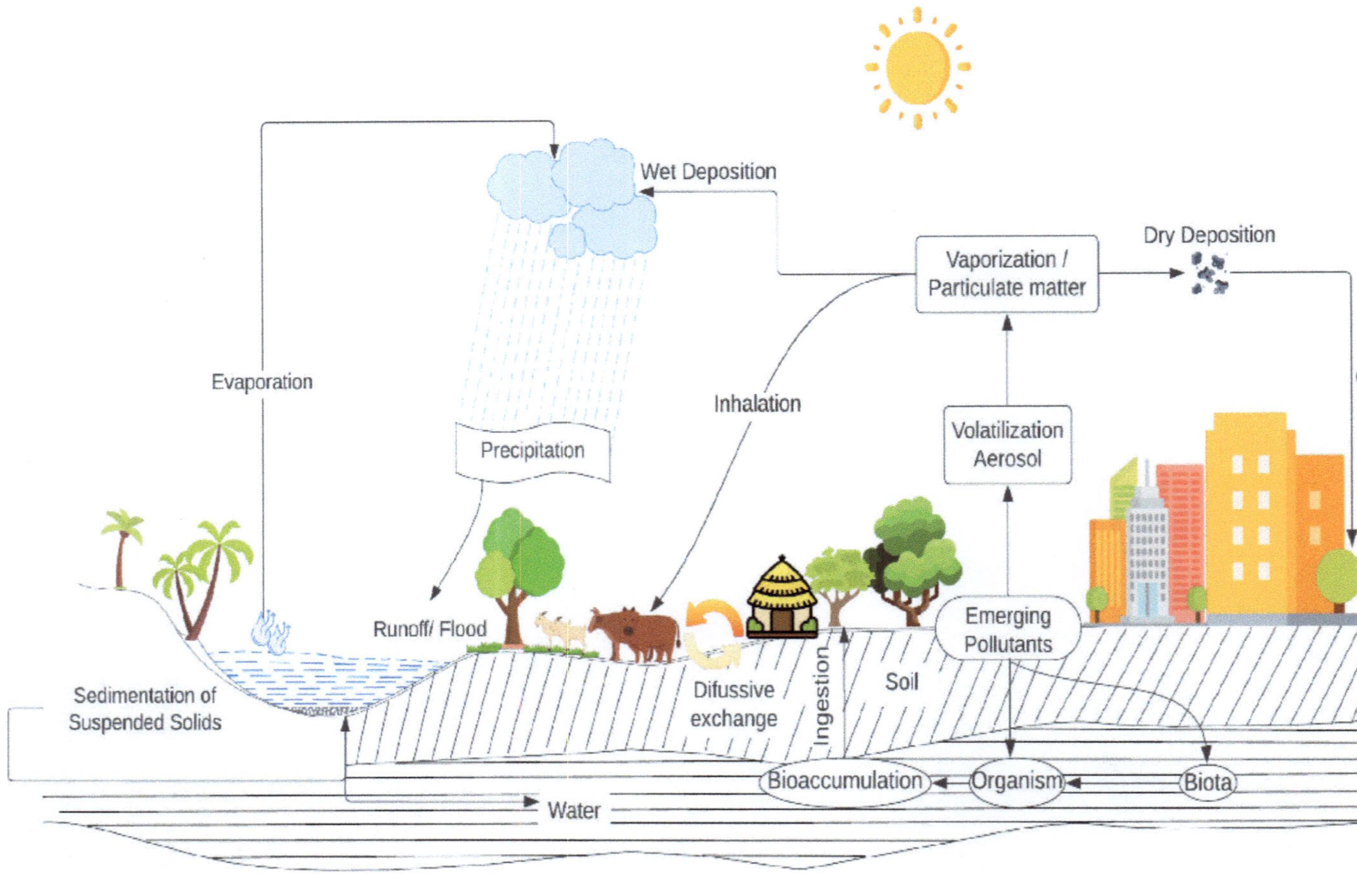

Figure 2: Providence and reciprocity of EC in the atmosphere.

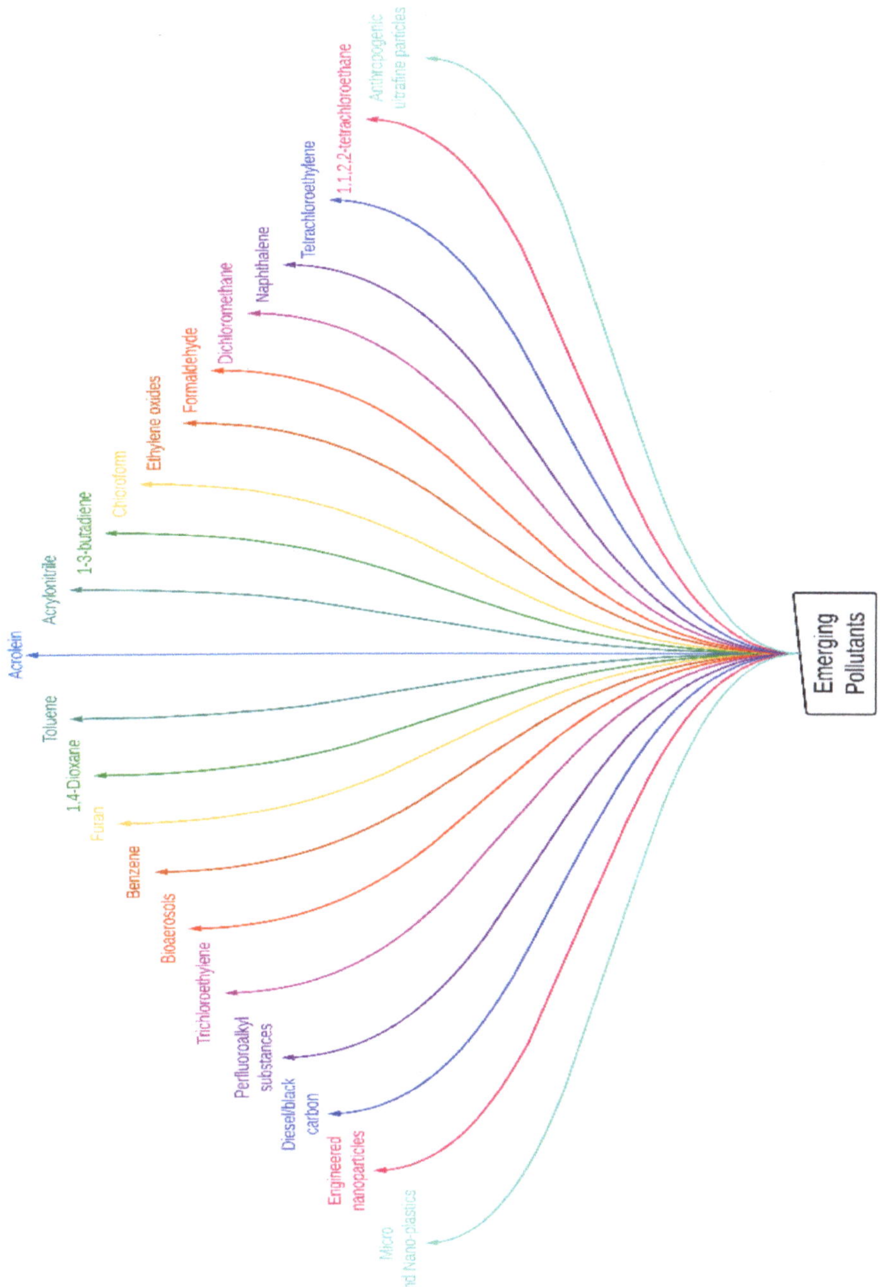

Figure 3: Selected environmental contaminants in the atmosphere.

Table 1: Arrangement of VOCs with consideration to their properties [5].

Characteristic	Class	Examples
Volatility	The boiling point is less than 100 °C	Alkanes
	The boiling point is less than 250 °C	Alcohol
	The boiling point is less than 390 °C	Polyaromatic hydrocarbon
	The boiling point is greater than 390 °C	Particulate organic matter
Biodiversity	Likely to form photochemical oxidants	Terpenes
	Very likely to deplete the stratospheric ozone layer	Halocarbons
	Extremely harmful to contemporary pathogens	methane
Mortal well-being	Harmful to mortal well-being	Oxygenated hydrocarbons
	Produced through photochemical oxidation, secondary hazardous waste	Aldehydes

Table 2: VOC components and its sources in the atmosphere.

S. no.	VOCs	Chemical structure	Source
1.	Ethanol	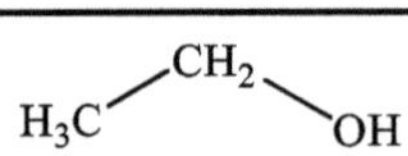	Industrial
2.	Propane	$CH_3—CH_2—CH_2—CH_3$	LP gas
3.	2-hexanone	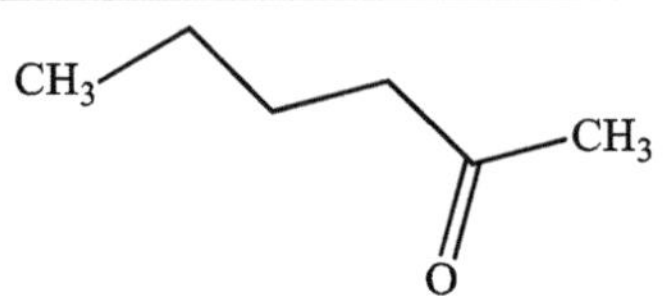	Industrial
4.	Acetone	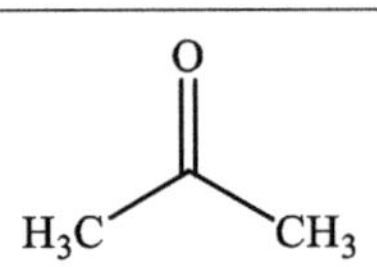	Industrial
5.	4-methyl-2-pentanone	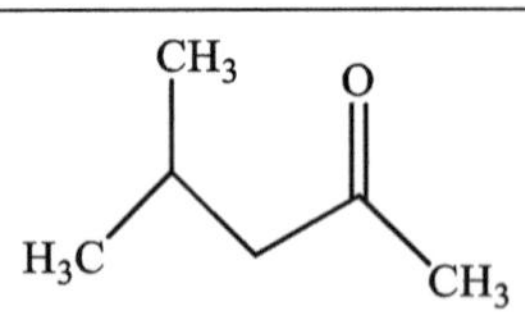	Industrial

Table 2 (continued)

S. no.	VOCs	Chemical structure	Source
6.	Hexachloro-1,3-butadiene	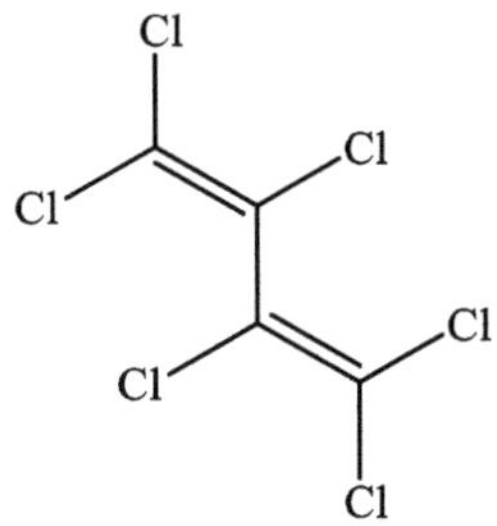	Industrial
7.	1,4-dichlorobenzene	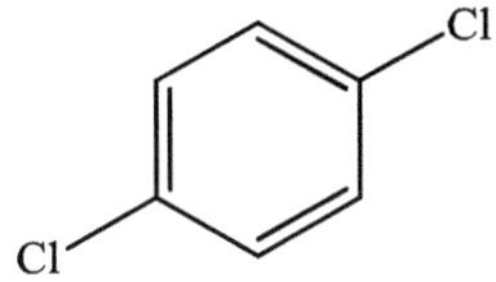	Industrial
8.	1,2,4-trimethylbenzene	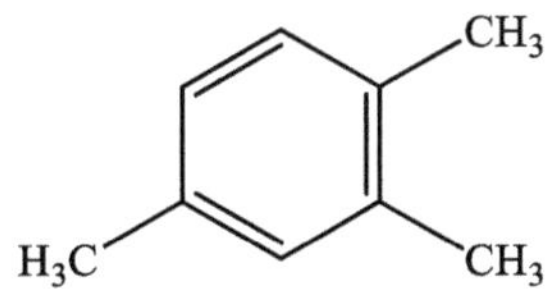	Industrial
9.	n-butane	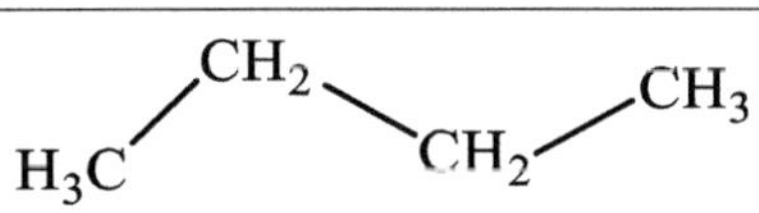	LP gas
10.	Methylcyclo-pentane	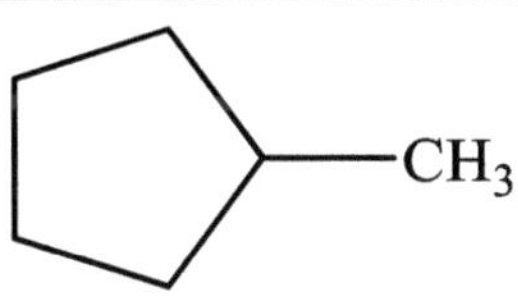	Industrial
11.	2-propanol	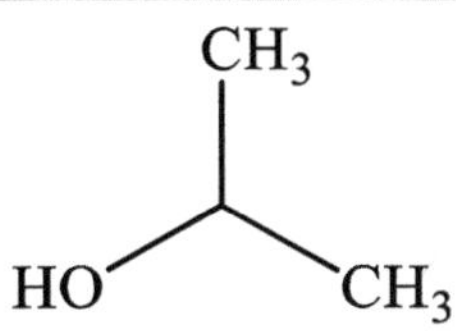	Industrial
12.	Toluene	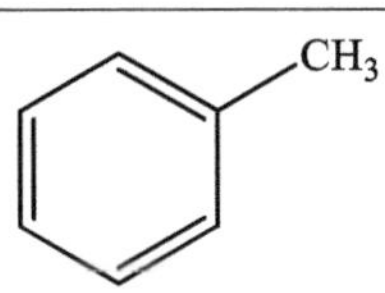	Domestic product

Table 2 (continued)

S. no.	VOCs	Chemical structure	Source
13.	Isobutane	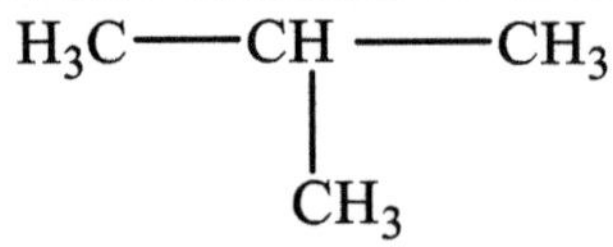	Industrial
14.	Acrolein	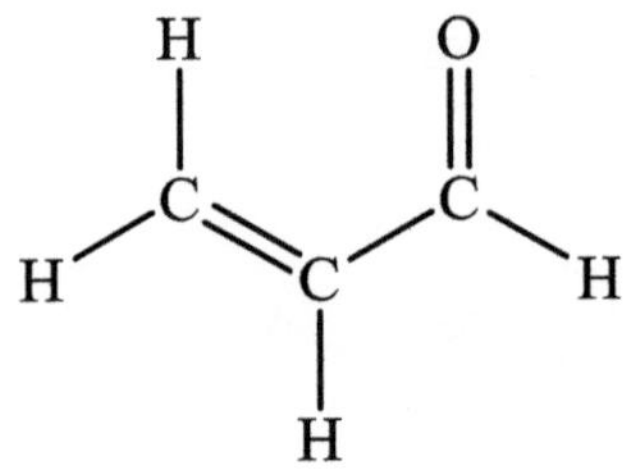	Industrial
15.	Acrylonitrile	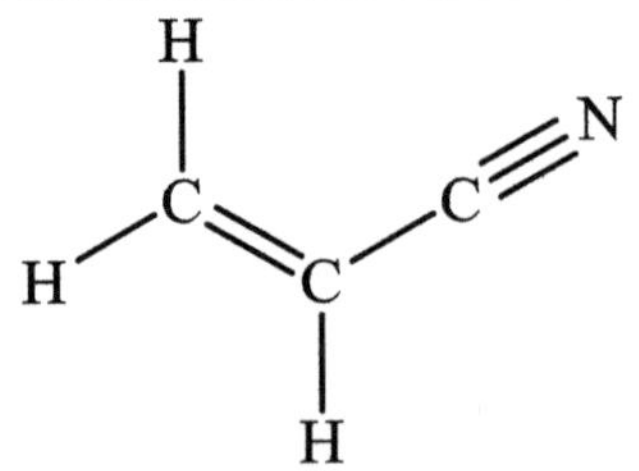	Plastic Material
16.	1,3-butadiene	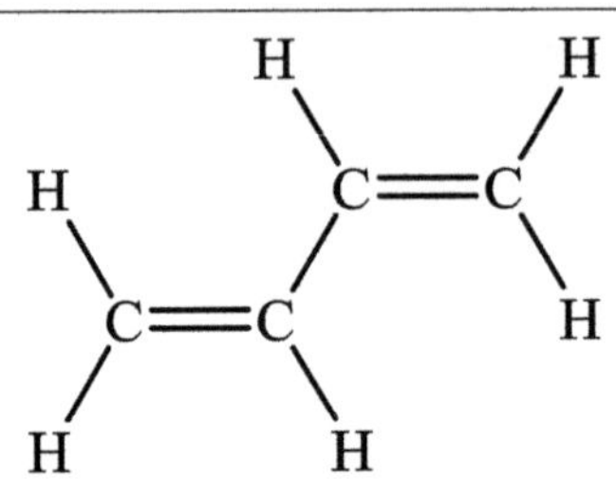	Petroleum refining
17.	Chloroform	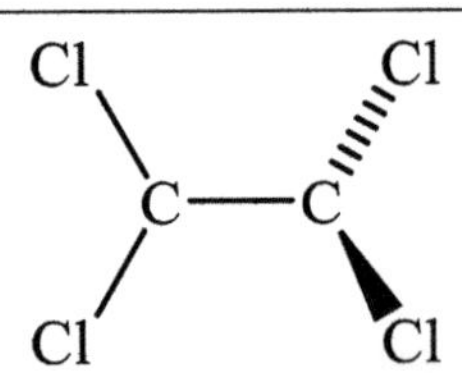	Humic materials
18.	Naphthalene	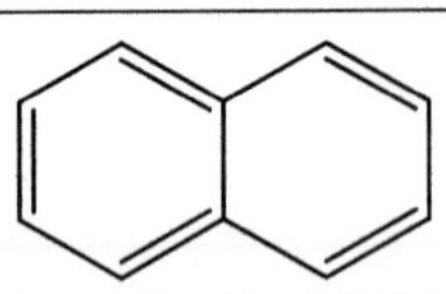	Diesel combustion

Table 2 (continued)

S. no.	VOCs	Chemical structure	Source
19.	Tetrachloroethylene	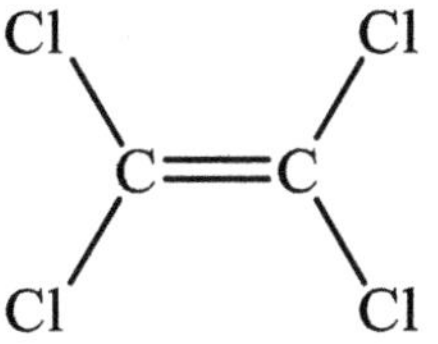	Consumer products
20.	1,1,2,2-tetrachloroethane	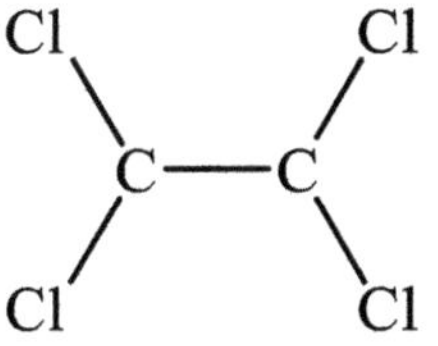	Discharges stemming
21.	Formaldehyde	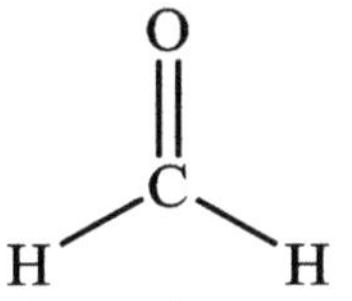	Household products
22.	Ethylene oxide	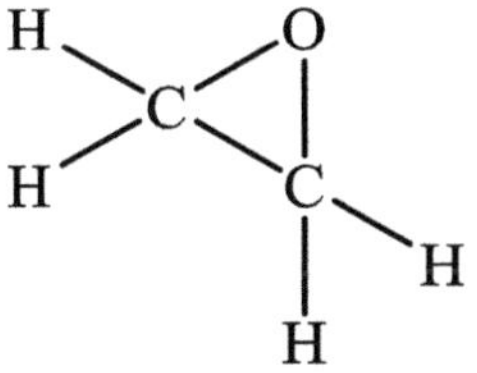	Plants
23.	Dichloromethane	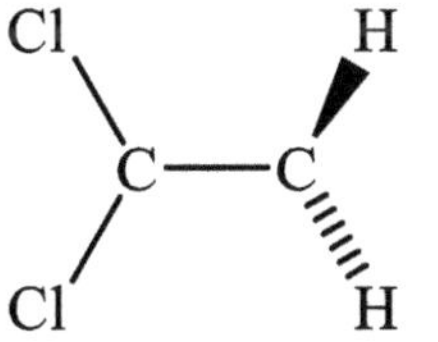	Ocean
24.	1,4-dioxane	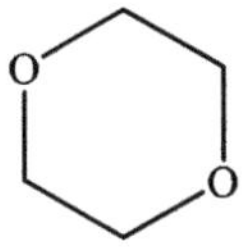	Shampoo
25.	Trichloroethylene	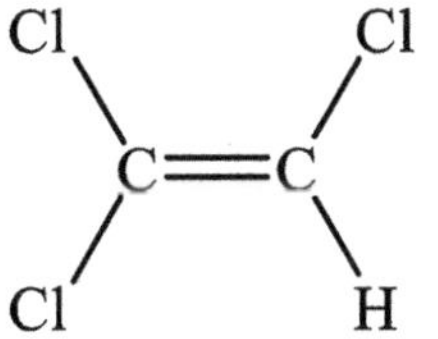	Resin workers

Table 2 (continued)

S. no.	VOCs	Chemical structure	Source
26.	Methyl chloride	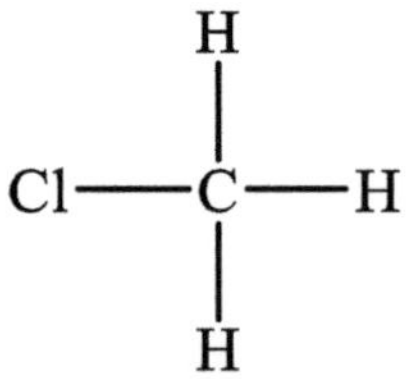	Ocean emission
27.	Vinyl chloride	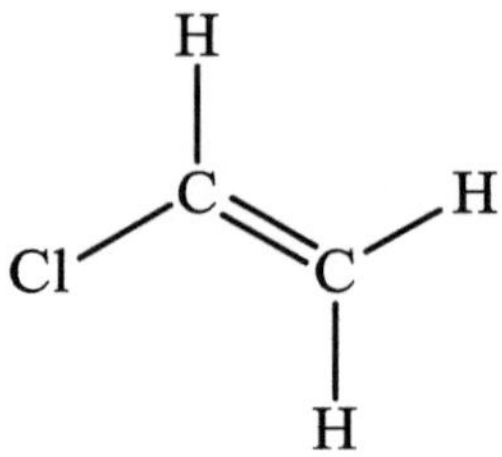	Industrial
28.	Carbon disulfide	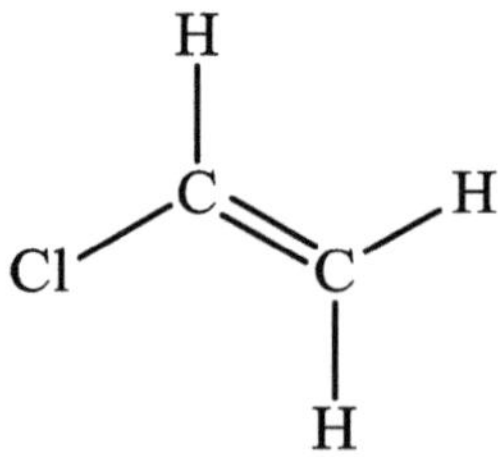	Geothermal
29.	d-Limonene	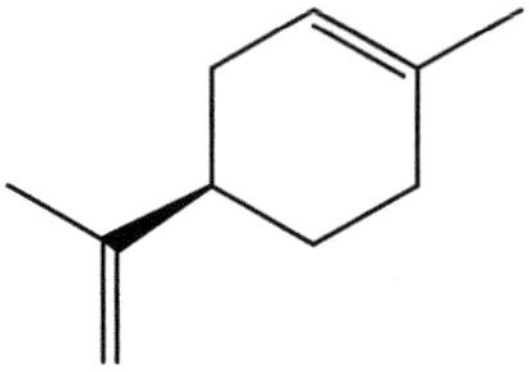	Oils of orange
30.	DDT	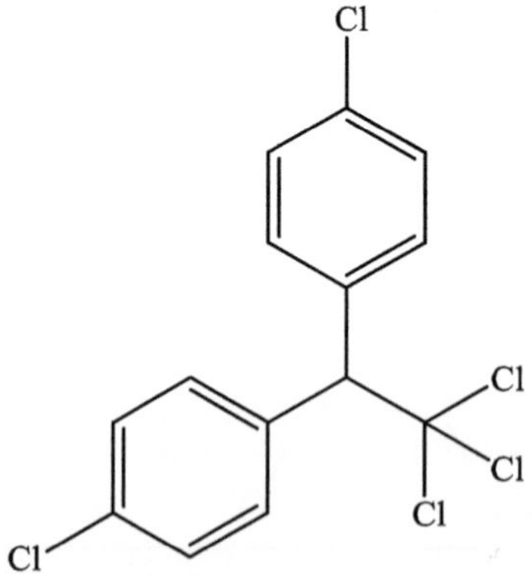	Pesticide

Table 2 (continued)

S. no.	VOCs	Chemical structure	Source
31.	Chlordane	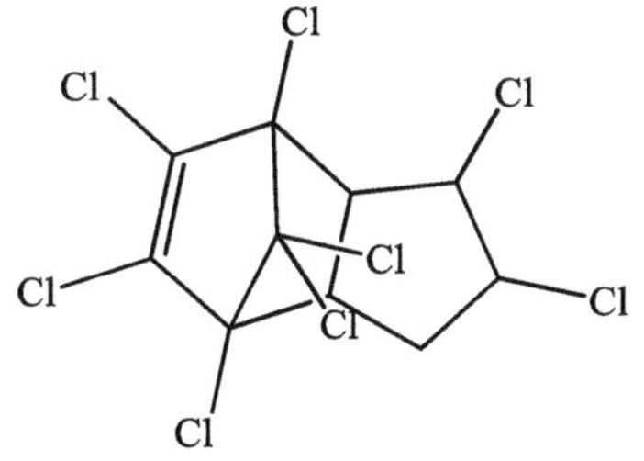	Industrial
32.	Phthalate	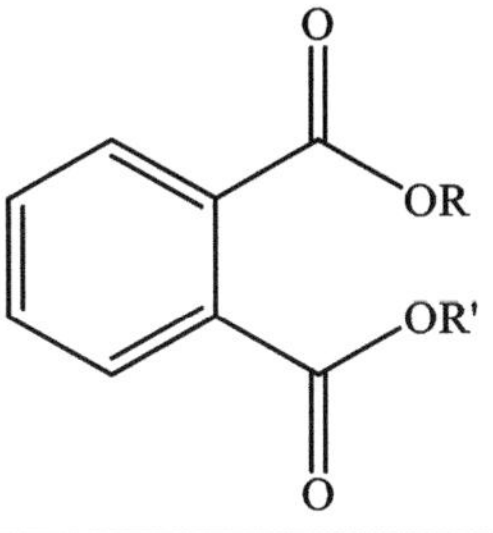	Plastics
33.	Polychlorinated biphenyl	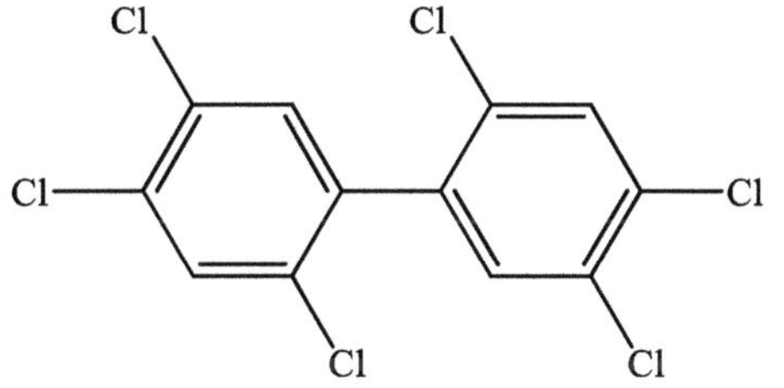	Cable insulation
34.	Polybrominated biphenyl	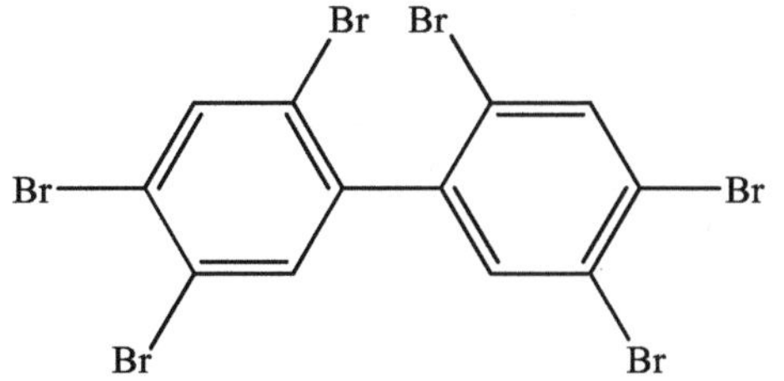	Consumer good

Table 3: Shows the information on the concentration of VOCs in the air in several countries.

Country	Compounds	24 h Average concentration (μg/m³)	Ref.
Spain	Benzene	1.33	[6]
	Toluene	8.74	[6]
	Ethyl benzene	1.30	[6]
Cyprus	Benzene	0.37	[7]
	Toluene	0.19	[7]
	Ethylene	0.35	[7]
Rome	Benzene	3.29	[8]
	Toluene	10.58	[8]
	Ethyl benzene	0.43	[8]

2.1 Formaldehyde (Methanal)

The results of a thorough assessment of formaldehyde (HCHO) levels globally in various indoor environments have been presented. The concentrations mentioned in some reports are combined. WHO proposed a guideline for 100 μg/m^3. However, the studies reported the highest formaldehyde concentration in Egypt, i.e., 96.6 μg/m^3 [9]. In the current scenario, further studies are needed on this EP world widely.

The Consumer Product Safety Commission (CPSC) has faced more than 3,000 lawsuits concerning purchasers' risk of vulnerability to formaldehyde condense emitted through organizing substances. Among such, 2,000 comprise urea-formaldehyde foam insulation (UFFI). The rest includes hardwood, paneling, chipboard, including particular wooden goods in adaptable houses. Some are because of another source of formaldehyde, which consists of

1. furnishings, inclusive of textiles, furniture, and carpets,
2. fueled home equipment, inclusive gas stoves, unvented house heaters, and ovens,
3. consumer products, consisting of paper products, air fresheners, adhesive gums, etc.,
4. humanity lifestyle, like a cigarette.

The methanal range in the interior atmosphere is usually more extensive than in outdoor air. Despite the degree of 0.2 μg/m^3 as soon as the recorded previous application of automobile discharge powers, the contents of methanal doors are frequently much less than 0.02 μg/m^3. The interior range of formaldehyde could also vary in mass from much less than 0.13 μg/m^3 to 3.9 μg/m^3 or higher. Facts on methanol levels in homes, including unique estimations in a home using different National institute for occupational Safety and Health (NIOSH) chronotropic approaches or the same approach, were complied with the assistance of the CPSC's Health Science directorate. UFFI-insulated homes have surprisingly better methanal convergences ($p < 0.02$) compared to non-UFFI residences. Conditional details were selected and scrutinized on UFFI managed below absolute states, and it concluded that UFFI was present in the residences' methanal charge by 0.07–0.4 mg/m^3 at 25 °C. UFFI consoles were analyzed by temperatures 24, 35, and 41 °C. The quantity of methanal emitted from the console extended six-fold by 35 °C and thirteen-fold near 40 °C compared to around 23 °C. UFFI installation in the cavity wall can comes across comparable degrees. CPSC is now investigating the beneficiation of methanal from more assets, which include textiles, firewood, and different consumer commodities. The exceeding transmission range was acquired from desiccator experimental analysis along relative humidity of 100–70 and ambient determining at 100–70 relative humidity. The methanal emitted from those origins may be from the product containing free formaldehyde or from the resin hydrolysis applied within the outcome. The formaldehyde emission range is close to wax, temperature, loading component, curing method, humidity, fabric age, and background formaldehyde concentration [10]. Outdoor air contains formaldehyde, which has been linked to cancer. Exposure to a suggested HCHO level of 1gm^3 over a lifetime will result in up to 13 cases of nasopharyngeal and lung cancer in every million individuals [11]. The EPA has identified 187 hazardous air pollutants (HAPs) as the primary cause of most cancers or other severe health impacts in ambient outdoor air, including formaldehyde. Given that it is to blame for more than 50% of all cancer risks associated with HAPs in the United States, it is without a doubt the HAP that poses the most health risks. The chemical scrubber showed an efficacy of 95.7–97.2% in removing formaldehyde from crude gas [12].

2.2 Benzene

Benzene (C_6H_6) is the second-most vital HAP. It may come from both natural and artificial sources. It is used in the manufacturing of plastics, resin, pesticides, etc. It is a proven carcinogen. The skin and lungs are entrance points for benzene into the human body. There are also several known non-cancer dangers, including tremors and headaches. In the atmosphere, it continues to exist as vapor. In the process of manufacture, it is also released into the atmosphere. Cigarette smoke constitutes one of the most popular origins of benzene in indoor air. The recommended benzene airborne limit in India is 5 g/m^3 (annually) [13]. Only New Zealand has a benzene standard of 5 g/m^3 (annual). Their standards were set based on EU and UK policies [14]. HPA is no longer associated with other anthropogenic hotspot but is still in the United States. The by-products released from the oxidation of VOC are specific biogenic isoprene [15] The information regarding the effect of human exposure to HAPs in monitoring locations has been released by the HAP sampling unit in the United States; however, it is confined to city/commercial sites [16]. Over there, they use 12 years of benzene observations with the help of an ozone monitoring instrument (OMI) satellite instrument [17] having 5×5 km^2 spatial selection enabled through the way of an oversampling approach [18] to track formaldehyde concentration in surface air over the contiguous United States and leads to the risk of cancer on a national scale. The predicted benzene emission concentration levels that are practical to achieve are: (a) Refrigeration-absorption (1,000 ppm), carbon adsorption (10 ppm), and thermal incineration (10 ppm). The technologies were evaluated using the above emission levels. The economic penalty for installing and operating a thermal incineration at 10 rather than 1,000 ppm is small. This is not the case with carbon adsorption and a meaningful economic comparison of this technology can only be made when it and competing technologies are evaluated at the same emission concentration level. Carbon adsorption is not as cost effective as thermal incineration when both are compared at 10 ppm. Thermal incineration offers the best means for control of benzene vapor to levels of 10 ppm benzene. Various techniques, including adsorption, photo-catalytic oxidation, and biofiltration, were developed to eliminate benzene from the air because of its threat to human health [19]. Its occupational exposure guidelines include a TLV of 0.5 ppm (8 h workshift).

2.3 Naphthalene

Only a few studies have been done on specific areas mentioned in several reviews [20]. Studies conducted on Portuguese college students demonstrated that concentrations were higher inside than outside the environment. Furthermore, an evaluation of work on naphthalene ($C_{10}H_8$) in Europe countries from 2007 to 2017 concludes that outdoor environments contain elevated concentrations [20]. There is a requirement for further research on airborne naphthalene, such as in Africa, which will allow for large-scale assessments of the substance's global distribution.

The EPA classifies naphthalene, the fundamental polycyclic aromatic hydrocarbon (PAH), as the most vaporous pollutant in the semi-volatile organic compounds (SVOC) category. Naphthalene was categorized may be human-hazardous by international agency for research on cancer (IARC). EPA prescribed limit of naphthalene in the air is 3 $\mu g/m^3$. A prolonged possibility of unusual nasal cancers in male rats supports this likelihood of human carcinogenicity [21]. The WHO is contemplating the progress of instruction for naphthalene in indoor air. It moves between the gaseous and particulate levels in the air depending on climate, rainstorms, and atmospheric features. Because naphthalene is photodegraded through hydroxyl radicals, its half-life in the environment is 8 h. High exposure to this compound involves being close to naphthalene-emitting resources [22].

Its primary anthropogenic sources are the burning of biomass, oil and gas burning, the chemical industry, and insect repellents. Emissions from motor vehicles indicate the most vital source in town areas. The production of mothballs, creosote impregnation, oil refineries, the cooking of plant life, and the production of pesticides and surfactants all result in occupational exposure. Cigarette smoking significantly increases micro-environmental exposure to naphthalene [23]. Naphthalene occupational exposure guidelines include a threshold value (TLV) of 50 $\mu g/m^3$ and a reference exposure limit (REL) of 75 $\mu g/m^3$ for a 15 min vulnerability. Utilizing metallic packaging bags and sorbent enrichment are just a few methods that can be used to pattern risky and SVOC in the atmosphere. These methods are effective, but they require many steps. Solid phase micro-extraction (SPME) is one immensely environmentally friendly method for exemplifying hazardous materials in the air. SPME consolidates examining and pre-convergences in one part and then desorption straight hooked on the analytical instrument. In the most recent discovery, a new choice machine for cold fiber (CF) SPME specimens in atmosphere air was used to generate naphthalene standards through permeation [24]. Determinations were made using GC/MS following straight desorption into the GC injector. The optimized method for evaluating naphthalene in environmental air samples is more straightforward and sufficiently sensitive.

Because transmissions are from numerous sources, including the chemical and essential mineral commercials, gas and lubricant burning, biomass burning, mothballs, smoking, fumigants, deodorizers, and many others. It is a poisonous atmosphere contaminant that can be present in both indoor and ambient air. It is frequently accustomed halfway in the manufacture of pesticides (14,000 tons), surfactants (27,000 tons), and cyclic dicarboxylic anhydride (66,000 metric heaps) [25]. It is too present in different conditions; e.g., 40% of national precedence catalog sites constitute naphthalene in soils [26]. In addition, naphthalin, naphthene, naphthaline, tar camphor, white tar, moth flakes, and mothballs are all names for Naphthalene. Decorator, also carbon, mighty 150, and mighty RD1 are all trade names for naphthalene [26]. It is the most hazardous member of the PAH family because of its bicyclic aromatic structure. In studies of VOCs, SVOCs, and PAHs in the environment, naphthalene has served as a target compound.

2.4 Perchloroethylene

Perchloroethylene (PEC), also called tetrachloroethylene (TEC), is an extraction procedure (EP) found in the atmosphere with a lifetime of 70 to 250 days. WHO has comprehensively reviewed 20 studies regarding the presence of perchloroethylene (C_2Cl_4) in interior and outside atmosphere cubicles. The highest concentrations were found, ranging from 0.4 to 355,000 μg/m^3. In general, concentrations vary depending on the season and the area of human activities. With increased knowledge of EP toxicology, the US Environmental Protection Agency (USEPA) suggested that the maximum perchloroethylene in the air is now 40 μg/m^3. It is a 60% reduction from the 1997 recommendation of 100 μg/m^3. PER is significantly applied in cleansing liquid for material in a dry-washing store. Many humans stay after or more than these dry-washing stores. Thus, the query emerged even if or not the PER transmitted would possibly attain certain subjects. Experimental learning is about to be taken on understanding the submission at calculating the convergences of PER in blowout air. This easy process maybe supplies higher records on the current direction than the PER convergences in the interior atmosphere. Various reviews confirmed a particular success after several hours of publicity to a consistent PER assembly, the concentration of PER in a blowout is around 0.4 ~ the attention in breath in the air, and the air sac concentration is half its close-transmission range after regarding fifteen minutes of like as an exposure [27]. It is a chlorinated liquid typically utilized in pharmaceutical manufacturing and solvent extraction in numerous industrial strategies. Metal degreasing is, therefore, the primary source of TEC's release into the environment. Nausea, weakness, headache, dizziness, loss of appetite, and tremors have been symptoms of occupational exposure to trichloroethylene. Nerve damage cannot be repaired, and death has occurred due to acute exposure to high concentrations. Long-term exposures to TEC have damaged the kidneys and liver [28]. IARC has determined that TEC has the potential to cause cancer in humans [29].

It is a vapor solvent utilized in business solvents for dry washing and removing grease. At low concentrations, severe breathing of tetrachloroethylene vapor causes central nervous system (CNS) despair, ranging from light-headedness and lack of muscle coordination to unconsciousness and diaphragmatic paralysis as quantity increases. Minor, reversible episodes of hepatic defect appear many days after unintentional personal vulnerability to tetrachloroethylene calmative convergences. Overexposure is typically thought to lead to CNS slowdown and the possible risks of weekend judgment in work environments. There is less concern about the prolonged consequences of frequent liability to significantly fewer concentrations. One potential issue within commercial persistent vulnerability circumstances is carcinogenicity. Although, when administered orally or inhaled, tetrachloroethylene no longer caused cancer in rats in long-term repeated publicity studies and routes, nor was it previously mutagenic in microorganisms. However, extended verbal management in rats was previously associated with a raised risk of primary liver cancer (hepatocellular carcinoma). Nevertheless, oral transmission to a large amount of a composite nearly inhaled in the habitat would consume the better manifestation of probable danger. Changes in uptake specimen or drench of detoxing and removing trace at exceptionally extreme doses may outcome previously unobservable lethality at convergences firmly akin to commercial emission. Determining tetrachloroethylene's uptake issue in trial wilds can also improve comprehension of the toxicity application and serve as a valuable resource for deciding exposure limits. The study's primary objectives were to look at the pharmacokinetic variables of tetrachloroethylene publicity in mice as a feature of the quantity or direction of the administration. One plausible mechanism for carcinogenesis is the covalent binding of reactive agents to tissue polymers. Thus, it was possible to ascertain the extent and dose dependence of tetrachloroethylene consumption binding to the target polymers. It is a volatile artificial chemical primarily utilized as an intermediate in synthesizing various chlorinated hydrocarbons, even though its use has decreased significantly. Emissions to ambient air, where the chemical is likely to remain for many weeks, have the highest environmental impacts. It is no longer anticipated that 1,1,2,2-Tetrachloroethane will contribute to global warming or stratospheric ozone loss. It disappears unexpectedly from marine structures and will not bioaccumulate. The majority of human vulnerability to 1,1,2,2-tetrachloroethane is through breathing [30]. Its occupational exposure guidelines include a TLV of 1 ppm.

2.5 Toluene

Toluene (C_7H_8) is an aromatic hydrocarbon compound with seven carbons that makes up only 5% of gasoline. It smells sweet and pungent, like benzene. In the environment, vapor-phase toluene will be degraded by photochemically produced hydroxyl radicals; according to USEPA, this response's half-life in the air is estimated to be two days. Due to its use in numerous household and industrial products, toluene significantly impacts the environment [31]. Concentrations, on average, can vary a lot [32]. In 1990, data collected from 11 US cities showed a total of 20 μg/m^3, with a value of 0.23–750 μg/m^3 [33], suggesting that the average time was never mentioned. Edgerton et al. [34] investigated 24-hour mean convergences of toluene in the air in US urban areas and they varied from 30 to 234 ppb and from 8 to 62 μg/m^3. Given the similarity between the pollution profile found in the cities studied and that found in vehicle exhaust, it was once thought that this was the primary source. There was a 24-hour mean of 17 μg/m^3, according to a screening study of 14 US city sites [35]. Toluene quantities of comparable importance were evaluated at locations in the United States, Australia, and the United Kingdom. At six city sites in Canada, mean concentrations (24-hour average) varied from 5 to 44 μg/m^3 [36]; for individual samples, the value of the mean concentration till 145 μg/m^3. At a rural location, the mean concentrations were observed between 3.5 and 5.0 μg/m^3.

It is frequently applied as a benzene substitute and as a solvent in producing products like leather and paints. Toluene vapor inhalation may have harmful consequences on people, including effects on the neurological system and muscular strength. Toluene must be removed from industrial workspaces to avoid cutaneous damage from breathing. Today, various treatment strategies have been used to eliminate toluene from air, including adsorption, equipment, thermal oxidation, and process control. It must be removed from manufacturing workspaces to avoid cutaneous damage from breathing. It is now removed from the air using various conduct techniques, such as adsorption, equipment, thermal oxidation, and process control, thermal oxidation. Its occupational exposure guidelines include a TLV of 20 ppm (8 hr workshift).

2.6 Trichloroethylene

The German Chemical Society's outdoor air research in Germany confirmed the trichloroethylene (TCE) concentration of 15 μg/m^3. The average convergence is 2.5 μg/m^3 in the United States [28], whereas, in Canada, the concentration varied from 0.96 μg/m^3 outdoors to 1.4 μg/m^3 indoors [28]. A more recent actual-time evaluation of TCE (C_2HCl_3) in the outside air surrounding a producing facility in Canada was carried out [37]. They suggested that the highest 30 min assembly of TCE estimated windward of the capability was once 1,300 μg/m^3 and that the most incredible immediate attention was 115,000 μg/m^3. Its occupational exposure guidelines include a TLV of 10 ppm (8 hr workshift).

During its production, use, and disposal, its releases into the environment are evident. Its primary function is as a degreaser during processes of metallic cleaning. Additionally, it is utilized in producing organic chemicals, as a solvent for adhesives, as an ingredient in varnishes and paints, and as a paint stripper. Even though it can be made by one red microalga and by temperate, subtropical, and tropical algae [38], releases from activities that humans do not cause are insignificant. Vapor degreasing operations release most of the TCE used in the United States into the environment. Water treatment facilities, landfills, sewage treatment, and disposal facilities all contain airborne emissions. It has been found in the stack emissions of municipal and hazardous waste incinerators. It may also release into groundwater through landfill leaching and industrial wastewater stream discharges [39].

2.7 1,4-dioxane

There is no current research on 1,4-dioxane ($C_4H_8O_2$) in the air; in any case, posted study has been conducted in the USA, and impacts demonstrated that concentrations went from 0.005 to 0.11 μg/m^3 [37]. Thus, additional research is required before significant conclusions can be drawn. Dioxins are chemical compounds of two C_6H_6 rings coupled through two O_2 atoms. The 8 C atoms presently not clung to oxygen on the rings can form bonds with H atoms or some more components. One of the additional polychlorinated dibenzodioxins has Cl atoms at places 2, 3, 7, and 8 and is a significant hazard. Chemically, this isomer of 2,3,7,8-TCDD is the more stable and harmful complex. Because it is indissoluble, dioxin cannot be dissolved by rainwater and remains in the earth, where it can puncture fatty tissues and build up. One-crystal XRD crystallography can be used to determine dioxin's molecular shape. The crystal form of TCDD is also determined by the molecular crystal of PAHs [40]. It is a contaminant from voluminous manufacturing utilized as a fluid stabilizer in producing and developing thread, papers, materials, automobile air filters, cosmetics, and shampoos. It is widespread in waters of the environment at ranges above aqua capacity standards and regulations. Succeeding the 2015 finding of excessive stages in the Cape Fear River Basin, a significant North Carolina water shed, dioxane has attracted considerable communication awareness in the United States. Dioxane is likely cancerous for humans, earlier categorized on the CCl_3 and now on the CCl_4. 1,4-Dioxane is employed as a solvent. Vertigo, headache, drowsiness, malnutrition, and irritation of the eyes, throat, nasal, and respiratory have all been reported following sharp acquaintance to the dominant capacity of 1,4-dioxane inhalation.

Additionally, it might irritate the body. Long-term (incurable) fragility to the kidneys and liver was discovered in mice water consumption. The number of malignancy cases found was not different from the anticipated number of cancerous failures in three epidemiologic researches on employees transmitted to polychlorinated dibenzodioxin. Animals that were exposed orally have been found to have tumors. The EPA has grouped polychlorinated dibenzodioxin as a possible humanoid carcinogen in Group B2. The EPA Air Toxics Hazards Summary states that polychlorinated dibenzodioxin is a transparent, neutral, and ethereal smell. The one in one million (10^{-6}) hazard level for its presence in drinking water is 0.35 μg/L. The flash point is 55°F (12.8 °C). It is soluble in water and slightly denser than water, and vapors are heavier than air. It is susceptible to peroxide formation by autoxidation. Its occupational exposure guidelines include a TLV of 90 μg/m^3.

It has been classified as an "emerging contaminant or contaminant of concern" following investigations into disease caused by 1,4-dioxane at legacy sites. In the past, closed dry-cleaning centers were re-examined for this poisonous, volatile, and water-soluble contaminant because 1,4-dioxane was utilized as a solvent stabilizer for chlorinated solvents like trichloroethane. Even though dioxane is labeled as a volatile compound with a boiling point of 101 °C, it is not well suited for standard VOC analyses like purge and traps. Because of its high polarity, GC-MS is difficult to extract using traditional liquid-liquid partitioning methods. The analytical chemist will have difficulty obtaining the regulatory authorities' sub-ppb reporting limits.

2.8 Acrolein

Acrolein, also known as propenal, is a rising chemical that belongs to the aldehyde family. With the chemical method C_3H_4O and a mean life expectancy of 15 to 20 h, it is the easiest unsaturated aldehyde. Many examine have measured acrolein in the air, probably because there needs to be a well-known evaluation method. It raises questions about the dimension's uniformity and dependability, and solutions. Nonetheless, the accessible studies are condensed and unveiled in a table, most of which are from the United States. Its convergences were typically more extraordinary and generated by the number of biogenic works in the field and the weather. According to Robichaud [41], research demonstrates that the magnitude in urban habitats exceeds the specified limit for humans 27% of the time. Oxidizing chemicals generate acrolein in the air, similarly 1, 3-butadiene, a crucial component of automobile exhaust, and the insufficient combustion of natural fibers. According to EPA, approximately 75% of the ambient acrolein comes from cells, the rest from agriculture, commercial processes, cigarette smoke, and conflagration. The EPA anticipated average acrolein concentrations of zero using 1996 emission data as a model for ambient air concentrations of 0.119 μg/m^3 (all over the country), and California has 0.157 μg/m^3, which is appreciably above the EPA reference awareness (RfC) of 0.02 μg/m^3. According to a recent theory, California's ambient air contains acrolein at 0.36 μg/m^3, more than twice as much as the EPA estimate. In 2004, 2005, and 2006, according to the California Air Resources Board (CARB), the statewide standard ambient concentrations were 1.21, 1.37, and 1.35 μg/m3, respectively, and based on instances of air monitoring of acrolein utilizing canister sampling method, it is 0.3 μg/m^3. Even though the modeling approximates and quantification aren't precisely the same, they all show ambient acrogenic ranges exceeding the environmental protection agency's request for correction (EPA RfC) [42].

It is the best but also the majority uncontrollable aldehyde with α, β- unsaturated construction. It has a horrible odor at room temperature and is a flammable liquid, colorless to yellowish. Because of its numerous transmission path and greater lethality, the EPA classified acrolein as a high-priority poisonous chemical. Acrolein's toxicity is often shown by reacting with histidine and lysine excess in as deoxyribonucleic acid (DNA) bases as well as proteins or peptides. These reactions lead to the depletion of glutathione (GSH), covalent changes, protein aggregation, denaturation, oxidative damage, and intervention with body cell signaling routes. It causes damage to the nerve cell and may play a part in the enlargement of Alzheimer's disorder, Parkinson's illness, mini-attack, apoplexy, and Sicca affliction. Acrolein can also lead to more diseases like coronary heart disease, juvenile-onset diabetes, respiratory inflammation, and testicular dysgenesis syndrome by programmed cell death, which reduces endoderm formation; WHO estimates that 7.5 μg/kg acrolein should be consumed daily. Nevertheless, acrolein from edibles, such as pastry, tuber products, cheese, alcohol, vegetables, and fruits, is frequently well above this maximum [43]. Its occupational exposure guidelines include a TLV of 0.1 ppm. Novoselov and Filatov [44] looked into how little acrolein vapor decomposed in air is exposed to pulsed electron beams.

2.9 Acrylonitrile

Acrylonitrile (AN) is a newly synthesized compound with the symbol CH_2CHCN that belongs to the amine group. It has a radical vinyl group attached to a nitrile in its molecular structure. It is toxic and reactive by less amount and stays in the atmosphere for a standard of 5.6 daytime. The research about CH_2CHCN needs to be updated, so there may be a need for more modern research. It is a neutral, vaporous, and combustible solvent with an inherent smell. It dissolves completely in water and can be mixed with most natural solvents. Its level is more than 99% pure, with some stabilizers and minor impurities. It is no longer listed as a natural substance. It is frequently used as a chemical intermediate and produces waxes, rubbers, acrylic, and modacrylic fibers. It could also enter the environment when waste is made, processed (completed), handled, stored, transported, and abandoned. In addition to emissions from manufacturing, there are too many environmental losses from injuries and inadequate protection. In 1974, it was anticipated that plants in the United States would emit 14,100 tonnes or 2.2% of the manufacturing total. Following the implementation of particular emission management in the United States, more recent estimates reveal a conventional reduction in emissions and a pattern shift: in 1981, 800 tons were used to make acrylonitrile, while 3,000 tons were used to make quit products. In 1981, 41 plants in 10 western European countries were expected to emit 4,970 tonnes of emissions, with Europe's proposed emission constituent for producing acrylonitrile 0.2–0.5% for processing acrylonitrile.

Under the aliases Acrylon, Carbacryl, and Ventox, Acrylonitrile is extensively utilized as a fumigant for wrap-up cereals and grains. Otherwise, this is used to make synthetic rubber, plastic fiber, and herbal syntheses, among other things. It is a carcinogen and mutagen and can be a severe eye or skin irritation that is very active. It is completely absorbed through the mouth, intact skin, or inhalation. A portion of the substance dissolved and liberated cyanide when inhaled. Its relative toxicity can be determined quickly using this method. The final component is said to have been altered into organic acid. The American conference of governmental industrial hygienists (ACGIH) recommended a lower limit charge is 2 ppm. Its high toxicity necessitates a quick method for restoring its strength at ppm levels. It has been determined using various forms, including polarographic, gas chromatographic, titrimetric, and spectrophotometric, each of which has its advantages and disadvantages [45]. TBAB is the one of the biofilteration technologies for AN control.

2.10 1-3-butadiene

1-3-butadiene (BD) is a new compound that belongs to the alkene group and has the components $(CH_2 = CH)_2$. Before 2000, butadiene distributions in ambient air were measured worldwide in indoor and outdoor booths. Preferably, butadiene concentrations were below occupational ranges of 2 ppm. Research regulated an eight-year observing work from 2008 to 2015 and recorded a yearly mean of BD that was identical to 0.33E-9 ppb [46]. The outcome acquired through research [47] established that the BD was 0.17 ppb, higher than the E-5 lifetime maximum cancer level set by the EPA at 0.14 ppb. The excessive degree was previously attributed to the range's extreme density of petrochemical services. To comprehend the chemical's spatial distribution, additional research is required to quantify the concentrations of BD in ambient air from numerous global cities. The problem with BD isn't a current one when it comes to environmental network situations. BD was added to the dangerous air pollution list maintained by the EPA. It is in the top 33 in the harmful liberate list and 36th out of the 50 most manufactured chemical additives in the Occupational Safety and Health Administration (OSHA). Biomass has been widely used in manufacturing since the 1930s, even though the utterly natural structure of BD has derived from biogas burning at some point throughout wildfires in wooded areas. Although BD is explicitly used in the production of polymers, this can occur in abundance in cooking oils, fungicides, and smoke. Outside convergence typically scales as little as 1–10 ppb in village places, with better culmination nearer transmitting commercial offerings, even though referred-to emissions vary and may benefit suitable 3,000 tons per month in the United States [48]. Its occupational exposure guidelines include a TLV of 2 ppm (8 hr workshift).

BD, a colorless gasoline, is a critical petrochemical manufactured in massive amounts and is essentially utilized to supply thermoplastic resins and synthetic rubber. Because it is formed through the imperfect combustion of biogas and non-renewable energy, it is a ubiquitous environmental pollutant. It is mechanically monitored VOCs in the air. It was identified as a human cancerous in 2002 by the EPA and in 2008 by the IARC. It causes lymph hematopoietic cancers in people exposed to it in the workplace. EPA states this is specific to 187 HAPs, also called air toxics. They once thoroughly examined it in volume 97 of the IARC articles on assessing human carcinogenic threats published in 2008. They noted almost every BD factor in exceptional detail in the files. In 2012, IARC issued volume 100 F and made a modification available on BD based solely on newly accessible documents.

Furthermore, in 2007 and 2010, they examined BD's metabolism, oncology, pharmacology, nontoxicity, and carcinogenic modes of action. Even so, numerous new studies, particularly epidemiological studies involving institutions with several childhood illnesses, came to be issued, yet cannot be reviewed. As a result, we will discuss the most significant developments in BD-related research over the past decade in this evaluation. However, we only attempt to include some published BD-related literature. In addition, since benzene is one of the essential components of tobacco smoke, we dispute its potential contribution to tobacco-related diseases, with lung and laryngeal cancer as the primary focus [49]. The biotreatment of 1,3-butadiene in air stream using a reactor technology was studied by Chou and Lu [50].

2.11 Chloroform

Chloroform, also known as trichloromethane, is a new chemical that belongs to the halocarbon family and has the formula $CHCl_3$. Its concentrations in the air were found to frequently exceed or approach regional levels in a tracking study conducted between 2009 and 2013 [51]. More contemporary research on chloroform in the atmosphere of six Norwegian swimming baths revealed a range of 127.9 ± 33.1 to 272.5 ± 77.2 μg/m^3 [52]. This point out that chloroform was widely recognized in the air of swimming pools. There hasn't been much research done on this growing pollutant. The production of pharmaceutical products, dyes, insecticides, CHCl, and fluorocarbon 22 (CHClF) are all examples of business strategies that use chloroform (CHCl). Additionally, it is accidentally produced in paper generators, landfills, and sewage treatment plants. According to an EPA study conducted by the National Response Center (NRC), chlorination treatment may also prevent excessive levels of chloroform in drinking water, which were found in 80% of the towns sampled. In addition, the NRC report emphasized the danger to human health posed by prolonged exposure to CHCl in water, foods, and the air. Due to the numerous identified most significant and secondary anthropogenic sources, the presence of CHCl in the environment has frequently been regarded as artificial. Evidence suggests that some or all of the chloroform in the ecosystem may have come from nature. Discussing the air quality and the health effects that come with them has prompted several studies into the prevalence and immensity of VOCs in interior environments. For instance, it finalized a study on the concentrations of numerous VOCs, including chloroform, in 12 Ontario homes. The concentration of chloroform in the drawing room air was frequently as high as five parts per billion (ppb), although they had not identified the specific source. The Total Exposure Assessment Method (TEAM) research executed by the USEPA between 1979 and 1985 revealed that chloroform concentrations in personal air are five to twelve conditions greater than those in outdoor air [53]. In the past, they expected one zero. The inhalation route accounts for 33% of all exposure to chloroform [54]. Although previous research has established that notable interior sources of chloroform must exist, they cannot thoroughly investigate these sources. TBAB is the one of the biofilteration technologies for chloroform control. Its occupational exposure guidelines include a TLV of 10 ppm (8 hr work shift).

2.12 Dichloromethane

Dichloromethane (DCM), also known as methyl dichloride, methylene bichloride, and methylene chloride, is a natural compound with an open chain that is halogenated. It shows that concentrations have typically been reported at shallow levels, in contrast to the recommended limit. In North American homes, the typical interior air concentration ranged from 0.4 to 3.5 μg/m^3 between 1990 and 2005 [55]. There were no discernible patterns between the interior and outside compartments. The anthropogenic intensity and seasonal deviations influence the number of dichloromethanes in the air, but concentrations were higher in the icy conditions than in the summer [56]. Its occupational exposure guidelines include a TLV of 100 ppm.

It is an organic solvent that has been abundantly utilized in both consumer goods and industrial settings. They liberated the amount of 3.4 million pounds of methylene chloride gas into the habitat in the United States in 2012. Due to its volatility, it is mainly present in the air, and most people are exposed to it through inhalation. According to the EPA, in 1996, outside air had a mean concentration of 0.47 μg/m^3. Utilizing consumer goods containing dichloromethane, like adherent, spray shoe furbishes, adhesive removers, building matters, and furniture equipment furnishings can result in exposure to dichloromethane through inhalation in the indoor environment. Inside air concentrations collected from densely populated, village, and rural homes in the middle of 1990 and 2005 ranged from 0.4 to 3.5 μg/m^3 [57]. It has a high solvent for lubricants, fats, and a few plastic materials. Due to its volatility and boiling point of 40.1 °C, it is extensively applied as an industrial solvent in the manufacture of chemicals and pharmaceuticals and, to a lesser extent, as a food extraction solvent for cleaning metal and paint. It has also been used in polyurethane (PU) foam blowing, aerosols, paint strippers, and as a laboratory agent. It is also a component of special adhesives. Many of these uses can cause a lot of the DCM used to be released into the environment, called "emissive uses." The production of HFC-32 (CH_2F_2, difluoromethane) has recently included the use of smaller quantities of DCM as a chemical feedstock; however, the use of DCM as a feedstock does not result in the emission of DCM, except for small losses (fugitive emissions).

Anthropogenic emissions have significantly augmented the natural flux. Analysis of the first air from Antarctica suggests that atmospheric cognizance was once relatively normal at 1.5 pmol/mol in the early twentieth century, a cease result of the natural sources noted above. However, it proliferated from 1950 to reap a southern hemisphere maximum of 9.4 pmol/mol in 1990. Its model is compatible with changes in manufacturing and discharge estimated from trade data in the latter part of the twentieth century. It has been investigated by many groups using a variety of catalysis in oxidative environments.

2.13 Ethylene oxides

Ethylene oxide is one of the epoxides and has a 69- to 149-year lifespan. Its chemical formula is C_2H_4O. They are antiquated, and additional contemporary research on this EP and a re-evaluation are in order. The simplest cyclic ether is ethylene oxide (oxirane). It smells sweet and etheric and is a colorless gas or liquid. It depicts the ethylene oxide molecule with its strained angles and short C–C bond. The structure's theoretical calculations and experimental findings are in good agreement. Ethylene oxide is very responsive because its tightly packed can be easily opened. It may be the most-flexible compound center. It is also a dangerous compound involved in numerous severe incidents due to its toxicity and reactivity. To safely handle it, one must have a good understanding of its properties. Its occupational exposure guidelines include a TLV of 1 ppm (8 hr workshift).

WURTZ first described ethylene oxide in 1859. To make it, he used a potassium hydroxide solution to remove hydrochloric acid from ethylene chlorohydrin. Based on WURTZ's discovery, the chlorohydrin process began industrial production in 1914. It production and significance have steadily increased. The direct catalytic oxidation of ethylene was discovered in 1931 by LEFORT, gradually replacing the chlorohydrin method. Presently, ethylene oxide is produced by oxidizing ethylene directly with oxygen or air; its worldwide production capacity is approximately 15,106 t, which makes it a significant industrial chemical. It is utilized as a fumigant, disinfectant, and sterilizing agent. Ethylene glycol, its most integral derivative, is found in antifreeze (car radiators). An interstellar source of ethylene has been found to have some endogenous formation in humans. Natural gas, cigarette smoke, and diesel exhaust contain it, making it an atmospheric pollutant. Even though it is rarely found in nature, its action in biology may significantly affect the evolution of DNA and RNA in entire body systems. Catalytic oxidizers and scrubbers are the mostly widely accepted control technologies available to control ethylene oxide emissions.

2.14 Furan

Furans are a poisonous organic material. Furan (C_4H4O) is a colorless, volatile liquid with a shallow boiling point, the same as at room temperature. Furans are aromatic heterocyclic compounds with a five-membered ring, four carbon atoms, and one oxygen atom. The worldwide ecological influence of hazardous compounds is based on several assets, including atmospheric pressure, heat condition, and climate and topography factors that influence the accumulation. Understanding the reactions that produce furan's derivatives and the speed with which these viable products are eliminated from the environment is fascinating. Response with the OH radical, which primarily begins reactions with volatile organic compounds via hydrogen extraction from a C-H bond, will likely be one of the five most oversized sinks of furan derivatives. Its occupational exposure guidelines include a TLV of 2 ppm.

One of the main human-caused origins of dioxin emissions is incineration. The combustion process throughout the procedure is strongly tied to the design itself. According to studies, fly ash and flue gas from incineration and combustion operations, as well as polluted soils from industrial and other origins, are the significant ways furans enter the environment. Carbon adsorption is one of the furans elimination technologies now in use for treating flue gases coming from incineration plants. Furans in stack gas were subsequently eliminated using the fixed bed adsorption technique.

3 Bioaerosols

Bioaerosols, also called organic soot, are tiny substances from living or dead plants, animals, and microorganisms involving hazards from pathogens, viruses, parasites, and mycotoxins, and their lipopolysaccharide (LPS), allergens, antigens, and so on. Coughing, sneezing, talking, and other human behaviors can release bioaerosols into the air. Annually 1,000 Tg of bioaerosols is released, according to estimates. Not many studies have discovered that whole infectious cell convergences of bioaerosol calculated utilizing unconventional procedure mean around 5 × 107 cells/m^3 [58]. In containment modern homes, audits have demonstrated certain negligible and most capable culture convergences of bioaerosols are for bacilli:104 and 107 CFU/m^3, for gram-negative bacteria:102 and 104 CFU/m^3, respectively, and fungus:102 CFU/m^3 and 105 CFU/m^3 consequently [59]. Nonetheless, additional research is required before significant conclusions can be drawn.

4 Particles

Particles indicate that particulate matter has been a pollutant in the air since the beginning of time. These particle sizes and behavior have been extensively studied. In any case, many wellsprings of particulates in the air deliver unprecedented fragments, generating them to be viewed as rising granules and grouped as EPs.

4.1 Ultra-fine particles

Usually, ultra-fine particles (UFPs) are clouds of dust, but they differ from added fragments because of their dimensions and effects. UFPs have an aerodynamic diameter of fewer than 0.1 μm (100 nm), an entire area per mass of 100, and an ambient air matter integer of 1,000,000. Because of their small size, UFPs can persist and accumulate in the air for a period and may be carried above lengthy spaces. UFPs are excellent air adsorbers of poisonous natural compounds or transition metals due to their high surface area. As a result, UFPs may pose a double threat to humans. The majority of essential products of combustion, including those from printing or photocopier, cooking, biofuel combustion, and mobile supply emissions, comprise ultrafine particles. Ultrafine particles in the air do not have any established limits or regulations. However, depending on the work of the newest science literature, a blank paper for policymakers subject to authority point of view organized by the "Thinking Outside the Box" group concurred on what comprises below versus above concentrations. They proposed that ultrafine particles need to be less than 1,000 particles/cm^3 (24 h mean) in an environment that is smooth and unaffected by anthropogenic emissions and with a city background, 10,000 particles.cm^{-3}/24 h.

4.2 Microplastics and nanoplastics

Microplastics (MPs) and nanoplastics (NPs) are tiny plastic particles with a dimension range between 0.001–0.1 μm and 0.1 μm – 5 mm, correspondingly, primarily obtained by the breakdown of significant polymers are the primary challenge and have garnered enlarging interest from researchers and the publishers. In the atmosphere, Microplastics and nanoplastics are firstly brought from direct manufacturing for a variety of commercial and commercial needs or secondly from the dissolution of large polymers materials, which can also present in the surrounding environment and can also be flattered into the atmosphere by the windblown in the dirt, among other things. Matter size and weather aspects like air pressure, temperature, wind (flow and point of the compass), rainstorms, and precipitation control the issuing and destiny of microplastics and nanoplastics in the atmosphere.

Microplastics and nanoplastics in the air do not currently have any regulations. Generally, research on the presence of nanoplastics in the atmosphere has yet to be done, but a few studies on microplastics have shown that they are present in high concentrations. All convenient examinations have articulated significant microplastics groupings in placed highway soot connected to human activities (commercial, business, residential, and heavy traffic).

4.3 Black carbon/diesel exhaust particles

The atmospheric aerosols known as black carbon (BC) and diesel exhaust particles (DEP) are the by-products of combustion processes and significantly affect human health and Earth's climate. These substances range in size from little nanometers for air aerosols to little centimeters for charcoal pieces from combustible vegetation products. The convergences and consequences of BC are the subjects of in-progress research, though hardly any works have been recorded in recent years. It examined the concentration of BC in the air in the Central Himalayas, India. Occasionally, BC mass concentrations had been more noteworthy over spring, which is credited to the uplift of contamination starting at the basin (underneath the mountaintop) by utilizing the relocation limit and raised normal discharges. Due to more extensive natural gas utilization and the below partition surface of the atmosphere, they received the most excellent attention during the winter. Generally, the measured average concentration was 9.5 μgm^3 and negatively correlated with air current and degrees. The evaluated studies typically demonstrated that, in addition to human activities, seasonal variations and meteorological factors control the concentration of carbon black in the atmosphere. Diesel exhaust particle convergences in ambient air are poorly reviewed compared to black carbon.

Using diesel-powered tools in underground mines suggest that diesel exhaust particles reached a concentration of 5,570 μgm^{-3} over 8 h of time-weighted. Although new modifications to diesel engines have changed exhaust characteristics [60], resulting in significantly lower emissions of diesel exhaust into the ambient air. In general, studies on diesel exhaust particles are used to measure exposure to diesel exhaust.

5 Source of EC

There are various types of sources, which are given below:

5.1 Incineration sources

Reducing the number of waste hospitals produce is the best strategy for addressing the issue. An incinerator is an essential approach for limiting the quantity of waste produced. However, the system's flue gas emissions of furans and dioxins are a significant drawback. The combustion processes release contaminants like dioxins and furans into the environment. Dioxins were deposited in soil, sediment, and clay before the industrial revolution due to the plant material combustion during the forest and varying fires. Industrial red-hot materials such as steel, ceramic, and coke are one of many post-industrial sources. The municipal stable waste incinerator produced many dioxins after the predecessors and de novo synthesis protocol. The emergence of dioxin and its compounds from municipal stable waste incinerators has been measured at approximately 50 ng 1-TEQ/kg for the past few decades [61]. Except for high-quality technologies, hospital waste is also dealt with by incineration. As a result, the ecosystem was contaminated by many dioxins when chlorine-containing products were burned. Dioxin is produced by burning hazardous waste, which is oxidizing, explosive, extremely inflammable, infectious, and mutagenic. Sewage sludge is the solid waste from treating wastewater with toxic natural pollutants. The limitation of the landfill disposal method is also dealt with through incineration. As a result, it also has a lot of dioxins [62].

5.2 Industrial sources

The use of harmful material as a gas contributes significantly to releasing dioxin and furans into the air when clinker is burned. Wood burning was found to be the primary source of Furan emissions in the air in a wide range of emission studies [63]. One appraisal that gives alluring records is dioxin discharge after wood copying was found as 945 g I-TEQ in one year [64]. There are a lot of phenolic and chlorinated compounds in waste goods from pulp and paper industries, as wood pulps release dioxin into water, land, and composition. According to a Chinese document, a paper company produced 300 pg/11-TEQ [65]. Metallurgical businesses, including several processes, like the recovery of scrap metal and the smelting process, are the primary origin of dioxin and related congeners. These toxic congeners are also obtained through iron ore sintering. Global dioxin production is estimated to be between 500–400 g I-TEQ annually.

5.3 Reservoir sources

Dioxin and furan are water-insoluble, stable compounds that accumulate excessively in sediment, soil, waste, and organic and vegetation materials. These materials can be recycled and distributed again in the environment. Dioxin and furans also have sources from biological processes. Dioxins and furans, two highly toxic compounds, are released into the atmosphere when chlorinated phenolic compounds are catalyzed by microorganisms [66]. Thermal oxygen-slicing steel at destruction sites, the Kraft Liquor Boiler, the combustion of workshop waste tires, and carbon activation amenities are all excluded [67].

6 Control techniques

The two approaches most often used to cut down on VOC emissions to the environment are as follows:
- Installing “add-on” control technology to capture or eliminate off-gas pollution. In addition to the add-on equipment, equipment is often needed to catch the emission.
- Modifications to a method and raw material to stop or lessen pollution production through the operation.

6.1 Add-on control equipment

The four main kinds of control equipment for preventive emission of VOC are as follows. Their foundations include condensation, adsorption, absorption, and combustion.

6.1.1 Combustion

Flares, process heaters, boilers, and catalytic and thermal incinerators are examples of combustion control equipment. VOC will be quickly and exothermically oxidized to produce H2O and CO2. Fuels and VOCs comprise C and H, wholly converted to CO2 and H2O when burnt with oxygen.

- Thermal incineration

By oxidizing volatile organics to CO2 and H20, incineration eliminates them. Any VOC heated to a high adequate temperature while exposed to oxygen will burn or oxidize. According to studies on thermal incineration performance, new incinerators employing modern methods may destroy VOCs by 98% or have an exit concentration of 20 ppmv [68]. Almost all producers of volatile organic emissions, such as solvent operations, reactor vents and kilns, may be reduced using this technique.

- Catalytic incinerators

In catalytic incinerators, catalysts make the oxidizing process possible at a temperature lower than that needed for thermal oxidation. Ninety-five percent of VOC destruction effectiveness may be attained by a catalytic proceed at around 840 F with only a catalyst bed volume from 0.014 to 0.057 m3 and off-gas running by the apparatus at a rate of 0.47 scm/sec [69]. Various stationary waste streams from different sources have been subjected to it.

– Industrial boilers and process heaters

It is capable of destroying VOCs. Like an incinerator, a process heater or boiler's combustion performance is determined by the average furnace temperature and residence duration. Both technologies are used in situations with a high potential for vent stream heat recovery. They are great control tools because, in most cases, they can destroy VOCs by at least 98%.

– Flares

A flare is a combustion control apparatus that provides a cost-effective solution to dispose of vast volumes of gas released suddenly safely. They are often used to burn waste and purge materials from the chemical and refinery industries. The EPA has determined that 98% combustion performance can be attained based on many flares' combustion empirical studies [70].

6.1.2 Adsorption

When a scrubber is unproductive or uneconomical, it may be utilized on waste-gas streams with low VOC concentrations. The ability of carbon to adsorb different organics is not constant. Typically, the relationship between volatility and adsorption capacity is inverse. All organic vapors are first evenly adsorbed. Typically, organics that have been desorbed from the carbon are condensed and either recycled right away or reconditioned. For this aim, moving bed adsorbers have been used in Japan and Europe [71]. Carbon adsorption can remove VOCs with an effectiveness of greater than 95%.

6.1.3 Absorption

Long contact durations and much absorbent are needed to remove modest organic contents in waste streams efficiently. As a result, it is often more costly than incineration or adsorption. It is appealing if a substantial amount of VOC can be collected and the recovered VOC can be put to good use. Typically, it is not measured if the VOC is between 200 and 300 ppmv [72].

6.1.4 Condensation

It is frequently used in conjunction with other air pollution control technology as an emission control strategy. Activities for ample value organics, including the reclamation of hydrocarbon vapors from gasoline marketing activities, are highly suitable for refrigeration systems. Condensers are often used to effectively lower coal tar and organic pollutants from refining petroleum. Condensers are run differently to help from 50% to 90% [72].

6.2 Various control techniques

6.2.1 Modifications to the procedures or raw materials

The diversity of modifications that may be performed to a procedure or its raw materials to minimize the emission rate is the controlling "option" that is the most challenging to talk about. According to reports, this has increased print quality and decreased ink use, adding value to the printer and positively impacting the environment. The opportunities are limitless, and a creative examination of the manufacturing process must be conducted to identify the potential for improvement.

6.2.2 Substitution of organic compounds with less photochemically reactive ones

According to studies conducted after the Los Angeles regulation was passed, almost all organic chemicals react photochemically to generate ozone according to the policy. Halogenated organics, which have very little photochemical reactivity, were most of the exclusions listed in the regulation.

7 Episode of elementary carbon

The analysis of elemental carbon in trace amounts in atmospheric matrices has been made possible by technological advancements in analytical instrumentation. As a result, more research has been done on evaluating and distributing these compounds in the environment in the last ten years. In studies adopting passive and active sampling, the air has been analyzed for over 170 distinct EC. Plasticizers, surfactants, phencyclidine (PCP), illicit drugs, and licit are among the most frequently studied rising contaminants. Most of this research describes concentration points as the entirety of the particular matter (PM) and vapor concentrations, while the rest of the studies refer to the PM or vapor concentrations separately. The current analysis differentiates between PM and vapor phase concentrations to provide a detailed evaluation of the atmospheric concentrations. Most studies on the prevalence of elemental carbon in ecosystems have been conducted in the northern hemisphere, specifically on the continents of Europe and the United States. Only two measures have been implemented in Africa: one in Algeria and the other in South Africa. Oceania also needs more information. Just two works expressed fixations in Australia and New Zealand. The research was made on maritime expeditions in one-of-a-kind seas and oceans, while various researchers have learned low pollution in very far remote zones, like mountain levels in the United States and the Alps [73].

8 Appearances of acquaintance valuation

An extensive condition of environmental fitness is preventing or reducing exposure of airborne chemicals that commit to increased disorder or disability. Evaluating mortal hazards to airborne VOCs has to establish an appraisal approach, selecting convenient sampling capacity, and excellent analytical manner. The assessment of mortal exposure to environmental VOCs combines an approximate explanation and an assessable estimate of exposure among a body and the synthetic assistant, authentic as exposure and enter the body, supposed as dosage.

Quantifiable evaluation of airborne VOC exposures is frequently a fundamental character of natural vulnerability assessment studies, bridging the gap with well-being possessions. There are two methods of measurable exposure assessment: Direct determination, along with biomarker evidence (biological monitoring); and indirect review, including airborne VOC evidence (environmental monitoring). This general binary access to commensurate exposure evaluation is autonomous and reciprocal. Individually trusts contrasting types of info and acquires distinct durability and deficiency. Therefore, we can check our results' flesh using multiple approaches. Although the eventual goal of the human liability schedule is to assess exposures from a significant perspective to concede the founding of dose–response communication, hypotheses based on preexisting mortality or preexisting mortality, there is room for subjective evaluation inside basic studies to formulate Based on the WHO Incidence Record.

Acquaintance assessment, including one and the other alien (air) and constitutional dosage (biomarkers), is an indispensable part of the inconclusive analysis. The exposure estimate for coincidental well-being research is well established, but this appraisal can be complicated and investigates contrasting attitudes toward people's exposure to toxicants. Precise and decent acquaintance-associated information is vital for making a knowledgeable agreement to protect and promote communal well-being. The demand for additional comprehensive and high-quality evidence on population exposure to coincidental chemicals is recognized by Sexton et al. [74] on the population's broad people and important subgroups. Exposure assessments should consider the disparate condition of human exposure to air pollutants:

- Exposure often includes various faculty or admixture rather than individual agents.
- The location or region of influence of the coincidental material in the air. It can be an anthropogenic or non-anthropogenic origin and can be a large operation or a clear-cut point that can be immobile or free. The outdoor habitat is a significant cause of toxic chemicals, but a person spends 90% of his time indoors, disclosed to chemical pollutants and poisons.
- Acquaintance to attenuated air can appear in employment or non-occupational, peopled, or non-residential settings. The consolidation (dose), extent, and prevalence of experience tell us the degree of exposure.
- Better compassion for improving eco-friendly exposures to contamination concerns requires considering the broad community, community subgroups, and individuals.

Evaluation of environmental exposures in community subcategories can advise classification differences in susceptibility. For case, children should consistently participate in an ecological investigation. These program desire adaptations of approaches presently used in adult research, advancement of tools, and acceptance of child-specific strategies. A parent-child learning unit-focused design provides adult evidence for connection and allows parents to assist in extra complex consideration construction. Child-friendly devices such as limited air pumps and identification displays are also essential.

There are many trails through which the environment is exposed to noxious materials, such as VOCs. Good human exposure assessment study composition is crucial for accurately estimating humanoid liability to ecological pollutants and associated health outcomes. It is fundamental for primary research. They have often established an attentive design to approve a particular hypothesis.

9 Measuring VOC acquaintance with biomarkers

VOCs protect various chemical grades with disparate biological and physicochemical possessions. The practice of biomarkers enables combined valuation of acquaintance built on interior quantities ascending from breath, cutaneous, and oral ways. It is valuable for beginning links amid physical revelations and potential properties. Ecological and occupational VOCs can arrive at the frame through intentional or accidental absorption, but inhalation acquaintance due to their volatility is the primary route.

It is essential to note that fewer biomarkers can be adequately recycled at the moment for visual assessment of past coincidental exposure of pollutants. In most essential facts, there is an approximate or semi-quantitative index to monitor such risk. Still, as the area advances and awareness of xenobiotic pharmacokinetics and pharmacodynamics progresses, so does the statistic of finitely accepted biomarkers of exposure.

Social exposure appraisal emphasizes the occurrence of pollutants in the atmosphere and their association with opposing well-being. The practice of biomarkers for acquaintance valuation introduces a different viewpoint than outdated ecological dimensions for evaluating humanoid introduction to VOC contaminants. There are unlike biomarkers, and they differ in thought and presentation. Exposure biomarkers signify xenobiotic materials or their metabolites unhurried in sections of products of connections amid mark molecules and xenobiotics. A biological consequence marker is a quantifiable biochemical, interactive, or other alteration in a creature that, contingent on its extent, contributes to a recognized or probable health disorder. Biological susceptibility markers can be identified as pointers of an organism's innate or developed aptitude to reply to an exact xenobiotic acquaintance encounter.

Biomonitoring in people is a device for experts and legislators and can be implemented for the most remarkable chemicals, with VOCs at the heart of discussions for global ecological well-being. Biomarkers have the exceptional advantage of being numerous acquaintance valuation devices. They can recover the risk valuation procedure as long as an essential connection between chemical revelation, interior amount, and well-being impairment. Biomonitoring deliberates altogether absorption ways and entirely applicable foundations, manufacturing it an exceptional tool for risk valuation and organization.

The convenience and circumspection of biomarker research were argued to identify and indicate convenient accurate benefits. Features of an excellent biomarker combine screen liability and the competence to demonstrate and relate holdings to construe the origin and extent of publicity and its emanation. While the biomarker is convenient for determining the well-being of exposure and, therefore, opportunity, more is needed to estimate the holdings. Still, the corporation amid copious biomarkers and the communication between these exposure and well-being conclusions is complex due to differences in susceptibility and perceptivity within individuals. Therefore, biomarkers of sensitivity are required. The functionally optimal biomarker can be approached by employing a set of biomarkers, exclusively providing a definitive property.

VOCs cast-off as outdoor and indoor air character markers include any chemical complexes that can be leisurely for ecological monitoring. Ecological and biological monitoring is not simultaneous; they complement each other, even as biomarkers play a cumulative part in assessing social acquaintance to these complexes. They hang on to the research target, ecological issues, and available analytical approaches. They may favor whensoever conceivable; together, procedures of monitoring and a mutual method are often required. The mainstream biomarker analyses currently obtainable are based on breath samples. The optimal average is contingent on numerous VOC features, such as physicochemical possessions, toxicology, and suitability of sample gathering.

10 Future challenges and directions

Although significant progress has been made in identifying, quantifying, and comprehending environmental issues and developing rehabilitating techniques for increasing environmental pollutants, numerous obstacles exist [75]. The environment contains multiple pollutants that must be identified. It is especially apparent in the recent class of PFAS instructions that are still being determined in the background. These include new PFAS that producers are switching due to the phase-out of perfluorooctanesulfonic acid (PFOS) and perfluorooctanic acid (PFOA), as well as pollutants that have existed in the atmosphere for so many years but have yet to be identified. Similarly, as existing flame retardants, such as polybrominated diphenyl ethers, are outlawed, new ones are added. The results from innovative environmental strategies focusing on cleansing the atmosphere and may continue to be a shifting goal in the future. To conclude, the chemical composition of water changes over time. Hence, it is to re-evaluate the engineering solutions and the impact of cleaning methods for such water. In addition, contamination from the primary source is not the only issue. These contaminants frequently change the environment, drinking water, or wastewater treatment into new chemical compounds, which are extra dangerous than their parental composites. As a result, a constant demand has been raised for non-target plans, such as high-determination mass spectrometry techniques that race up the authentication of unknown compounds by assigning empirical formulation data for the particles and their ions. Once discovered, original analytical methods will be required to quantify these novel transformation goods in the environmental matrices. It is an overwhelming undertaking, given the sheer figure of synthetic substances that ought to exist in the surroundings, which variables to essential to work on a method for focusing on this extensive posting of compound substances or lead to the chance of employing group (bunch) estimations to concede a supplementary significant complete component of their existence in the atmosphere.

This group measurement presently includes the total oxidizable precursors (TOP) analysis, which oxidizes the PFAS materials into PFAA, which can be identified and analyzed using LC-MS-MS. The total nitrosamine (TONO) analysis can reduce nitrosamines to nitric oxide which can be detected using chemiluminescent exposure, or bromine, iodine, and the total organic chlorine analysis is used to produce iodinated, brominated, and chlorinated disinfection by-products (DBPs) and also use carbon for combustion, sorption, and identification of halides using ICP-MS. Thus, there is a need to innovate new advanced oxidation processes (AOPs) and another technology that can contribute to removing those pollutants. In this aspect, membranes still grip promises, mainly with recent investigations into developing self-healing and anti-fouling membranes. Ensuring that introducing poisonous by-products from these processes does not affect the health of humans or the environment is an essential consideration for these treatment methods, particularly those utilizing AOPs. Combining toxicology research with various terminations that apply to possible effects on humanoid or ecology fitness is critical.

11 Conclusions and prospects

Air pollution is the leading contributor to illness and mortality in the modern world. The number of different kinds and forms of new pollutants has increased due to growing urbanization and industrialization, yet these pollutants are challenging to identify and classify due to their foul odors and complicated generation sources. Environmental research is now seriously concerned about emerging contaminants in urban, rural, or isolated air environments. It led to the publishing of various papers regarding the analysis, existence, and dispersion of chemicals in the atmosphere. The state of the knowledge regarding EC in the atmospheric compartment is covered in this paper. As far as we can tell, the issues listed below are of widespread interest and need further research.

There have been reports of microplastics in both indoor and outdoor settings, as well as atmospheric fallout. The study of atmospheric microplastics is still in its infancy; research indicates that the pollution has been dispersing in the air for many years. And experts believe these particles might all seriously impact the health of people, the ecology, and the climate. However, the field presently contains more questions than answers. Researchers attempt to comprehend the issue as quickly as possible, but the analytical methods required to examine air microplastics are time- and labor-intensive. Scientists are working to enhance these techniques and take more accurate measurements of the amount of plastic in the atmosphere, its composition, and its global movement to understand the possible effects of this pollution.

An exhaustive examination of the literature was carried out to fill the knowledge gap. The discussion includes the compound's toxicity profile, diffusion over space and time, destiny in the atmosphere, monitoring techniques, and health consequences it may have. The information will assist decision-makers in upcoming requirements to add or alter the regulatory obligations on developing pollution.

References

[1] Ahmad S, Singh S, Arfin T, Neeti K. Fluoride contamination, consequences and removal techniques in water: A review. Environ Sci Adv 2022, 1, 620–661.

[2] Richardson SD, Ternes TA. Water analysis: Emerging contaminants and current issues. Anal Chem 2014, 86, 2813–2828.

[3] Zhang Q, Jiang X, Tong D, Davis SJ, Zhao H, Geng G, Feng T, Zheng B, Lu Z, Streets DG, Ni R, Brauer M, Donkelaar A, Martin RV, Fuo H, Liu Z, Pan D, Kan H, Yan Y, Lin J, He K, Guan D. Transboundary health impacts of transported global air pollution and international trade. Nature 2017, 543, 705–709.

[4] Arfin T, Kate P, Singh A, Kumari K. Alternatives to POPs for a healthy life. In: Kumari K (ed.). Persistent organic pollutants: Gaps in management and associated challenges. CRC Press, Boca Raton, USA, 2021, 201–226.

[5] World Health Organization, 1989. Indoor air quality: Organic pollutants. Report on a WHO Meeting, Berlin, 23–27 August 1987. EURO Reports and Studies 111. Copenhagen, World Health Organization Regional Office for Europe.

[6] Parra MA, Elustondo D, Bermejo R, Santamaria JM. Ambient air levels of volatile organic compounds (VOC) and nitrogen dioxide (NO2) in a medium size city in northern spain. Sci Total Environ 2009, 407, 999–1009.

[7] Debevec C, Sauvage S, Gros V, Sciare J, Pikridas M, Stavroulas I. Origin and variability in volatile organic compounds observed at an Eastern Mediterranean background sife (cyprus). Atmosperic Chem Phys 2017, 17, 11355–11388.

[8] Fanizza C, Incoronato F, Baiguera S, Schiro R, Brocco D. Volatile organic compound levels at one site in rome urban air. Almos Pollut Res 2014, 5, 304–314.

[9] Khoder MI, Shakour AA, Faran SA, Hameed AAA. Indoor and outdoor formaldehyde concentrations in homes in residential areas in Greater Cairo. J Environ Monit 2000, 2, 123–126.

[10] Gupta KC, Ulsamer AG, Preuss PW. Formaldehyde in indoor air: Sources and toxicity. Environ Int 1982, 8, 349–358.
[11] Zhu L, Jacob DJ, Keutsch FN, Mickley LJ, Scheffe R, Strum M, Abad GG. Formaldehyde (HCHO) as a hazardous air pollutant: Mapping surface air concentrations from satellite and inferring cancer risks in the United States. Environ Sci Technol 2017, 51, 5650–5657.
[12] Dobslaw D, Engesser KH, Stork H, Gerl T. Low-cost process for emission abatement of biogas internal combustion engines. J Clean Prod 2019, 227, 1079–1092.
[13] Central Pollution Control Board, 2009. National Ambient Air Quality Standards.
[14] Ministry for the Environment and the Ministry of Health, 2002. Ambient air Quality Guidelines.
[15] Wang JL, Chew C, Chang CY, Liao WC, Lung SCC, Chen WN, Lee PJ, Lin PH, Chang CC. Biogenic isoprene in subtropical urban settings and implications for air quality. Atmos Environ 2013, 79, 369–379.
[16] Strum M, Scheffe R. National review of ambient air toxics observations. J Air Waste Manag Assoc 2016, 66, 120–133.
[17] Abad G, Liu GX, Chance K, Wang H, Kurosu TP, Suleiman R. Updated Smithsonian Astrophysical Observatory Ozone Monitoring Instrument (SAO OMI) formaldehyde retrieval. Atmos Meas Tech 2015, 8, 19–32.
[18] Zhu JL, Mickley DJ, Marais LJ, Cohan EA, Yoshida DS, Duncan Y, Abad BNG, Chance G. Anthropogenic emissions of highly reactive volatile organic compounds in eastern Texas inferred from oversampling of satellite (OMI) measurements of HCHO columns. Environ Res Lett 2014, 9, 114004.
[19] Hosseinabadi MB, Timoori S, Zarandi AF. Functionalized graphene-trimethoxyphenyl silane for toluene removal from workplace air sorbent gas extraction method. Anal Methods Environ Chem J 2019, 2, 45–54.
[20] Oliveira M, Slezakova K, Delerue-Matos C, Pereira S, Morais MC. Children environmental exposure to particulate matter and polycyclic aromatic hydrocarbons and biomonitoring in school environments: A review on indoor and outdoor exposure levels, major sources and health impacts. Environ Int 2019, 124, 180–204.
[21] Abdo M, Grumbein S, Chou BJ, Herbert R. Toxicity and carcinogenicity study in F344 rats following 2 years of whole-body exposure to naphthalene vapors. Inhal Toxicol 2001, 13, 931–950.
[22] Jia C, Batterman S. A critical review of naphthalene sources and exposures relevant to indoor and outdoor air. Int J Environ Res Public Health 2010, 7, 2903–2939.
[23] Charles SM, Jia C, Batterman SA, Godwin C. VOC and particulate emissions from commercial cigarettes: Analysis of 2, 5-DMF as an ETS tracer. Environ Sci Technol 2008, 42, 1324–1331.
[24] Amorim LCA, Carneiro JP, Cardeal ZL. An optimized method for determination of benzene in exhaled air by gas chromatography–mass spectrometry using solid phase microextraction as a sampling technique. J Chromatogr A 2008, 865, 141–146.
[25] Chaffin KA, Knutsen JS, Brant P, Bates FS. High-strength welds in metallocene polypropylene/ polyethylene laminates. Sci 2000, 288, 2187–2190.
[26] Masri EI H. Toxicological profile for naphthalene, 1-methylnaphthalene, and 2-methylnaphthalene. ATSDR, 2005.
[27] Karasek FW, Hutzinger O. Environmental chemistry in the Netherlands. Anal Lett 1976, 9, vii–xvii.
[28] Karellas SN, Chen Q. Real-time air monitoring of trichloroethylene and tetrachloroethylene using mobile TAGA mass spectrometry. J Environ Prot Sci 2013, 4, 99–105.
[29] Murnone SS, Lehocky AH, Owens PD. Odor thresholds for chemicals with established occupational health standards. 3rd edition. AIHA, Falls Church, VA, 1989.
[30] Hughes K, Meek ME. World Health Organization, 1, 1, 2, 2-Tetrachloroethane. WHO, 1998.
[31] Fishbein L. An overview of environmental and toxicological aspects of aromatic hydrocarbons II. Toluene Sci Total Environ 1985, 42, 267–288.

[32] World Health Organization Air quality guidelines for Europe, Copenhagen: World Health Organization Regional Office for Europe. WHO Reg Publ Eur Ser 2000, 91.
[33] Kelly TJ, Ramamurthi M, Pollack AJ, Spicer CW, Culpitt LT Ambient concentration summaries for clean air act Title III hazardous air pollutants. Final report for US Environmental Protection Agency Contract No 1993, 68–D80082.
[34] Edgerton SA, Holdren MW, Smith DL, Shah JJ. Inter-urban comparison of ambient volatile organic compound concentrations in US cities. JAPCA 1989, 39, 729–732.
[35] Helmig D, Arey J. Organic chemicals in the air at Whitaker's Forest/Sierra Nevada Mountains, California. Sci Total Environ 1992, 112, 233–250.
[36] Dann T, Wang D, Etlinger A. Volatile organic compounds in Canadian ambient air: A new emphasis. Pollution Measurement Division, Conservation and Protection, Environment Canada. PMD, 1989, 89–26.
[37] Enyoh CE, Verla AW, Qingyue W, Ohiagu FO, Chowdhury AH, Enyoh EC, Chowdhury T, Verla EN, Chinwendu UP. An overview of emerging pollutants in air: Method of analysis and potential public health concern from human environmental exposure. Trends Environ Anal Chem 2020, 28, e00107.
[38] IARC Working Group on the Evaluation of Carcinogenic Risks to Humans, International Agency for Research on Cancer, and World Health Organization. Dry cleaning, some chlorinated solvents and other industrial chemicals. World Health Organization 1995, 63.
[39] Johnson MW, Mandell D, Anderson E ATSDR toxicological profile for trichloroethylene 1997.
[40] Boer FP, Neuman MA, Van Remoortere FP, North PP, Rinn HW, X-ray Diffraction Studies of Chlorinated Dibenzo-p-dioxins 1973.
[41] Robichaud A. An overview of selected emerging outdoor airborne pollutants and air quality issues: The need to reduce uncertainty about environmental and human impacts. J Air Waste Manag Assoc. 70 2020, 341–378.
[42] Seaman VY, Bennett DH, Cahill TM. Origin, occurrence, and source emission rate of acrolein in residential indoor air. Environ Sci Technol 2007, 41, 6940–6949.
[43] Zou Z, Yin Z, Ou J, Zheng J, Liu F, Huang C, Ou S. Identification of adducts formed between acrolein and alanine or serine in fried potato crisps and the cytotoxicity-lowering effect of acrolein in three cell lines. Food Chem 2021, 361, 130164.
[44] Novoselov YN, Filatov IE. Removal of acrolein vapor from air by a nanosecond electron beam. Tech Phys Lett 1998, 24, 638–639. Technical physics letters 24 (1998) 638–639.
[45] Koerselman W, Graaf MV. Acrylonitrile: A suspected human carcinogen. Int Arch Occup Environ Health 1984, 54, 317–324.
[46] Thongkum W, Wibuloutai J, Thitisutthi S. Benzene and 1, 3 Butadiene Concentration and Its Potential Health Impact in Chiang Mai, Thailand. Eur J Sustain Dev 2017, 6, 187–192.
[47] Yacovitch TI, Herndon SC, Roscioli JR, Floerchinger C, Knighton WB, Kolb CE. Air pollutant mapping with a mobile laboratory during the BEE-TEX field study. Environ Health Insights 2020, 9, 7–13.
[48] Doyle M, Sexton KG, Jeffries H, Bridge K, Jaspers I. Effects of 1,3-Butadiene, isoprene, and their photochemical degradation products on human lung cells. Environ Health Perspect 2004, 112, 1488–1495.
[49] France L 1,3-Butadiene, ethylene oxide and vinyl halides (vinyl fluoride, vinyl chloride and vinyl bromide) IARC Monographs on the Evaluation of Carcinogenic Risks to Humans 2008, 97.
[50] Chou MS, Lu SL. Treatment of 1,3-butadiene in air stream by a biotrickling filter and a biofilter. J Air Water Manage Assoc 1998, 48, 711–720.
[51] Elisabeth G, Wang D, Zlotorzynska ED, Siu M, Celo V, Tardif M, Harnish D, Jiang Y. Air toxics in Canada measured by the National Air Pollution Surveillance (NAPS) program and their relation to ambient air quality guidelines. J Air Waste Manag Assoc 2016, 66, 184–200.

[52] Nitter TB, Svendsen K. Modelling the concentration of chloroform in the air of a Norwegian swimming pool facility-A repeated measures study. Sci Total Environ 2019, 664, 1039–1044.
[53] Chan CC, Vainer L, Martin JW, Williams DT. Determination of organic contaminants in residential indoor air using an adsorption-thermal desorption technique. J Air Waste Manag Assoc 1990, 40, 62–67.
[54] Wallace LA, Pellizzari E, Hartwell T, Rosenzweig M, Erickson M, Sparacino C, Zelon H. Personal exposure to volatile organic compounds. Direct measurements in breathing-zone air, drinking water, food, and exhaled breath. Environ Res Lett 1984, 35, 293–319.
[55] Dawson HE, McAlary T. A Compilation of statistics for VOCs from post-1990 indoor air concentration studies in North American residences unaffected by subsurface vapor intrusion. Ground Water Monit Remed 2009, 29, 60–69.
[56] Lei T, Liao X, Chen J, Xiao H, Xu L, Zhang F, Niu Z, Yu J. Pollution characteristics of ambient volatile organic compounds (VOCs) in the southeast coastal cities of China. Environ Sci Pollut Res 2013, 20, 2603–2615.
[57] Schlosser PM, Bale AS, Gibbons CF, Wilkins A, Cooper GS. Human health effects of dichloromethane: Key findings and scientific issues. Environ Health Perspect 2015, 123, 114–119.
[58] Ching CM, Li C. Analysis of bioaerosols from chicken houses by culture and non-culture method. Aerosol Sci Technol 2006, 40, 1071–1079.
[59] Venter P, Lues JF, Theron H. Quantification of bioaerosols in automated chicken egg production plants. Poult Sci 2004, 83, 1226–1231.
[60] Hesterberg TW, Long CM, Bunn WB. Non cancer health effects of diesel exhaust: A critical assessment of recent human and animal toxicological literature. Crit Rev Toxicol 2009, 39, 195–227.
[61] Abad E, Adrados M, Caixach J, Rivera J. Dioxin abatement strategies and mass balance at a municipal waste management plant. Environ Sci Technol 2002, 36, 92–99.
[62] Fullana A, Conesa JA, Font R, Sidhu S. Formation and destruction of chlorinated pollutants during sewage sludge incineration. Environ Sci Technol 2004, 38, 2953–2958.
[63] Quass U, Fermann MW, Broker G. Steps towards a European dioxin emission inventory. Chemosphere 2000, 40, 1125–1129.
[64] Lavric ED, Konnov AA, Ruyck D. Dioxin levels in wood combustion – A review. Biomass Bioenergy 2004, 2, 115–145.
[65] Rappe C, Andersson R, Bergqvist PA, Brohede C, Hansson M, Kjeller LO. Overview on environmental fate of chlorinated dioxins and dibenzofurans. Sources, levels and isomeric pattern in various matrices. Chemosphere 1987, 16, 1603–1618.
[66] Anthony E, Jia L, Granatstein D. Dioxin and furan formation in FBC boilers. Environ Sci Technol 2001, 35, 3002–3007.
[67] Petroff BK, Roby KF, Gao X, Son DS, Williams S, Johnson D. A review of mechanisms controlling ovulation with implications for the anovulatory effects of polychlorinated dibenzo-p-dioxins in rodents. Toxicology 2001, 158, 91–107.
[68] Memo and attachments from Farmer JR. EPA to distribution. August 22, 1980. 29 Thermal incinerator performance for NSPS.
[69] Key JA (IT Environscience) Control device evaluation catalytic oxidation. In: US Environmental Protection agency. Organic chemical manufacturing, Vol. 4: Combustion control devices. Publication no. EPA-450/3-80-026, Research Triangle Park NC, December 1980. 1–1. Report 3.
[70] Federal Register 51 FR 2699.
[71] Kenson RE. KPR system for VOC emission control from paint spray booths. Paper presented at air pollution control association annual meeting, June 1985. Detroit Michigan.
[72] Distillation operations in synthetic organic chemical manufacturing background information for proposed standards. Draft EIS. EPA-450/3-83-005a. Office of air planning and standards, US Environmental Protection Agency, December 1983.

[73] Richardson SD, Kimura SY. Emerging environmental contaminants: Challenges facing our next generation and potential engineering solutions. Environ Technol Innov 8, 2017, 40–56.

[74] Sexton K, Callahan MA, Bryan EF, Saint CG, Wood WP. Informed decisions about protecting and promoting public health rationale for a national human exposure assessment survey. J Expo Anal Env Epid 1995, 5, 233–256.

[75] Arfin T, Bhaisare DA, Waghmare SS. Development of a PANI/Fe $(NO_3)_2$ nanomaterial for reactive orange 16 (RO16) dye removal. Anal Methods 2021, 13, 5309–5327.

Tijo Cherian*, Wajih Jamal

Use of Green Inspired Nanomaterials for the Biological Remediation of Pollutants

Abstract: The rising environmental contaminants brought on by human activity are having a negative impact and posing a threat to many biotic life forms on the earth. By virtue of their hazardous endurance in the environment, heavy metals and some organic pollutants are regarded as important pollutants all over the world. Pollutants have been present in the environment for some time now, but traditional technologies have difficulty getting rid of them due to a variety of limitations. This has caused research to focus toward the creation of effectual and affordable technologies for the removal of pollutants. The current study for reducing environmental contaminants with minimal environmental concerns uses the adsorption phenomena from old technologies in combination with modified adsorbents at the nanoscale. In recent years, fast research for eco-friendly pollution clearance has been stimulated by the unexplored capacity of biomass entities for the bioengineering of nanomaterials. The biofabricated nanomaterials exhibit unique phenomena such as electrostatic contact, surface active site interactions, and photo- and enzymatic catalysis that aid in the detoxification of various contaminants. In this context, with an emphasis on the reduction of pathogen-derived toxins and air pollutants, the current chapter focuses on the creation of biofabricated nano-based materials and their mechanisms underlying for the cleanup and remediation of persistent organic pollutants in the environment.

Keywords: Green chemistry, nanomaterials, environmental pollution, bioremediation

1 Introduction

Due to the extensive consumption of contaminants during the past years, contamination by both inorganic and organic pollutants has significantly increased. Persistent organic pollutants, or POPs for short, are a group of hazardous chemicals that are created, utilized, and dumped as a result of human activities [1]. Because they are easily transported by wind and water, they can have a negative effect on other living things as well as the environment. Every day, the amount of dangerous persistent organic con-

***Corresponding author: Tijo Cherian**, Department of Ocean Studies and Marine Biology, Faculty of Life Sciences, Pondicherry University, Puducherry, India, e-mail: tvarghese891@gmail.com
Wajih Jamal, Department of Zoology, Faculty of Life Sciences, Aligarh Muslim University, Aligarh, Uttar Pradesh, India

https://doi.org/10.1515/9783111016825-015

taminants in the environment rises. POPs constitute a significant threat because they are easily transported from one source to another and discharged into the environment through groundwater and drinking water. POPs have the capacity to accumulate in the body and are known to be exceedingly hazardous even at low concentrations. They also persist in the environment for a long time [2]. They can quickly move up the food chain from one organism to another, harming organisms that are far away from the source of contamination. POPs have become a global issue because of their persistence, long-range transportability, and propensity of bioaccumulation [3]. In accordance with the Stockholm Convention, 22 nonpolar organic compounds are POPs. OCs (organochlorines), which include aldrin and dichlorodiphenyltrichloroethane (DDT), are the most common class of POPs. Organophosphorus (OPs) pesticides are next, followed by PAHs (polyaromatic hydrocarbons), flame retardants, and other compounds (including PCBs [polychlorinated biphenyls]) [4, 5].

Because POPs have a propensity to bioaccumulate and biomagnify throughout the food chain, their release into water bodies raises concerns for both the environment and public safety. POPs may enter human bodies in excess through drinking water due to the leaching of contaminants [6]. Technological innovation for the cleanup of such contaminants that is extremely effective, economical, and low maintenance is difficult to implement and develop. Researchers from all over the world have found advancements in nanotechnology to be appealing among the many wastewater treatment solutions. Due to the fact that unique nanomaterials can be used for the effective degradation of such contaminants, nanotechnology has emerged as an intriguing and revolutionary technology. The present chapter aims to highlight recent advancements in POP removal using nanotechnology while simultaneously emphasizing the environmental effects of POPs. The present chapter provides a critical evaluation of the many published articles on the use of nanoparticles for the removal of POPs from contaminated water sources in order to further the creation of novel nanomaterials for such applications.

2 Overview of pollution, source, and effects

POPs have drawn a lot of interest globally and are classified as xenobiotic compounds of various etiologies, but they all have a number of traits, including high toxicity, hydrophobicity, bioaccumulation, and environmental persistence [7]. POPs are carbon-containing compounds that prefer to accumulate among fatty tissues due to their increased solubility in lipids. Because of their ability to interfere with an organism's endocrine system, they are frequently submitted to as ED (endocrine disruptors) [8]. The physicochemical characteristics of POPs include their high adsorption, low water solubility, hydrophobic nature, and slow breakdown causing their diffusion and distribution in the environment [9].

The usage and consumption of pesticides suddenly increased during the green revolution; at the time, POPs' potential for harm was unknown. The hazardous consequences of pesticides have recently come to light on a global scale. The general people began to become aware of the dangers of pesticides and other organic pollutants. POP was divided into three groups following the Stockholm Convention: pesticides, by-products, and commercial chemicals [10]. POPs are a group of dangerous man-made compounds that can be generated intentionally or accidentally [11]. Pesticides are purposely made chemicals that are used in pest control in homes and farms; DDT is one well-known case that was outlawed around the globe due to its severe toxicity [12]. Others include industrial goods or chemicals that were mistakenly produced such as dioxins. Perfluorinated compounds (PF), PBDE (polybrominated diphenyl ethers), and a long list of other contaminants that are being added every day are examples of developing contaminants that are included in the new class of POPs. OCP (organochlorine pesticides), HCB (hexachlorobenzene), and other PCBzs (polychlorinated benzenes), PCN (polychlorinated naphthalenes), PCDD (polychlorinated dibenzo-p-dioxins) and -dibenzofurans (PCDF), and OCEC (other contaminants of emerging concerns) are some of the categories for POPs [13].

POPs have a lengthy half-life in the environment and can take decades or even centuries to totally disintegrate. Due to their physicochemical properties, POPs have a tendency to travel across greater distances and withstand disintegration (both chemically and biologically). This enables them to bioaccumulate to a higher level through biomagnification, and their release can have a major negative impact on the environment as well as human health [14]. Given that most POPs are semi-volatile chemicals, which rapidly soak onto atmospheric particles and travel into air, soil, and water media [15–17], numerous studies have suggested a spectrum of unfavorable impacts caused by POPs. POPs are prevalent in every medium and are known to be found present in every medium across the environmental spectrum if analyzed [18]. Pharmaceutical waste, garbage from the chemical and electrical industries, and waste from the agricultural sector all include POPs.

POPs are so poisonous that even minute amounts of them are discovered to be extremely lethal to organisms. POPs have a low solubility, which makes them resistant to biological, chemical, and photo-degradation. It is also challenging to break down POPs using conventional wastewater-treatment methods [19, 20]. In the recent past, POP cleanup was accomplished using cutting-edge wastewater-treatment technology or by combining one or more techniques. POPs typically demonstrate lipid solubility, and as a result, they frequently build up in the fatty and adipose tissues of an organism. Furthermore, due to the non-reactivity of C-Cl bonds, halogenated compounds exhibit excellent durability toward hydrolysis and photolytic destruction [21]. Because of their resistance to degradation and lipophilicity, POPs compounds are particularly dangerous.

2.1 Causes of environmental pollution

Global industrialization has increased in every industry as a result of population growth, urban sprawl, and the need for survival. Rapid advancements in the agricultural, scientific, and industrial fields – including those of textiles, paints, paper pulp, chemical fertilizers, tannery, automobiles, biomedical wastes, petroleum mining, electric and electronic devices, pharmaceuticals – have resulted in an increase in inorganic and organic pollutants in water, soil, and the atmosphere in recent years. These are endangering the living biota's health seriously [22]. The foundation for advancing the global economy is provided by the industry and agricultural sectors [23]. The largest contributors to intricate inorganic and organic pollutant generators are industrial production units because of poor management and a lack of adherence to environmental legislation or defined standards for the proper disposal of wastewater into the environment, including nearby land regions and aquatic bodies at various levels [24, 25]. Numerous toxins have entered the earthly and aquatic habitats as a result of quick industrialization, which has led to real natural issues [26]. Water and land are precious natural resources that are reliant on human civilization and agricultural management [27]. The contaminants include effluents, significant metal waste from mining, vehicle wear, insecticides, bug sprays, and naturally occurring salts mixed with soil and water that pose serious risks to the environment [28]. Municipal waste management techniques, human and agricultural activities, and the industrial sector all play a role in the discharge of pollutants into different ecosystems, including soils and water bodies. These residues have the ability to remain in the environment for an extended period of time without going through any sort of degradation or detoxification processes, which has detrimental impacts on the ecosystem. Providing safe and wholesome environmental conditions has grown to be difficult in many nations or globally.

2.2 Sources of inorganic pollution

The ubiquitous, naturally occurring, extremely poisonous, non-biodegradable inorganic pollutants have a high atomic density and are made up of radionuclides, heavy metals, and metalloids. The main causes of inorganic pollution are anthropogenic activities and the biogeochemical cycle. Radioactive elements like Uranium, Radon, Thorium, Polonium-210 naturally develop as a result of nucleogenesis reactions at different atmospheric levels. Nuclear waste is being produced from a variety of sources, including oil extraction, nuclear fuel generation, volcanic eruption, nuclear power plants, manufacturing facilities, other mineral extraction, use of radioactive materials in anthropological, forensic research, and medical care [29].

The raw materials for ore mining and ore smelting, which yield high-purity metals including gold, silver, copper, mercury, zinc, and cadmium, are from the earth's crust and rock sediments. Metalworking techniques have been used since the dawn of civilization to create materials for jewelry, building materials, and the manufacture of stainless steel, among other things. Industries are utilizing heavy metals in a variety of processes to produce the desired output. In conventional drug formulations, medications, thermometers, sphygmomanometers, dental amalgams, and aesthetic preparations, heavy metals, particularly Hg, are used [30]. Heavy metals and metalloids are pesticide residues that are responsible for contaminating adjacent water sources and agricultural land sites due to overuse. Electroplating, smelting, incorrect battery disposal, tannery effluents (Cr), oil refinery products (Hg, Pb), paint and textile (metal complexed dyes), landfill leachates, and acid mine are some of the other significant causes of contamination (Pb, Cd, Zn, As, Cr). These heavy metals and metalloids are contaminants that do not degrade and have a tendency to persist in the environment. This causes bioaccumulation/ biomagnification in a variety of organisms, including plants, microbes, aquatic, and terrestrial organisms. These organisms then enter the food chain, impair physiological processes, and, ultimately, cause toxicity-induced ailments [30].

2.3 Sources of organic pollutants

Chemical fertilizers and pesticides are highly necessary in the current global environment to match up the need for agricultural production and to reduce insect-pest plant maladies. POPs, which accumulate in soil and water as a result of the unrestrained and unscientific use of pesticides in agricultural activities, have a considerable negative influence on human health and the environment. POPs are being spread, which has resulted in the phenomena of bioaccumulation and biomagnification at various ecosystem levels, which has an impact on the physiological functioning of flora and fauna [31]. Numerous synthetic azo dyes are used in the fabrication of textiles. The percentage of dye lost during this process is higher. Dye and heavy metal poisoning of the environment is a result of untreated effluents from the tannery paint and textile industries. Its refractory character provided by its conformation and aromaticity confers the main concern with the pesticide and dye contamination. The majority of conventional dye remediation techniques were water remediation tactics intended to undermine resistant techniques [32]. The sources for inorganic and organic pollution are represented in Figure 1.

2.4 Effects on biota

Pesticides, heavy metals/metalloids, and fertilizers are only a few examples of the inorganic and organic contaminants that have affected plant growth, soil microbiota, and the nutrition of the soil. The harmful effects are directly related to the pollutant type and soil pH. Pollutant residue persistence has an impact on enzymatic processes such

Figure 1: Sources of inorganic and organic pollution in the environment.

as cellulose ammonification, nitrification, and breakdown [31]. The effects of over-fertilization have been noted in the publications, including yield loss, diminished inflorescence and fruit set, and decreased synthesis of secondary metabolites and vitamins [33]. Excessive pesticide runoff from manufacturing, agricultural, or industrial effluents, dyes, and POPs builds up in local water bodies, causing bioaccumulation in biological systems. As a result, these substances move up the food chain and have an intentional impact on both aquatic and terrestrial organisms. The ecosystems have been affected to varying degrees by the persistence of contaminants both in parental or converted state. The negative effects of pesticides, organophosphates, organochlorines, and herbicides on algae, bacteria, and other phototropic microbes have also been brought to light by studies [34]. Pollutants, particularly herbicides, have a strong inhibitory effect on the mitochondrial systems and the building blocks of chlorophyll synthesis in phototropic microorganisms, reducing the efficiency of the photosynthetic process. A variety of immunological reactions, including acute and chronic toxicity, can result from the interaction of pesticides and other pollutants, such as heavy metals, with non-specific target signals. The kind of contaminant concentration and the source of exposure determine how strong the effects are. They mostly irritate the skin and the eyes, and trigger allergic reactions that affect central nervous system (CNS). Breathing issues, vomiting, lightheadedness, anxiety, depression, headaches, and muscle aches are a few symptoms. The signaling mechanisms or enzymes are the main targets of high doses, which function as enzyme inhibitors, particularly for cholinesterase, obstructing, or reducing the conductivity of neural signals. Prolonged exposure can result in epigenetic changes through the early onset of amyotrophic lateral sclerosis, cognitive decline, dementia, Alzheimer's disease, Parkinson's disease, immunological suppression, hormonal imbalance, and endocrine-disrupting hormones [35].

The oxidative stress, neurotoxicity, ROS (reactive oxygen species), reproductive issues like infertility and congenital abnormalities, dental deformities, cancer, and liver malfunction/ failure are all brought on by ingesting or inhaling hazardous pollutants like pesticides and dyes from numerous sources [36]. In agricultural workers at higher risk of cardiovascular dysfunction after pesticide exposure, modified erythrocyte acetylcholinesterase and other hematological characteristics were discovered, changing the nitric oxide signaling pathway. Above all, environmental problems brought on by different organic and inorganic contaminants are incalculable. Further investigation is necessary to understand how contaminants affect the microbial community. Due to the various levels of toxicity caused by the pollutants, scientists, environmental law makers, food producers, and responsible citizens are now concentrating on limiting the use of pollution-levying compounds and constructing sustainable methods for degradation and detoxification [30]. Figure 2 shows the effects of pollutants on biota.

3 Conventional technology

The literature has amply demonstrated that the existence of POPs and other emergent pollutants has significantly reduced the usage of conventional technologies for wastewater treatment. The range of POPs that are highly toxic cannot be treated with the currently available treatment techniques, such as activated sludge, activated carbon, reverse osmosis, and oxidation; and the by-products generated during the degradation process may prove to be more dangerous than the parent compound [37]. As a result, there should be efficient treatment technologies for POPs that do not require secondary or tertiary care. Numerous studies revealed that countries with restricted POP application have greater levels of organic pollutants in their water sources. The traditional methods do not operate sustainably and leave contaminants for future generations. To overcome the limitations of traditional treatment technologies, an efficient alternative strategy should be sought to completely breakdown the POPs into environmentally benign or less damaging end products [38].

Physical, chemical, and biological approaches are used in every wastewater treatment process. These methods are divided into three fundamental stages, namely, primary, secondary, and tertiary treatment, and are then followed by more developed treatment technologies [39]. Simple water filtration, decoloration, deionization, reverse osmosis, disinfection, ion exchange, coagulation and flocculation, sedimentation, steam distillation, and biological treatment utilizing microorganisms are examples of such techniques [40]. The materials utilized in old processes can result in secondary pollutants

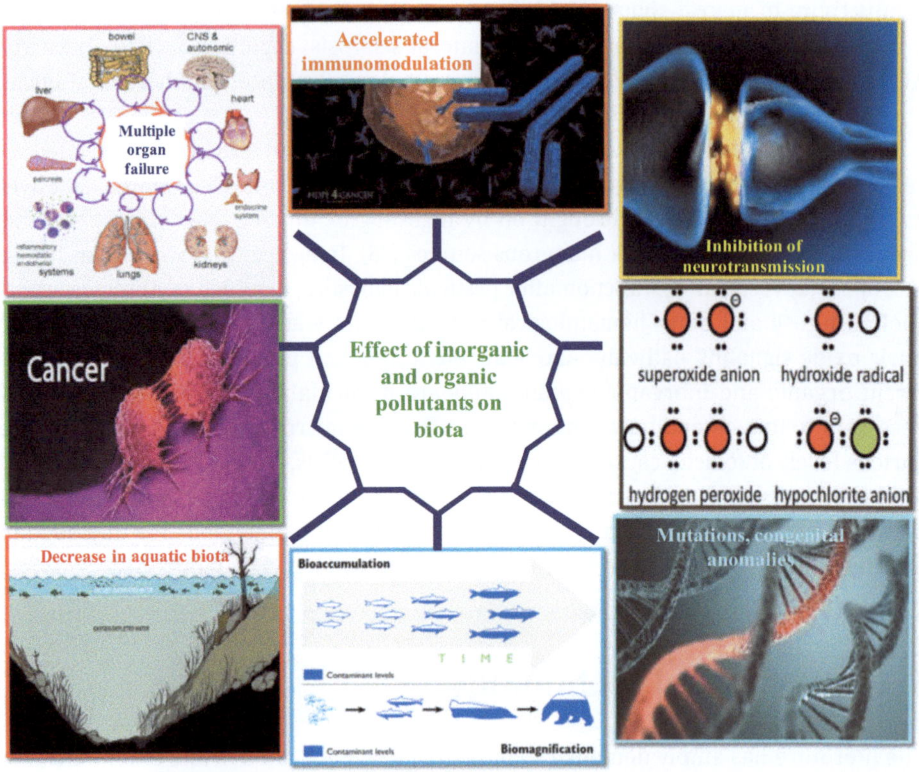

Figure 2: Effect of pollutants on biota.

and are highly expensive. Studies showing the inadequacy of frequently used traditional methods of treatment have led to the need for more sophisticated and effective techniques for removing POPs from water and wastewater. With satisfactory performance and cost-effective solutions, nanotechnology enables an incredibly effective, resilient, resourceful, and stable process that can offer a potential option to traditional technologies [38].

As a result of the above discussions, it is concluded that traditional treatment techniques can, to a limited extent, treat the majority of organic pollutants. However, in order to overcome the limitations associated with these technologies, it is necessary to search for efficient advanced treatment technologies while taking into consideration the characteristics of pollutants and all other environmental variables. The substitute treatment strategy must be effective, durable, and provide a one-step resolution to the existing problem without generating additional pollutants [38].

4 Advanced nanotechnological methods for removal of POPs

Before choosing the best treatment technique, economic and technological variables must be carefully taken into account because they have a significant impact on the removal of POPs. POPs' fate, transit, and degradation entail complicated processes. One of the most promising ways to modify contaminated management technologies is through the use of nanotechnology, which appears to be the biggest blessing for environmental challenges. A novel technique for the long-term elimination of organic contaminants is nanoremediation, defined as the use of nanomaterial for the cleanup of environmental contaminants [41, 42]. Nanomaterials are advantageous for an assortment of potential relevance due to their distinctive optical, mechanical, thermal, electromagnetic, and structural properties. Physical, chemical, or biological methods can be used to create nanomaterials, which can then be used to make nanoparticles, nanosensors, nanoadsorbents, nanomembranes, or nanocatalysts for wastewater management [43]. A sustainable option to chemical-zero processes with zero waste generation is now widely used for the production of nanomaterials for ecological applications based on green chemistry [38, 41–42].

Gibbs free energy (ΔG), the quantity of energy needed for each chemical reaction to occur, is crucial in defining how POPs transform into different intermediate and end compounds. Using DG values, one may calculate the amount of energy released from any biological reaction and assess if the reduction process is in spontaneity or not, which determines the outcome of the reaction. POPs are known to be resistant to the majority of environmental degradation processes, although some molecular modification that results in the creation of an even more hazardous intermediates or end-products is still a possibility [44]. The following sections will conduct in-depth study on the application of nanomaterials to the management of POPs in a number of sectors.

4.1 Nanofilteration

By employing nanofiltration membranes, nanotechnology has pioneered the door for improved water treatment systems. The pressure-driven filtration techniques known as membrane processes, including nanofiltration (NF), ultrafiltration (UF), reverse osmosis (RO), and microfiltration (MF), are thought to be very successful for the treatment of wastewater [45]. They are regarded as alternatives to traditional remediation techniques for organic micropollutants in water bodies. Although membrane systems for wastewater treatment are expensive, they are the ideal alternatives to traditional methods due to their extremely high removal efficiency. For the development of membranes with desired functionality and features, the nanoparticle can be introduced into the membrane either through surfacial immobilization, surface grafting, or blending [46].

By employing materials like biomass wastes, ceramics, metals, or polymers in the range of 10–1,000 nm, the electrospinning method can be used to create composite/polymeric nanofibrous membranes that are used to create ultrafine nanofibers [47, 48]. Out of all the membrane approaches, RO and NF demonstrated their effectiveness in the cleanup of micro/trace organic contaminants by efficient filtration. In contrast to RO, where filtration is primarily driven by charge, diffusion, and convection, NF is far more effective at cleaning up contaminants. This is at the expense of higher energy and maintenance costs. Because of its tiny pore sizes, great effectiveness, and ease of usage, NF is useful for the cleanup of micropollutants. Due to their efficiency in removing POPs, a variety of polymers (both natural and synthetic) including cellulose acetate, polypropylene, polyvinyl fluoride, and polyacrylonitrile have been employed to create nanofiltration membranes [38].

The pressure-driven method of nanofiltration, which is based on hydrodynamics between membrane nanopores and membrane surface, may successfully remove low molecular weight molecules with a size range between 1 and 10 nm [49]. Nanofilters contribute to a decrease in the ionic strength of the solution by reducing the hardness of organic contaminants. The membrane's surface area, porosity, and charge have a significant impact on how well it filters. High-quality nanofibrous membranes are made using the electrospinning process. Nanofiltration is extremely effective at treating arsenic in drinking water because it efficiently eliminates practically all dissolved salts and discards univalent, divalent, and multivalent ions. The safest technology is determined to be membrane filtration, and NF is very effective at removing low-molecular-weight contaminants. The only filtration method that has been proven to effectively remove organic pollutants and pesticides is NF. The disadvantage of filtration technology, however, is membrane fouling and blockage, which can be remedied by using hybrid technologies [50].

4.2 Nanocatalysis

The necessity to create a green, inventive, and maintainable technology that can totally breakdown and sequester the organic contaminants with considerably less energy and chemical use emerges from the inability of traditional technologies to do so [51]. As a result, due to the strong oxidizing radicals produced by sophisticated oxidation methods, the science world has been searching for low-cost, efficient ways to oxidize and mineralize a variety of contaminants, including POPs. Nanocatalysis is the conversion of pollutants into environmentally beneficial molecules using semiconducting wide-band gap nanoparticles. The semiconductor metal and metal-oxide nanoparticles have drawn a lot of interest in the sustainable treatment of POPs. For efficient elimination of POPs from wastewater, a variety of nanocatalysts are used, including photocatalyst, electrocatalyst, Fenton-based, and doped multifunctional nanocatalyst [52]. One well-known AOP is photocatalysis/nanocatalysis, which is used to increase the biodegradability of POPs by oxidizing organic pollutants and releasing highly ROS in order for the chemical reaction to take place. Photocatalysis depends on the generation of strong radicals like H_2O_2, O_2^-, O_3, OH^- radicals, which effectively destroy all organic molecules, in order to initiate the catalytic process in the presence of light. By generating free radicals, photocatalysis is even useful for remediating VOCs (volatile organic compounds) like PCBs, dioxins, and PHA. In context of either natural or artificial light source and oxygen, wide bandgap nanocatalyst (such as ZnO, TiO_2, or WO_3) employed to break down POPs, photocatalysis begins. TiO_2 and ZnO are currently the most frequently used semiconductors in the degradation of POP. When it comes to eliminating extremely hydrophobic POPs from the environment, the heterogeneous photocatalyst is effective. When choosing the right catalyst, it is important to take into account the semiconducting material's surface, pore volume, and structural characteristics [53]. The degrading effectiveness of the photocatalyst can be increased by adjusting the material's surface characteristics and crystal structure. The elimination of nanomaterial from the reaction fluid after photocatalysis is complete is one of the main drawbacks of this method. Unquestionably, the most popular method for treating sewage is photocatalysis. For the degradation of a variety of inorganic and organic contaminants, photocatalysis is a preferred technology due to its excellent efficiency, sustainability, and effective outcomes.

4.3 Nanoadsorption

Due to their exceptionally large surface area, numerous sorption sites, variable pore size, high surface activity, and significantly reduced intraparticle-diffusion distance, nanoadsorbents have remarkable sorption effectiveness for a wide variety of organic and other contaminants [54]. The use of nanoadsorbents has the benefit that any contaminants can be easily functionalized making them extremely selective. For the cleanup of POPs like phenols, hydrocarbon, detergents, medicines, dyes, pesticides, and biphenyls, it has been discovered that the adsorption procedure is effective. An easy and secure method for removing POPs from water sources is nanoadsorption. Nanoadsorption has so far proven to be one of the most effective techniques for POP remediation among a variety of technologies. Since > 90% elimination efficiency was attained in the majority of tests for up to ten cycles, many researches demonstrate the effectiveness of nanomaterials for the absorption of different POPs from wastewater. Creating a composite with the surface of metallic oxides and seeing if it can withstand a one-electron oxidation reaction while being exposed to visible light is the fundamental procedure for evaluating the adsorption efficiency of nanomaterials. Hydrogen bonding, electrostatic interactions, hydrophobic interactions like van der Waals, electron donor-acceptor, etc. are the foundation of nanoadsorption [55].

The process of nanoadsorption makes considerable use of nanomaterials such as alumina, zeolite, clay, carbon-based nanomaterials, activated carbon, metal/metal oxides, nanocomposites, nanotubes, nanosheets, graphene-based nanomaterials, and chitosan-based polymers [56]. The utilization of magnetic nanoparticles, particularly iron oxide, has resulted in a new category of magnetic-separation techniques for the efficient removal of POPs. The microporous features in activated carbon improve POP adsorption effectiveness [57]. Because of their hydrophobicity, hydrogen bonds, covalent contacts, and electrostatic and covalent interactions, carbon-based nanoadsorbents frequently interact with pollutants. Nanoadsorbents are frequently employed as a low-cost, successful, and long-lasting therapy procedure. Due to their recycling capacity and low operating and maintenance costs, they are particularly effective at removing heavy metals from wastewater. One of the biggest benefits of employing nanoadsorbents is how simple it is to separate magnetic nanoadsorbents from the reaction system by using an external magnet [58].

4.4 Nanobiotechnology and hybrid technologies

Through the use of bioremediation and nanoremediation techniques, nanobiotechnology is emerging as an environmentally friendly and sustainable approach for the elimination of POPs. Microbes + NP (micro-nanoremediation) and plant + NP (phyto-nanoremediation) are two different types of nanobioremediation, which uses microorganisms and nanoparticles to eliminate contaminants [59]. The sorption process, which includes both adsorption and absorption, is what drives the nanoremediation process most of the time. Understanding the behavior of nanomaterials and interaction of contaminants requires doing thermodynamic, mechanistic, and kinetic research. Even diverse nanomaterials are employed to boost microbial degradation effectiveness.

Bioremediation is a time-consuming and ineffective technique. Therefore, it is extremely effective and efficient due to the combination of nanoremediation and bioremediation. The focus has switched to hybrid technologies, which involve the fusion of many different technologies for the efficient removal of environmental pollutants. The most efficient hybrid technology combines membrane filtration with photocatalysis; nanobiotechnology methods also pave the way for the one-step elimination of a variety of organic contaminants [60]. Since each method has its own advantages and disadvantages as well as a certain spectrum of removal efficiency, it is also necessary to design integrative or composite nanomaterials to enhance remediation processes. While nanocatalysts can efficiently convert POPs into environmentally acceptable molecules, nanoadsorbent materials have remarkable qualities for the remediation of a wide variety of heavy metals.

5 Future outlook

There is a critical need for advanced POP treatment technologies to guarantee a sustainable future. Due to the inefficiency of current technologies in achieving a zero-elimination rate, POP treatment has not been completely successful. According to numerous reports on the elimination of POPs, using nanotechnology in lab-scale experiments has successfully resolved the majority of the problems encountered during the remediation of POPs. The study findings on different nanomaterials used to remove POPs from water sources are critically evaluated in this chapter. The following recommendations have been made after reading numerous pieces of literature: first, it is possible to modify nanomaterials' surface qualities so that they may be used quickly to remove POPs; second, the material cost can be reduced by using waste materials and renewable resources; and third, the material can be reused. Green synthesis techniques and green nanotechnologies should receive more attention for environmental cleanup; nevertheless, further in-depth research is required to close knowledge gaps by comparing conventional and cutting-edge treatment approaches. Furthermore, the general people should be made aware of the potential risks and effects of POPs by the government and nongovernmental organizations (NGOs).

In order to remove POPs in an environmentally favorable way in the future, additional research will be required in the development and design of sustainable and green nanomaterials. This will encourage the further creation of nanomaterials that are more successfully in the elimination a variety of POPs. Future studies in the topic must concentrate on how POPs are handled, absorbed, and handled as well as how nanomaterials interact with POPs. For a sustainable society, scientists must develop cutting-edge biobased nanomaterials that are ecologically friendly and more effective than existing technologies, rather than combining diverse technologies.

6 Conclusion

In this chapter, we have covered the effects, destiny, and treatment options for POPs. Due to their extremely low concentrations, POPs are difficult to sufficiently remove using conventional procedures. Based on the physicochemical properties of the contaminants, an advanced treatment must be selected. Due to the special characteristics of nanoparticles, which are excellent for diverse environmental applications, nano-based technologies have transformed the field of wastewater treatment. We have talked about the many nano-based procedures for the efficient removal of POPs, including photocatalysis, nanoadsorption, and nanofiltration. Although many nano-based removal methods are still in the research phase, several have reached the pilot scale. To address the shortcomings, governments, industry, and research experts must work together to find a workable solution to the problems posed by the management of POPs.

References

[1] Alharbi OM, Khattab RA, Ali I. Health and environmental effects of persistent organic pollutants. J Mol Liq 2018, 263, 442–453.

[2] El-Shahawi MS, Hamza A, Bashammakh AS, Al-Saggaf WT. An overview on the accumulation, distribution, transformations, toxicity and analytical methods for the monitoring of persistent organic pollutants. Talanta 2010, 80, 1587–1597.

[3] Fernández P, Grimalt JO. On the global distribution of persistent organic pollutants. CHIMIA Int J Chem 2003, 57, 514–521.

[4] Stokke OS, Thommessen OB. Stockholm convention persistent organic pollutants (Stockholm Convention: POPs). In yearbook of international cooperation on environment and development 2001–02. Routledge, London, UK,, 2013, 122–123.

[5] Alam J, Yadav VK, Yadav KK, Cabral-Pinto MM, Tavker N, Choudhary N, Shukla AK, Ali FAA, Alhoshan M, Hamid AA. Recent advances in methods for the recovery of carbon nanominerals and polyaromatic hydrocarbons from coal fly ash andtheir emerging applications. Crystals 2021, 11, 88.

[6] Kelly BC, Ikonomou MG, Blair JD, Morin AE, Gobas FA. Food web specific biomagnification of persistent organic pollutants. Science 2007, 317, 236–239.

[7] Lefebvre T, Fréour T, Ploteau S, Le Bizec B, Antignac JP, Cano-Sancho G. Associations between human internal chemical exposure to Persistent Organic Pollutants (POPs) and In Vitro Fertilization (IVF) outcomes: Systematic review and evidence map of human epidemiological evidence. Reprod Toxicol 2021, 105, 184–197.

[8] Adithya S, Jayaraman RS, Krishnan A, Malolan R, Gopinath KP, Arun J, Govarthanan M. A critical review on theformation, fate and degradation of the persistent organic pollutant hexachlorocyclohexane in water systems and waste streams. Chemosphere 2021, 271, 129866.

[9] Jacob J. A review of the accumulation and distribution of persistent organic pollutants in the environment. Int J Biosci BiochemBioinform 2013, 3, 657.

[10] Xie L, Du T, Wang J, Ma Y, Ni Y, Liu Z, Wang J. Recent advances on heterojunction-based photocatalysts for the degradation of persistent organic pollutants. Chem Eng J 2021, 426, 130617.

[11] Kanzari F, Syakti AD, Asia L, Malleret L, Piram A, Mille G, Doumenq P. Distributions and sources of persistent organic pollutants (aliphatic hydrocarbons, PAHs, PCBs and pesticides) in surface sediments of an industrialized urban river (Huveaune), France. Sci Total Environ 2014, 478, 141–151.

[12] Venier M, Salamova A, Hites RA. How to distinguish urban vs. agricultural sources of persistent organic pollutants?. Curr Opin Environ Sci Health 2019, 8, 23–28.

[13] Guo W, Pan B, Sakkiah S, Yavas G, Ge W, Hong HZ. Persistent organic pollutants in food: Contamination sources, health effects and detection methods. Int J Environ Res Public Health 2019, 16, 4361.

[14] Göktaş RK, MacLeod M. Remoteness from sources of persistent organic pollutants in the multi-media global environment. Environ Pollut 2016, 217, 33–41.

[15] Adeola FO. Boon or bane? The environmental and health impacts of persistent organic pollutants (POPs). Hum Ecol Rev 2004, 11, 27–35.

[16] Abelsohn A, Gibson BL, Sanborn MD, Weir E. Identifying and managing adverse environmental health effects: Persistent organic pollutants. CMAJ 2002, 166, 1549–1554.

[17] Qing L, Loganath A, Seng Chong Y, Tan J, Philip Obbard J. Persistent organic pollutants and adverse health effects in humans. J Toxicol Environ Health Part A 2006, 69, 1987–2005.

[18] Gaur N, Narasimhulu K, PydiSetty Y. Recent advances in the bio-remediation of persistent organic pollutants and its effect on environment. J Clean Prod 2018, 198, 1602–1631.

[19] Sun B, Li Q, Zheng M, Su G, Lin S, Wu M, Meng B. Recent advances in the removal of persistent organic pollutants (POPs) using multifunctional materials: A review. Environ Pollut 2020, 265, 114908.

[20] Nguyen VH, Smith SM, Wantala K, Kajitvichyanukul P. Photocatalytic remediation of persistent organic pollutants (POPs): A review. Arab J Chem 2020, 13, 8309–8337.

[21] Karthigadevi G, Manikandan S, Karmegam N, Subbaiya R, Chozhavendhan S, Ravindran B, Awasthi MK. Chemico-nanotreatment methods for the removal of persistent organic pollutants and xenobiotics in water – A review. Bioresour Technol 2021, 324, 124678.

[22] Sharma A, Kumar V, Shahzad B, Tanveer M, Sidhu GPS, Handa N, Kohli SK, Yadav P, Bali AS, Parihar RD, Dar OI. Worldwide pesticide usage and its impacts on ecosystem. SN Applied Sciences 2019, 1, 1446.

[23] Alston JM, Pardey PG. Agriculture in the global economy. J Econ Perspect 2014, 28, 121–146.

[24] Singh AK, Chandra R. Pollutants released from the pulp paper industry: aquatic toxicity and their health hazards. Aquat Toxicol 2019, 211, 202–216.

[25] Satapute P, Kamble MV, Shivakantkumar SA, Jogaiah S. Influence of triazole pesticides on tillage soil microbial populations and metabolic changes. Sci Total Environ 2019, 651, 2334–2344.

[26] Luo F, Chen Z, Megharaj M, Naidu R. Simultaneous removal of trichloroethylene and hexavalent chromium by green synthesized agarose-Fe nanoparticles hydrogel. Chem Eng J 2016, 294, 290–297.

[27] Gowdy J. Our hunter-gatherer future: climate change, agriculture and uncivilization. Futures 2020, 115, 102488.

[28] Wang B, Sun Z, Sun Q, Wang J, Du Z, Li C, Li X. The preparation of bifunctional electrospun air filtration membranes by introducing attapulgite for the efficient capturing of ultrafine PMs and hazardous heavy metal ions. Environ Pollut 2019, 249, 851–859.

[29] Bharagava RN, Saxena G, Mulla SI. Introduction to industrial wastes containing organic and inorganic pollutants and bioremediation approaches for environmental management. In: Saxena G, Bharagava R (eds.). Bioremediation of industrial waste for environmental safety. Springer, Singapore, 2020.

[30] Bhavya G, Seema AB, Raja M, Nagaraja G, Huntrike SS, Shashikant SU, Sudisha J. Remediation of emerging environmental pollutants: A review based on advances in the uses of eco-friendly biofabricated nanomaterials. Chemosphere 2021, 275, 129975.

[31] Tripathi S, Srivastava P, Devi RS, Bhadouria R. Influence of synthetic fertilizers and pesticides on soil health and soil microbiology. In: Agrochemicals detection, treatment and remediation. Butterworth-Heinemann, 2020, 25–54.

[32] Nyankson E, Kumar RV. Removal of water-soluble dyes and pharmaceutical wastes by combining the photocatalytic properties of Ag_3PO_4 with the adsorption properties of halloysite nanotubes. Mat Tod Adv 2019, 4, 100025.

[33] Sun Y, Liang J, Tang L, Li H, Zhu Y, Jiang D, Song B, Chen M, Zeng G. Nano-pesticides: a great challenge for biodiversity?. Nano Tod 2019, 28, 100757.

[34] Satapute P, Paidi MK, Kurjogi M, Jogaiah S. Physiological adaptation and spectral annotation of Arsenic and Cadmium heavy metal-resistant and susceptible strain *Pseudomonas taiwanensis*. Environ Pollut 2019, 251, 555–563.

[35] Aloizou AM, Siokas V, Vogiatzi C, Peristeri E, Docea A, Petrakis D, Provatas A, Folia V, Chalkia C, Vinceti M, Wilks M. Pesticides, cognitive functions and dementia: a review. Toxicol Lett 2020, 326, 31–51.

[36] Kumar A, Luxmi V. Novel green photo-catalyst 'turmeric roots' for pesticides degradation: preparation and characterizations. Mater Lett 2020, 262, 127030.

[37] Tijani JO, Fatoba OO, Madzivire G, Petrik LF. A review of combined advanced oxidation technologies for the removal of organic pollutants from water. Water Air Soil Pollut 2014, 225, 2102.

[38] Boulkhessaim S, Gacem A, Khan SH, Amari A, Yadav VK, Harharah HN, Elkhaleefa AM, Yadav KK, Rather S-U, Ahn HJ, *et al.* Emerging trends in the remediation of persistent organic pollutants using nanomaterials and related processes: A review. Nanomat 2022, vol 12, 2148.

[39] Seow TW, Lim CK, Nor MHM, Mubarak MFM, Lam CY, Yahya A, Ibrahim Z. Review on wastewater treatment technologies. Int J Appl Environ Sci 2016, 11, 111–126.
[40] Crini G, Lichtfouse E. Advantages and disadvantages of techniques used for wastewater treatment. Environ Chem Lett 2019, 17, 145–155.
[41] Ganie AS, Bano S, Khan N, Sultana S, Rehman Z, Rahman MM, Khan MZ. Nanoremediation technologies for sustainable remediation of contaminated environments: Recent advances and challenges. Chemo 2021, 275, 130065.
[42] Rane NR, Tapase S, Kanojia A, Watharkar A, Salama ES, Jang M, Jeon BH. Molecular insights into plant–microbe interactions for sustainable remediation of contaminated environment. Bioresour Technol 2022, 344, 126246.
[43] Chauhan HA, Rafatullah M, Ali KA, Umar MF, Khan MA, Jeon BH. Photocatalytic activity of graphene oxide/zinc oxide nanocomposite derived from rice husk for the degradation of phenanthrene under ultraviolet-visible light. J Water Process Eng 2022, 47, 102714.
[44] Zhang Y, Li J, Liu F, Yan H, Li J, Zhang X. Reduction of Gibbs free energy and enhancement of Methanosaeta by bicarbonate to promote anaerobic syntrophic butyrate oxidation. Bioresour Technol 2018, 267, 209–217.
[45] Musteret CP, Teodosiu C. Removal of persistent organic pollutants from textile wastewater by membrane processes. Environ Eng Manag J 2007, 6, 175–187.
[46] Fujii S, Polprasert C, Tanaka S, Hong Lien NP, Qiu Y. New POPs in the water environment: Distribution, bioaccumulation and treatment of perfluorinated compounds–a review paper. J Water Supply Res Technol AQUA 2007, 56, 313–326.
[47] Tibi F, Charfi A, Cho J, Kim J. Fabrication of polymeric membranes for membrane distillation process and application for wastewater treatment: Critical review. Process Saf Environ Prot 2020, 141, 190–201.
[48] Sarkar S, Chakraborty S. Nanocomposite polymeric membrane a new trend of water and wastewater treatment: A short review. Groundw Sustain Dev 2021, 12, 100533.
[49] Marchetti P, Jimenez Solomon MF, Szekely G, Livingston AG. Molecular separation with organic solvent nanofiltration: A critical review. Chem Rev 2014, 114, 10735–10806.
[50] Shon HK, Phuntsho S, Chaudhary DS, Vigneswaran S, Cho J. Nanofiltration for water and wastewater treatment–a mini review. Drink Water Eng Sci 2013, 6, 47–53.
[51] Modi S, Yadav VK, Gacem A, Ali IH, Dave D, Khan SH, Jeon BH. Recent and emerging trends in remediation of methylene blue dye from wastewater by using zinc oxide nanoparticles. Water 2022, 14, 1749.
[52] Li K, Yan L, Zeng Z, Luo S, Luo X, Liu X, Guo Y. Fabrication of $H_3PW_{12}O_{40}$-doped carbon nitride nanotubes by one-step hydrothermal treatment strategy and their efficient visible-light photocatalytic activity toward representative aqueous persistent organic pollutants degradation. Appl Catal B Environ 2014, 156, 141–152.
[53] Valizadeh S, Lee SS, Baek K, Choi YJ, Jeon BH, Rhee GH, Park YK. Bioremediation strategies with biochar for polychlorinated biphenyls (PCBs)-contaminated soils: A review. Environ Res 2021, 200, 111757.
[54] Yadav VK, Ali D, Khan SH, Gnanamoorthy G, Choudhary N, Yadav KK, Manhrdas S. Synthesis and characterization of amorphous iron oxide nanoparticles by the sonochemical method and their application for the remediation of heavy metals from wastewater. Nanomat 2020, 10, 1551.
[55] Ghosh S, Malloum A, Bornman C, Othmani A, Osagie C, Esfahani Z, Dehghani MH. Novel green adsorbents for removal of aniline from industrial effluents: A review. J Mol Liq 2022, 345, 118167.
[56] Sadegh H, Ali GA. Potential applications of nanomaterials in wastewater treatment: Nanoadsorbents performance. In: Research anthology on synthesis, characterization, and applications of nanomaterials. IGI Global, Hershey PA USA, 2021, 1230–1240.

[57] Subbaiah MP, Kalimuthu P, Jung J, Jeon BH. Recent advances in effective capture of inorganic mercury from aqueous solutions through sulfurized 2D-material-based adsorbents. J Mater Chem A 2021, 9, 18086–18101.
[58] Sahoo SK, Padhiari S, Biswal SK, Panda BB. Hota G. Fe3O4 nanoparticles functionalized GO/g-C_3N_4 nanocomposite: An efficient magnetic nano adsorbent for adsorptive removal of organic pollutants. Mater Chem Phys 2020, 244, 122710.
[59] Yadav VK, Khan SH, Choudhary N, Tirth V, Kumar P, Ravi RK, Godha M. Nanobioremediation: A sustainable approach towards the degradation of sodium dodecyl sulfate in the environment and simulated conditions. J Basic Microbiol 2022, 62, 348–360.
[60] Ahmed SF, Mofijur M, Nuzhat S, Chowdhury AT, Rafa N, Uddin MA, Show PL. Recent developments in physical, biological, chemical, and hybrid treatment techniques for removing emerging contaminants from wastewater. J Hazard Mater 2021, 416, 125912.

Tijo Cherian*, Wajih Jamal, Shibin E., R. Mohanraju

Green Technologies for the Removal of Heavy Metals

Abstract: Environmental habitats are contaminated by industrial and mining practices, as well as the ongoing discharge of industrial effluents. Heavy metals (HMs) are among the contaminants that are causing the most worry due to their widespread pollution and toxicity for biota, especially the native microbial community, which plays a crucial part in maintaining ecosystem health by nutrient cycling and contaminant elimination. Many economic activities produce heavy metals like lead and cadmium, which are seriously neurotoxic, notably plating and those producing batteries, pigments, and armaments. HMs can persist in the ecosystem through the processes of biomagnification and bioaccumulation, which are particularly significant when it comes to edible species (like fishes), posing a toxicological danger to both humans and other biota. Conventionally, the most widely used techniques for removing metals from water include both physical and chemical processes However, the majority of these techniques have drawbacks, including high energy requirements, use of chemicals, non-selectivity, high operational and maintenance costs, and the requirement for proper disposal of the toxic waste produced. Keeping in mind the green remedial approach as an alternative to disposal and incineration has been encouraged by the safeguarding of natural resources and the optimization of the use of non-renewable energy. In this context, the present chapter focuses on the development of novel green remediation technologies utilizing naturally occurring biological sources for the decontamination heavy metal contaminated varied environmental domains.

Keywords: Green technologies, environmental pollution, heavy metals

1 Introduction

The public concern over environmental deterioration, particularly water pollution, has always existed as it has posed a hazard to both environment and human health by metal-containing wastewater resulting from anthropogenic activity in a variety of

***Corresponding author: Tijo Cherian,** Department of Ocean Studies and Marine Biology, Faculty of Life Sciences, Pondicherry University, Puducherry, India, e-mail: tvarghese891@gmail.com
Shibin E., R. Mohanraju, Department of Ocean Studies and Marine Biology, Faculty of Life Sciences, Pondicherry University, Puducherry, India
Wajih Jamal, Department of Zoology, Faculty of Life Sciences, Aligarh Muslim University, Aligarh, Uttar Pradesh, India

https://doi.org/10.1515/9783111016825-016

industries, including agriculture, mining, batteries, nuclear, textile dyes, tanneries, and others. The degradation of soil, which reduces crop output, and the absorption of heavy metals from contaminated water by plants, which enter food chains and have harmful effects on humans and other animals, are some of the most immediate environmental impacts [1, 2]. Heavy metals, including Cu, Cd, Pb, Ni, Zn, Hg, Fe, Co, and Mn, are typically present in trace concentrations in wastewater effluent, but they are known as the most dangerous and pervasive elements [3]. As a result of the body's inability to metabolize certain heavy metals, they build up in soft tissues, endangering human health [1]. According to the United Nations, 80% of the world's wastewater is occasionally discharged to the ecosystem without treatment or with minimal treatment. Heavy metals are traces of elements that have an atomic density >4 g/cm^3 and can be found in both natural and anthropogenic wastewater. The primary human-caused sources of heavy metal pollutants are metal finishing, metal electroplating, fuel combustion, extraction activities, street runoffs, and landfills. The primary natural sites of heavy metal pollution are soil erosion, aerosol particulates, urban runoffs, and volcanic eruptions. Heavy metals have negative effects on human health as well as the environment. For instance, copper (Cu^{2+}) ions can harm the hepatic system, cause sleeplessness, and prevent soil enzymatic activity. While chromium may cause nausea, headache, and diarrhea; mercaptan causes rheumatoid arthritis, neurological and circulatory diseases, as well as effects on aquatic systems. In contrast to lead (Pb) poisoning, which contributes to kidney failure, muscle weakness, and baby brain damage, copper poisoning may cause liver disease and anemia [4]. Anemia, skin irritation, reduced voluntary muscle performance, and numerous organ diseases are only a few of the harmful impacts of heavy metals on human health. Due to the stable, non-biodegradable, highly soluble, and migratory nature of heavy metals, contaminated wastewater effluents have a negative impact on human health as well as adverse effects on plants, including inhibition of photosynthetic process, reduced enzyme activity, decreased seed germination, and decreased chlorophyll synthesis and function. The propensity of elements to form complexes, which in turn contribute to physiologic and biochemical activity, is influenced by the shape and array of their atomic electronic shells, which determines their reactivity [2].

There are many methods that have been suggested to effectively remove heavy metal from aqueous medium, including solvent extraction, coagulation, membrane filtration, ion exchange, chemical precipitation, and electrochemical methods. The selection of appropriate method depends on the costs, efficiency, dependability, feasibility, environmental effect, usability, and operational challenges that are typically taken into account [5]. But these technologies do have certain limitations, though, such as low effectiveness, insufficient elimination, delicate operating environments, large energy needs, and expensive disposal processes.

Among the removal techniques, the adsorption approach stands out due to its adaptability in terms of operation and design procedures and apparent significant impact on the toxicity, transport of heavy metals in wastewater, and biological availability. Transit of contaminant from aqueous solution to sorbent surface, transport within sorbent particle and adsorption onto solid surface are the three primary processes of adsorption onto solid sorbent. Through the electrostatic force of attraction, the charged pollutants have a tendency to adsorb on the oppositely charged adsorbents. Heavy metals have a high affinity for functional groups like surface hydroxyls. Adsorbents can be regenerate for multiple uses because adsorption is frequently reversible because it is convoyed by desorption making it an economical and highly efficient process in the treatment of high-quality effluents. The metrics of Freundlich adsorption isotherm (K_F) and Langmuir maximum adsorption capacity (q_{max}) are frequently employed to assess the metal adsorption of diverse materials. Physical adsorption and active adsorption (also known as chemisorption) are the two types of adsorption. The former involves the formation of ionic or covalency connections by chemical reactions, whilst the latter involves the interaction of adsorbent and adsorbate surface. Physical adsorption is less specific and reversible than chemisorption, which is irreversible and more specific. The efficiency of the adsorbent is affected by a number of variables, including pH, temperature, stirring time, and initial concentration. With the increase of the aforementioned parameters, the adsorption rate of heavy metal normally increases [6]. Additionally, to successfully remove the pollutants, adsorbents with high adsorption interactions with the target heavy metal ions should be utilized. Activated carbon, agricultural waste, biomass, NMOs (nanosized metal oxides), and others are a few examples of typical adsorbents. Because of increased adsorption capacity due to chemical alterations, the activated carbon is the most popular and widely utilized of these adsorbents in water purification. However, the powdered form of activated carbon is expensive and difficult to separate from the solution [7]. Additionally, studies show that using bioadsorbent in wastewater treatment can be both cost-effective and environmentally friendly.

Recently, given the advantages of nanotechnology, there is currently considerable interest in using highly porous nanostructures, such as fullerenes, graphene, metal organic frameworks (MOFs), and graphitic carbon nitride, as an option to remove and eliminate heavy metals from wastewater. Due to this, such nano-modified adsorbents may present opportunities for providing potential solutions to address the problem of water pollution. This chapter provides a concise overview of the theory behind the use of both traditional and cutting-edge adsorbents for the elimination of heavy metallic ions from wastewater. Additionally, a comparison of adsorptive performance of activated carbon with these materials can be appropriated as a benchmark, and a conclusion is drawn regarding the technical difficulties now faced by innovative materials and the prospects for using nano-based materials in the future.

2 Sources, consequences, and detection

2.1 Sources of heavy metal pollution

Heavy metal contamination has affected about 40–45% of the planet's water bodies like rivers and lakes [3]. These contaminants might have both natural and man-made sources. Natural sources include volcanic eruptions and interactions with naturally occurring metal-containing materials. Volcanoes can contribute to climate change by constant low emissions, such as degassing and geothermal activity, or through explosive volcanic activity [8]. Industrial (such as the burning of fossil fuels and the processing of metals), agricultural (such as the use of pesticides), and home activities are examples of anthropogenic sources (e.g. garbage, detergents) [9]. One of the important industries to take into account is mining, in part because it is essential to the financial system of both industrialized and developing nations. Due to the emission of heavy metals during mineral extraction and transportation through streams and rivers, this activity produces a significant amount of heavy metals which can be a part of sediments or soluble in water. Owing to their propensity to leak into the groundwater, these metal species can also contribute to water scarcity and inhibit crop growth due to soil erosion, posing major health risks to both local humans and animals [10]. Figure 1 abridges the major sources of heavy metal pollution in the environment.

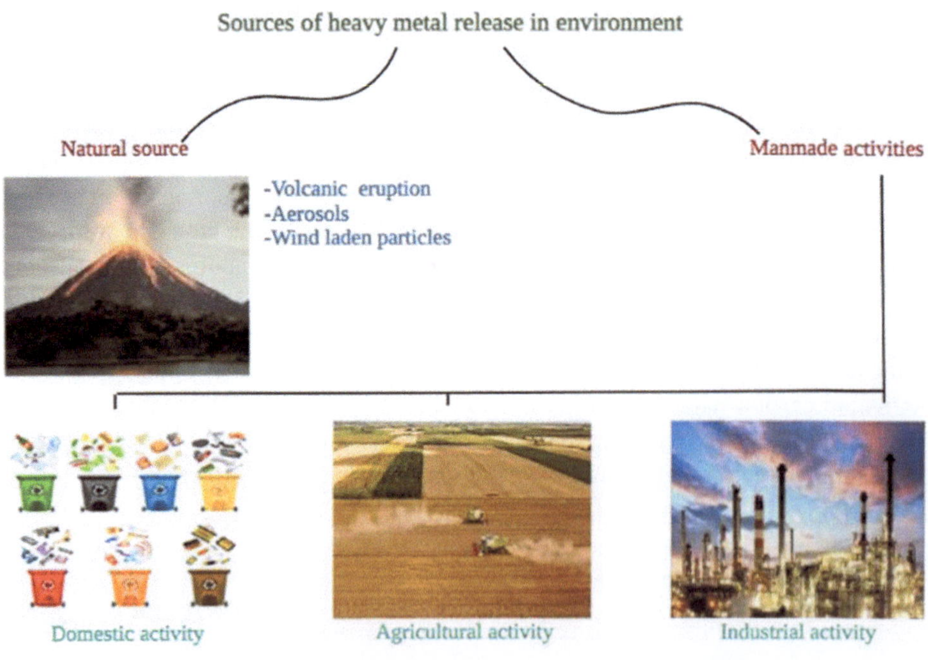

Figure 1: Sources of heavy metal pollution.

2.2 Toxic consequences of heavy metals

It is challenging to eliminate heavy metals from water as they can accrue in their specific reactive form or in conjunction with other metallic elements. These metals must be present in trace amounts for metabolic processes to function properly and prevent human illnesses and syndromes [7]. But, an excessive quantity of these heavy metals can cause major health issues in people, such as degenerative processes in the muscles, body, and nervous system [11]. Multiple sclerosis, Alzheimer's disease, muscular dystrophy, several forms of cancer has been linked to the long-term exposure to hazardous heavy metals [12–14]. In addition to gross injuries to the kidneys, blood components, liver, lungs, and other important organs, impairment to the central nervous system and mental functions has also been noted. In addition to producing changes in variety, densities, community structure, and species composition of populations, the disposal of heavy metals into water bodies can also result in physicochemical, and pathological diseases [15]. The type and intensity of alteration, which may potentially stop enzymatic operations, will principally rely on the concentration of heavy metal ions [16].

2.3 Detection

The amount of heavy metal pollution in given water must be found and measured before choosing an effective restoration strategy. The method chosen for this job must be economical, ecologically friendly, sensitive, and selectively adequate to find traces with reasonably high accuracy [17]. In spite of the challenges of fulfilling all these requirements, the modern technology allows for the reliable identification of heavy metals and quick determination of their quantities. This makes it possible to learn more about the scope of the problem of water contamination throughout the world.

3 Mechanisms for the removal of heavy metal ions

It's an open issue that how heavy metal ions can be eliminated from solutions through interaction mechanisms [18]. Most contemporary remedial techniques incorporate sorption-reduction, sorption, or photocatalytic degradation in a means or another. In this situation, determining the physical and chemical characteristics of these pollutants and combining sophisticated characterization (such as spectroscopic estimations with numerical simulations at the molecular level) with elimination route definition seems to be necessary to comprehend their potential effects on the environment [19]. One of the most fundamental methods for obtaining heavy metal ions from solutions on the sorption is chemical interaction between functional groups and metal ions, particularly those with different carbonyl groups, hydroxyl groups, or oxygen-containing groups.

Strong surface complexes are formed through hydrophobic contacts, π-π donor-acceptor interactions, hydrogen bonding, or electrostatic attractions [20]. Similarly, the sorption-reduction process is another method for the conversion of high valency metallic ion to low valency metallic ion and their immobilization. The high valency metallic ion undergoes initially a reduction step, after which they develop into denser particles or clusters that precipitate with ease. Therefore, in the natural environment low valency metallic ions are far more biofriendly. It is generally acknowledged that the high valency metallic ions are far more transportable than the low valency metallic ions. The degradative photocatalysis method under visible light irradiation has been widely used to remove low quantities of heavy metal ions from liquids. POPs (persistent organic pollutants) and other organic contaminants are typically eliminated via this approach [21]. This stratagem is based on photocatalytic reactions, which are influenced by the topologies of catalyst, visible light absorption, mass transport mechanisms, and the distribution of active sites on the surface [20]. In order to increase the photocatalytic efficiency, it is quite common to create customized composites or nanoparticles where the band energy gap can be regulated.

4 Conventional methods

This section explains some common heavy metal removal techniques as well as the underlying concepts behind them (Figure 2).

4.1 Chemical precipitation

Chemical precipitation is an efficient way to get rid of heavy metals, especially from the waste water from paper and electroplating units. In this procedure, heavy metals found in wastewater react with chemical precipitants such as iron salts, lime, alum, certain polymers to form insoluble precipitates [22]. Metals are easier to remove by this reaction. Optimizing variables like temperature, pH, starting dose, and ion charge will increase the removal capacity and efficiency. Sulfide and hydroxide precipitation are two types of chemical precipitation. Coagulants may enhance the removal of heavy metals by filtering or sedimentation during hydroxide precipitation. On the other hand, filtration or gravity settling is used to remove the sludge that results from metal sulfide precipitation. Because of the toxic nature of the sulfide ion and H_2S, this procedure necessitates pre- and post-treatment, along with careful monitoring of the amount of chemicals.

4.2 Coagulation/flocculation

A very effective physicochemical approach for eliminating heavy metals is coagulation-flocculation. NOM, turbidity, and other wastewater pollutants are decreased as a result of the process, which causes colloids and fine particles to aggregate into larger particles. In initial stages, a coagulant is added to water which encourages colloidal particles to coalesce into tiny aggregates known as flocs. Aluminum sulfate, ferric sulfate, PACI (polyaluminum chloride), polyacrylamide (PAM), and polymeric ferric sulfate (PFS) are the most commonly used coagulants [23]. Flocs gather together under gentle agitation in the second step, settle, and finally are discarded as sludge. Due to its adaptability, this procedure is employed as a pre- and post-treatment/primary wastewater treatment.

4.3 Ion exchange process

The mechanism of reversible ion exchange takes place between the solid and liquid states. The ions of an electrolytic solution are specifically removed by an insoluble substance, which then releases chemically comparable amounts of additional ions with the same charge. Inorganic 3D matrices, synthetic organic resins, and new band hybrid materials are the most widely used ion exchange materials [24, 25]. A suitable ion exchange resin can provide an effective and affordable solution to pollution management requirements.

4.4 Membrane-based technology

A membrane is a partition that allows the movement of a particular group of substances while obstructing the transit of others. The charge-charge repulsion effect (also known as Donnan exclusion), steric hindrance/size exclusion mechanism, and the adsorption capacity of particular pollutants play a vital role in this technology. Organic and inorganic pollutants, as well as suspended particles, can be removed by this treatment. Membranes can be categorized as: organic (constructed of man-made organic polymers like cellulose acetate and polyethylene) and inorganic (constructed of metals, zeolites, silica, ceramics) [26]. Microfiltration, ultrafiltration, and distillation are examples of low-pressure membrane processes; Nanofiltration, reverse osmosis, and electrodialysis are examples of high-pressure membrane processes and, direct osmosis, electrodialysis, and liquid membrane processes are examples of osmotic pressure-driven membrane processes [27]. The primary techniques employed in the removal of heavy metals from wastewater include reverse osmosis, ultrafiltration, microfiltration, and nanofilteration. The exclusion performance of a membrane is greatly influenced by a number of variables, including the surface charge, dimension, and distribution of pores, solution flow, degree of hydrophilicity, and the presence of functional groups [27].

4.5 Electrochemical-based technology

Heavy metal ions from bodies of water can be effectively removed using electrochemical treatment techniques. These methods recover metals in their elemental form by using anodic and cathodic reactions in an electrochemical cell [28]. Electrochemical procedures such as electroflotation, electrodeposition, and electrocoagulation are available. In traditional chemical coagulation, electrocoagulation takes place where both oxidation and reduction reactions occur at electrodes serving as anode and cathode, respectively. The anion in the effluent reacts with the charged ionic metals, and coagulant is produced by the electrolytic oxidation of an appropriate anode. By depositing pollutants on the cathode or by flotation, the concurrent cathodic reaction enables the elimination of contaminants. This method produces less sludge, requires no chemicals, and is simple to use.

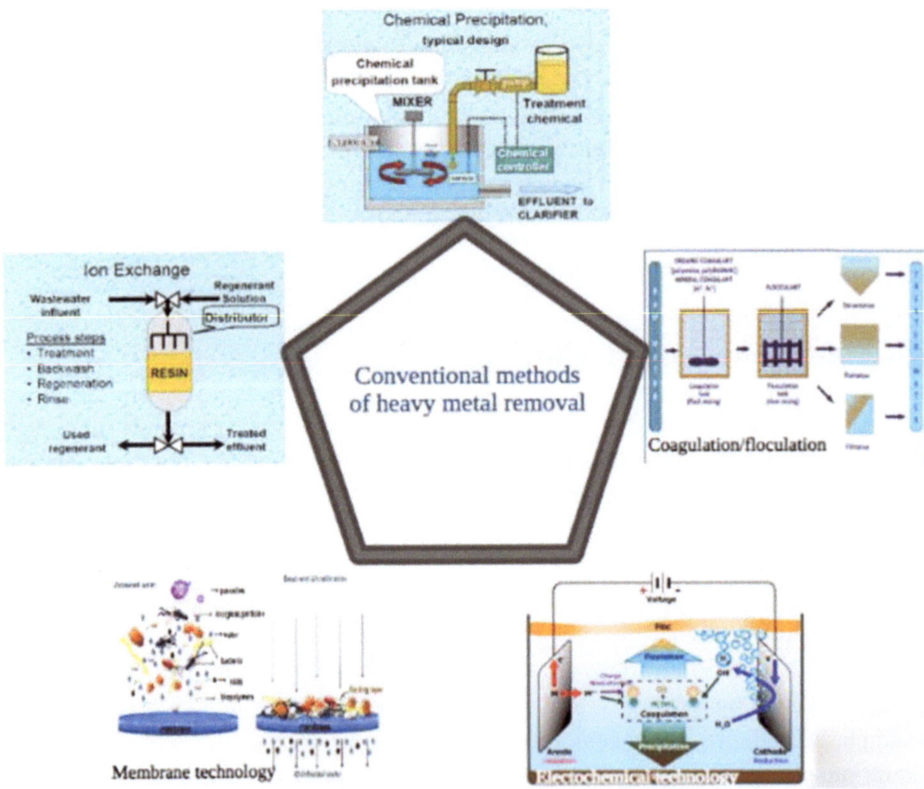

Figure 2: Conventional treatments for removing heavy metals.

5 Non-conventional methods

This section discusses the most contemporary and advanced treatment techniques, as follows:

5.1 Adsorption

One of the finest ways to eliminate heavy metals and other impurities is adsorption. The possibility of preventing substantial secondary pollutants, fairly low energy consumption, excellent removal performance, and technical operational specifications are some of its advantages [6]. Adsorbents should possess a number of advantageous qualities, such as high mechanical strength, sizable specific surface area, strong thermal stability, predictable morphology, and ecologically benign processing. Given the virtues of selectivity, high adsorption efficiency, low cost, and reusability, the reaction proceeds with high performance. Some of the most popular adsorbents are polymer-based materials, activated carbon (AC), magnetic materials, biomaterials, and industrial and agricultural wastes.

Due to their enormous surface area, complicated chemical makeup, and microporous structure, ACs (activated carbons) are often utilized as adsorbents in water treatment facilities to remove heavy metal ions. They have a variety of functional groups on their exterior, including carboxyl, carbonyl, phenol, lactone, quinone, etc. It has been demonstrated that pore size distribution and surfacial functional groups affect the adsorption efficiency of ACs [29]. Usually, the manufacture of ACs begins with the use of coal, coconut shells, and wood. Due to the high cost of coal-based ACs, more research has been done on less-priced ACs made from a variety of available agricultural waste products.

Polymeric materials provide excellent stability, processability, and accurately bind to target pollutants when used as a matrix. The structure of these materials can be modified to considerably improve the adsorption capabilities, and they are easily functionalized and bendable. A substance that will be used to remove pollutants needs to have qualities including affordability, non-toxicity, biocompatibility, recyclability, biodegradability, and ease of synthesis, preparation, or extraction. Many natural polymers, including chitosan, alginate, lignin, silk, cellulose, have these essential properties [30].

Cellulose, hemicellulose, starch, and lignin and are the primary components of agricultural leftovers. These biopolymers have robust, exceptional capacities for adsorption and a variety of active functional groups that allow metal complexation. The capacity of biosorbents to be easily tweaked to improve their industrial scale adaptability and adsorption capabilities is another one of their primary advantages. These residues/by-products are more cost-effective and produce less chemical/biological sludge when compared to conventional cleanup procedures. The efficacy of residues is influenced by variables such temperature, pH, heavy metal content and adsorbent dosage [31]. Also, as industrial wastes are inexpensive, accessible, and effective, they have been utilized in the elimination of heavy metals. Ligin, fly ash, blast furnace slag, sludge, flue dust, red mud are some of these materials.

One of the reasons for studying the pollutant removal from water using various composites, nanoparticles, and magnetic particles is the ease with which magnetic adsorbents can be readily isolated from an aqueous solution using an external magnetic field, leading to more affordable and effective water treatment methods [8]. Iron oxides like hematite (α-Fe_2O_3), magnetite (Fe_3O_4), and maghemite (γ-Fe_2O_3), zero-valent iron and spinel ferrites are a few examples. These iron oxides are very useful when used as nanoparticles because of their large surface area, high redox activity, extraordinary surface charge, and abundance of active sites.

The benefit of composite adsorbents is their capacity to integrate two or more different types of materials to produce an adsorbent that outperforms the sum of its parts. When magnetite nanoparticles (MNP) were added to lignocellulosic sawdust residues (LR), the resulting adsorbents required fewer contact times to obtain the similar elevated elimination efficiencies as those accomplished with nanoparticles/residues alone. Furthermore, because the adsorptive removal performance of both composites was found to be equivalent, it was documented that relatively high volumes of the inexpensive LR could be combined with minimal volumes of expensive MNP to produce very effective adsorbents [7].

5.2 MFCs (microbial fuel cells)

MFCs are a practical technology because they can utilize the organic material in wastewater to produce electricity using biocatalysts like bacteria. In MFCs, bacteria and other microbes produce protons and electrons on an anode cell (anaerobic), which sends protons through a proton exchange membrane and electrons through an external circuit to the cathode cell (aerobic). The circuit is closed by the high redox potential of O_2 (oxygen), which is reduced to water at the cathode [10]. With MFCs, heavy metals (of higher redox potential) can be removed and recovered at cathode chambers rather than the anode electrode.

5.3 Nanotechnology-based approach

Nanomaterials, which are used in therapies based on nanotechnology, have drawn immense contemplation in recent years due to their distinctive electrical, optical, and magnetic properties and high surface-to-volume ratio [32]. One of the main nanotechnological method employed for the removal of heavy metals is nanofiltration. Alumina nanofiber–based systems are very efficient at chemisorbing away dissolved forms. Additionally, low-dimensional structures are most frequently utilized for the filtration, sterilization, and elimination of heavy metals from water, including magnetic nanoparticles, single/multi-metallic oxides, and non-metallic oxides, nanocarbon, nanoclays [33]. All these nanostructures have many reactive surfaces, and many of them may be made cheaply or using abundant natural resources. Similar technologies for wastewater treatment include nanoplates, nano-assemblies, hierarchical ZnO nano-rods, and microspheres with nanosheets. However, there is a considerable knowledge gap about the toxicity, environmental impact, and health consequences of nanomaterials, which limits their full potential [34].

5.3.1 Nanosized metal oxides

Metal oxides are one of the most promising groups of adsorbent materials due to their exceptional adsorption capability, simplicity in modification and synthesis, low cost, and ability of mass production. A significant advantage of adsorption over competing methods is the ability of nanosized metal oxides (NMOs) to eliminate the majority of pollutants to trace amounts even in low concentrations. Additionally, NMOs have larger adsorption rate constants than activated carbon, typically 10 to 200 times higher. This is due to their highly reactive surfaces, which enable adsorption to occur more quickly toward certain pollutants. Due to their abundance of surface-active sites, metal oxides can interact with a variety of pollutants [35]. Some multivalent heavy metal ions, such as Cr^{6+}, Cr^{3}+, Sb^{3+}, Sb^{5+}, As^{3+}, and As^{5+}, due to their low affinity for adsorption sites, have been hypothesized to undergo reduction or oxidation when in contact with metal oxides. Redox processes involve metal oxides as MnO_2, Fe_3O_4, CeO_2, and TiO_2. Electrostatic attraction causes an outer-sphere complex formation followed by the oxidation or reduction between the adsorbent and adsorbate [2]. Also, the process called surface precipitation is also used by metal oxides to absorb heavy metals. Desorption and regenerability is essential to prevent secondary pollution issues and to reuse the costly adsorbents, thereby reducing the costs. In order to desorb different organic and inorganic impurities, metal oxides are typically recycled chemically through a process called chemical regeneration. Acids, chelating agents, and organic solvents are the chemical solvents used to desorb metal oxides. Electrostatic repulsion is a common method for desorbing heavy metal cations from acidic solutions during pH-dependent adsorption [2]. Nanoscale metal oxides are also capable of firmly adsorbing metal ions from aqueous environments due to their large surface area and high adsorption capacity. Additionally, due to their reduced diffusion resistance, smaller particles have faster adsorption kinetics. On the other hand, nanoparticles frequently agglomerate. Adsorption capabilities of metal oxides are hypothesized to be enhanced by surface modification.

Zinc, titanium, silicon, magnesium, iron, and manganese oxides are widely used for heavy metal cleanup in the form of nanoparticles. High porosity silica nanoparticles have silanol groups on their surface that are simplified and modified using silane chemistry. By themselves, in combination with substances like GO, or when functionalized, silica nanoparticles may efficiently eliminate metal ions like Hg^{2+}, Pb^{2+}, and Ni^{2+} [36]. IONPs (magnetic iron oxide nanoparticles) are useful for easily removing and sorting heavy metals. The ferric oxide nanoparticles are the ones that are employed the most because of their strong redox potential, huge surface charge, and reusable nature. The hydroxyl groups on the surface of IONPs due to surface modification. Similar to silica, IONPs are capable of removing a range of heavy metals, and functionalization can enhance their adsorption ability. Compared to their bulk counterparts, TiO_2 nanoparticles exhibit better catalytic capabilities. The metals and metalloids Cd^{2+}, Zn^{2+}, Cr^{6+}, As^{5+}, Nd^{3+}, and La^{3+} can be adsorbed by these nanoparticles [37].

5.3.2 Nanocarbons

Nanocarbons are employed as very effective adsorbents and are functionalized in form of graphene, CNTs (carbon nanotubes), or their derivatives. Because of their extremely reactive surface area and carbonaceous makeup, CNTs may directly interact with water contaminants through electrostatic, hydrophobic, π-π interactions, hydrogen bonding, and covalent bond formation [38]. Additionally, they are great for water cleanup due to their hollow, rod-like form, which provides several opportunities for adsorption. Even if it is difficult, by developing composite materials in the form of fibers, films, reinforced polymers, aerogels, and gels, it is feasible to transfer the properties of CNTs to the macroscopic world [39]. Through acidic treatments, functional molecule/group grafting, and metal impregnation, the CNT surface can be easily altered. The most recent model enables control of the hydrophobicity, dispersion, surface charge, and surface area of nanotubes, as well as potential adsorption. When heavy metals like copper, cadmium, lead, and mercury have to be removed from wastewater, functionalization by sulfuric, nitric, hydrochloric acid introduces oxy functional groups [38].

Another material with promise for treating water is graphene. Its hybridized sp^2-structure, along with its flat, 2D (two-dimensional) honeycomb atom arrangement, and the related electrical band structure, are all easily modifiable for special functionalities [40]. Applications for heavy metal removal have utilized graphene oxide (GO)-based materials, including composites and those that have undergone chemical modification [41]. Other graphene-based derivatives, such its reduced form (RGO), are effectively used in the treatment of water [32]. As an illustration, 3D graphene-based macrostructures offer a sizeable surface area, a porous network, excellent electrical mobility, ultra-lightweight characteristics, and exceptional chemical active sites. These structures usually act as scaffolding to immobilize nanoparticles, greatly expanding the range of possible uses for them in water treatment [42].

5.3.3 Graphitic carbon nitride (g-C_3N_4)

As the most stable allotrope of carbon nitride, g-C_3N_4 is a 2D nano-adsorbent. It has exceptional thermal and chemical stability under optimal conditions and has stacked layer architectures. Tri-s-triazine (C_6N_7) units in g-C_3N_4 are organized through hydrophobic effect and π-π contact. Lone pair electrons on nitrogen are responsible for the negatively charged functions, and these negatively charged functional groups have a fair amount of potential for metal cation reactions. Additionally, the surfacial amino groups of g-C_3N_4 strongly participate in the adsorption of heavy metals [43]. It is envisaged that g-C_3N_4 will be a cheap adsorbent for contaminants. It has been reported that g-C_3N_4 nanosheets made by thermal and ultrasonic methods have outstanding metal ion and adsorption capabilities. The specific surface area of g-C_3N_4 is 418.28 m^2/g while the average pore size was found to be 9.762 nm. The ionic strength had no effect on the adsorption process, whereas pH and temperature had a significant impact. However, only a small number of the precursors utilized to create g- C_3N_4 are dangerous, and mass manufacture continues to be difficult. Even though g-C_3N_4 has certain limitations when it comes to its synthesis and use in the management of environmental pollution, it demonstrates potential as an adsorbent for extracting heavy metal cations and anions from wastewater. Future studies should concentrate on developing suitable reactors to attain the practicability in wastewater treatment.

5.3.4 MXenes

Ceramics known as MXenes are one of the most significant categories of 2D transition metal carbide/nitride materials. They have great electrical conductivity, high chemical stability, and eco-friendly qualities. When applied to wastewater treatment, 2D nanomaterials have a huge specific surface area due to their enormous lateral size and nanoscale thickness. Additionally, the surface of MXene is abundant with active chemical functional groups (hydroxyl, fluorine, oxygen, etc.), and its hydrophilic nature makes it an efficient adsorbent for a variety of ionic and molecular species. Both pure and functionalized MXenes efficiently adsorb heavy metal ions due to their profusion of functional groups, flexible surface chemistry, and capacity to respond to local electrical and chemical disturbances. However, there are just a few MXenes that have been experimentally synthesized, which have restricted research in this area. Given their effectiveness and high selectivity adsorption capability, MXene adsorbents are expected to be less expensive than other high-performance adsorbent materials to address the challenge of wastewater treatment.

5.3.5 MOF (metal organic framework)

The cage-like structure of MOFs, which are crystalline porous solids formed of a three-dimensional (three-dimensional) network of positively charged metal ions (metal nodes) that are held in place by multidentate organic molecules by coordination bonds [44]. MOFs have an impressively huge interior surface area (1,000–10,000 m^2/g) greater than activated carbons because of their hollow and highly organized structure. The majority of MOFs display excellent chemical stability even in challenging circumstances. Additionally, they are easily and cheaply generated on a big scale. With its highly organized porosity structure and adaptable physiochemical characteristics, the next-generation adsorbent material MOF is promising and effective at removing heavy metal ions [45].

The first anion examined with a heterometallic 3D framework 1-ClO_4 as an adsorbent was CrO_4^{2-}. The technique of ion exchange was used to remove 85% of the ions. Utilizing a novel 3D MOF, a elevated adsorption capacity (207 mg/g) was attained. However, during the recycling procedures, where the chromate ions were desorbed, its single crystallinity was also found to be destroyed [46].

Due to its developed porosity, which facilitates transfer of mass inside the structure of pore, ZIF-8, a MOF-based material, demonstrated excellent exclusion of As(V) (arsenic ions) (exhibiting 90.92 mg/g adsorption capacity) [47]. When hydroxyl and amine groups are present, As(V) and As(III) can be withheld at the same time [48]. Compared to pure ZIF-8 nanoparticles, MOF-based composites (140.27 mg/g) had a 1.6 times higher q_{max} for As (III). There is still little knowledge of their regeneration [48]. Due of sulfur's powerful binding capacity, MOFs have been extensively researched for Hg^{2+} adsorption [49].

The close fit pore size of MIL-101 to the metal ion radius is responsible for the excellent selectivity of MOF towards Pb^{2+} [50]. ZIF-8, a MOF, had the highest Cu^{2+} adsorption capacity among the MOFs (800 mg/g), outperforming GO and AC-based adsorbent (GO/Fe_3O_4: 18 mg/g). At low initial metal concentrations, ion exchange took place, whereas at high initial Cu^{2+} concentrations, Cu^{2+} coordinates with nitrogen groups [51]. The common characteristics of MOFs and MOFs-based materials were great selectivity, quick adsorption kinetics, and excellent adsorption capacity. Less detailed information is provided with regard to their reusability, MOF stability across a larger pH range, and less expensive alternatives to expensive ligands.

5.4 Phyto-remediation of heavy metals

Phytoremediation is one of the most widely used techniques for eliminating heavy metal pollution from ecosystems. It employs unprocessed or genetically altered plant species to lessen the toxicity of contaminants [52, 53]. In order to restore contaminated ecosystems, plants frequently work in conjunction with other living things (like microorganisms), as well as external mechanisms (physical, electrical, chemical, thermal,) or inputs (macro-micronutrients, soil improvement, plant-litter, nanocomposites, fertilizers, emulsifiers, sorbents, biosurfactants, gels, hydrogels, clays, ceramics, water, aeration) [54, 55]. To increase the effectiveness of the approach and overcome its existing limits, a number of recently developed complementary approaches and processes can be used. However, infrastructure, time, logistics, and prices are also important as well as technical aspects for determining the effectivity of phytoremediation process. Phytoremediation helps preserve the structure of soil and ecosystems, prevents erosion and metal leaching, and can be used on a wide scale with little investment in installation and maintenance. One of its benefits is the potential for phytomining, which involves the recovery of metals that have bioaccumulated. Phytoremediation can also aid in creating local job and employment [56].

There are four stages that can be used to organize a heavy metal phyto-remediation framework. These four stages are (a) preliminary considerations, (b) processes, (c) cutting-edge phytoremediation, and (d) ultimate heavy metal disposal. During the first stage, it is important to focus on the following main issues: (a) identification of the polluted ecosystem; (b) delineating the physical-chemical characteristics of the heavy metals, such as their surface, volume, and interface with other compounds and their surroundings; and (c) selecting association of plant and the living organisms [52, 53]. The root system and biomass of the plant, growth rate, environmental adaptation, toxicity of the contaminants to the plant, and the amount of time required to lower the concentration of contaminants to the target level are some elements to consider when choosing a phytoremediator plant. Additionally, the plant must be resistant to pests and diseases [57].

Choosing an economical remediation method is important during the second stage after the pollution's magnitude has been determined. Stage three associates with the implementation of advanced automations, the majority of which are reliant on nanotechnology. The plants may, for instance, take up the metals from the growing medium, break them down into nanoparticles by their metabolic processes and amass them in their stems, leaves, and roots. By avoiding the translocation or obliterating the pollutants, plants that are designated as phyto-accumulators or phytotolerant which lessen the toxicity. The last stage involves analyzing how heavy metals, their by-products, and wastes are disposed of. If the post-harvesting mechanisms used are appropriately selected and used, this management could be effective and financially viable. Owing to the advantageous technological, environmental, economic outcomes in the favor of idea of circular economy-sustainable growth, valorization of residual wastes/substances with elevated heavy metal concentration has drawn particular attention globally [7].

In earlier sections, the main mechanisms and techniques for eliminating heavy metal ions from water have been covered. Consequently, it is conceivable to claim that no technology currently available can guarantee straightforward, affordable installation, affordable, environmentally stable operations, and high efficiency in the wide variety of conditions that call for the removal of water pollutants. So, a description regarding the key elements of the methodologies evaluated, emphasizing their limitations and merits needs to be warranted. The traditional approaches typically involve a mix of physicochemical and biological processes. However, due to technological and financial constraints, only a part of them are presently engaged by the industrial sector [58].

The available research claims that chemical precipitation-based therapies have straightforward operational procedures and are quite simple to automate. One of this major downside of this method is the generation of a lot of poisonous sludge, which necessitates additional treatment. On the other hand, the increased implementation costs are one of the key drawbacks of non-conventional treatment and new elimination technologies, from adsorption to MFCs, incorporating nanobiotechnology advancements, and Fenton reactions. The hybrid organic/inorganic composites have drawn particular interest in recent years due to their physicochemical characteristics, which support sorption-reduction processes and higher performance during sorption. This is a crucial component of the majority of remediation strategies. Combining the best adsorbent with the precise molecule will lead to high performance, cost-effective, and reliable treatment approaches. In an effort to benefit from the advantageous economic, social and environmental characteristics of these materials, numerous attempts have been made in the development of sustainable remedial techniques employing agro-industrial and industrial wastes. However, there have been a number of challenges with these initiatives, including the high costs of the pretreatments needed to condition the residues, some locations for the water treatment and the huge distances between the sources of the industrial and agro-industrial waste. Fenton-like reactions provide selective efficiency progress, but to meet the needs of industrially scalable processes, they should be integrated with other technologies. Nevertheless, the utilization of living things and inputs like soil to clean up polluted ecosystems and lessen their toxic effects is a benefit of plant remediation methods. To be more effective and competitive, they must be used in conjunction with other technologies [54, 55].

6 Conclusion

Due to a sharp rise over the past few decades, heavy metal pollution is especially concerning. Therefore, it is essential to create initiatives, regulations, techniques, and materials to address this issue without having any negative consequences on the environment or human health. Through this chapter, we have demonstrated how important it is to choose the best detection and measurement methods, treatment options, and operating parameters in order to maximize the efficiency and effectiveness of heavy metal removal. Fast, inexpensive, and reliable procedures that are appropriate for in situ and spot quantification of heavy metal ions must be researched and developed. This will help us understand and evaluate the severity of the issues caused by heavy metal pollution in different ecosystems. Although there are a number of established heavy metal elimination technologies and methods, it is important to pay close attention to the context of usage and differentiation between the suitable for centralized and decentralized situations. The financial and technical obligations for configuration, execution, and preservation must also be assessed. The advantages of utilizing a wide variety of adsorbents have been investigated, and adsorption has been recognized as one of the more practical options. Low-cost and convenient materials for disposal as well as the creation of nanostructured materials and advanced composite appear to be of special interest. Future research opportunities include working at sizes above the laboratory level, analyzing the toxicity of advanced materials presently under investigation, and doing investigations utilizing actual rather than synthetic waters. The results of this research should enable the advancement of current water treatment techniques and the development of cost-effective, safe, cutting-edge, eco-friendly, and efficient technologies that make use of residual wastes, advanced materials, and natural resources.

References

[1] Sankaran R, Show PL, Ooi CW, Ling TC, Shu-Jen C, Chen SY, Chang YK. Feasibility assessment of removal of heavy metals and soluble microbial products from aqueous solutions using eggshell wastes. Clean Technol Environ Pol 2020, 22, 773e786.

[2] Chai WS, Cheun JY, Kumar P,S, Mubashir M, Majeed Z, Banat F, Ho S-H, Show PL. A review on conventional and novel materials towards heavy metal adsorption in wastewater treatment application. J Cleaner Prod 2021, 296, 126589.

[3] Zhou Q, Yang N, Li Y, Ren B, Ding X, Bian H, Yao X. Total concentrations and sources of heavy metal pollution in global river and lake water bodies from 1972 to 2017. Global Ecol Conserv 2020, 22.

[4] Cheng SY, Show PL, Lau BF, Chang JS, Ling TC. New prospects for modified algae in heavy metal adsorption. Trends Biotechnol 2019, 37, 1255e1268.

[5] Crini G, Lichtfouse E. Advantages and disadvantages of techniques used for wastewater treatment. Environ Chem Lett 2019, 17, 145e155.

[6] Burakov AE, Galunin EV, Burakova IV, Kucherova AE, Agarwal S, Tkachev AG, Gupta VK. Adsorption of heavy metals on conventional and nanostructured materials for wastewater treatment purposes: a review. Ecotoxicol Environ Saf 2018, 148, 702e712.

[7] Zamora-Ledezma C, Negrete-Bolagay D, Figueroa F, Zamora-Ledezma E, Ni M, Alexis F, Guerrero VH. Heavy metal water pollution: A fresh look about hazards, novel and conventional remediation methods. Environ Technol Innov 2021, 22, 2021, 101504.

[8] Ali H, Khan E, Ilahi I. Environmental chemistry and ecotoxicology of hazardous heavy metals: Environmental persistence, toxicity, and bioaccumulation. J Chem 2019.

[9] Vareda J, Valencye A, Durães L. Assessment of heavy metal pollution from anthropogenic activities and remediation strategies: A review. J Environ Manag 2019, 246, 101–118.

[10] Vélez-Pérez LS, Ramirez-Nava J, Hernández-Flores G, Talavera-Mendoza O, Escamilla-Alvarado C, Poggi-Varaldo HM, Solorza-Feria O, López-Díaz JA. Industrial acid mine drainage and municipal wastewater co-treatment by dual-chamber microbial fuel cells. Int J Hydrog Energy 2020, 45, 13757–13766.

[11] Azeh Engwa G, Udoka Ferdinand P, Nweke Nwalo F, Unachukwu MN. Mechanism and health effects of heavy metal toxicity in humans. In: Poisoning in the modern world- new tricks for an old dog. Intech Open, 2019.

[12] MMS C-P, Inácio M, Neves O, Almeida AA, Pinto E, Oliveiros B, Ferreira da Silva EA. Human health risk assessment due to agricultural activities and crop consumption in the surroundings of an industrial area. Expos Health 2020a, 12, 629–640.

[13] Cabral Pinto MMS, Ordens CM, Condesso de Melo MT, Inácio M, Almeida A, Pinto E, Ferreira DSEA. An inter-disciplinary approach to evaluate human health risks due to long-term exposure to contaminated groundwater near a chemical complex. Expos Health 2020b, 12, 199–214.

[14] Gerardo B, Pinto MC, Nogueira J, Pinto P, Almeida A, Pinto E, Marinho-Reis P, Diniz L, Moreira PI, Simões MR, Freitas S. Associations between trace elements and cognitive decline: An exploratory 5-year follow-up study of an elderly cohort. Int J Environ Res Public Health 2020, 17, 1–18.

[15] Pratush A, Kumar A, Hu Z. Adverse effect of heavy metals (As, Pb, Hg, and Cr) on health and their bioremediation strategies: a review. Int Microbiol 2018, 21, 97–106.

[16] Mokarram M, Saber A, Sheykhi V. Effects of heavy metal contamination on river water quality due to release of industrial effluents. J Clean Prod 2020, 277.

[17] Malik LA, Bashir A, Qureashi A, Pandith AH. Detection and removal of heavy metal ions: a review. Environ Chem Lett 2019, 17, 1495–1521.

[18] Crini G, Lichtfouse E, Wilson LD, Morin-Crini N. Conventional and non-conventional adsorbents for wastewater treatment. Environ Chem Lett 2019, 17, 195–213.

[19] Liu X, Pang H, Liu X, Li Q, Zhang N, Mao L, Qiu M, Hu B, Yang H, Wang X, Liu X, Pang H, Liu X, Li Q, Zhang N, Mao L, Qiu M, Hu B, Yang H, Wang X. Orderly porous covalency organic frameworks-based materials: Superior adsorbents for pollutants removal from aqueous solutions orderly porous covalency organic frameworks-based materials: Superior adsorbents for pollutants removal from aqueous solutions. Innovations 2021, 2, 100076.

[20] Alharbi NS, Hu B, Hayat T, Rabah SO, Alsaedi A, Zhuang L, Wang X. Efficient elimination of environmental pollutants through sorption reduction and photocatalytic degradation using nanomaterials. Front Chem Sci Eng 2020, 14, 1124–1135.

[21] Cheng L, Liu S, He G, Hu Y. The simultaneous removal of heavy metals and organic contaminants over a bi2wo6/mesoporous tio2 nanotube composite photocatalyst. RSC Adv 2020, 10, 21228–21237.

[22] Verma B, Balomajumder C. Hexavalent chromium reduction from real electroplating wastewater by chemical precipitation. Bull Chem Soc Ethiop 2020, 34, 67–74.

[23] Xu D, Zhou B, Yuan R. Optimization of coagulation-flocculation treatment of wastewater containing Zn (II) and cr (VI).IOP Conf Ser Earth Environ Sci 2019, 227.

[24] Bashir A, Ahmad L, Sozia M, Taniya A, Mudasir M, Bhat A. Removal of heavy metal ions from aqueous system by ion – exchange and biosorption methods. Environ Chem Lett 2019, 17, 729–754.

[25] Dula T, Duke TN. Removal methods of heavy metals from laboratory wastewater. J Nat Sci Res 2019, 9.

[26] Ezugbe EO, Rathilal S. Membrane technologies in wastewater treatment: A review. Membranes (Basel) 2020, 10, 5.

[27] Abdullah N, Yusof N, Lau WJ, Jaafar J, Ismail AF. Recent trends of heavy metal removal from water/wastewater by membrane technologies. J Ind Eng Chem 2019, 76, 17–38.

[28] Vardhan KH, Kumar PS, Panda RC. A review on heavy metal pollution, toxicity and remedial measures: Current trends and future perspectives. J Mol Liq 2019, 290, 111197.

[29] Ahmad A, Azam T. Water purification technologies. In: Grumezescu AM, Holban AM (eds.). Bottled and packaged water. Elsevier, 2019, 83–120.

[30] Naseer A, Jamshaid A, Hamid A, Muhammad N, Ghauri M, Iqbal J, Rafiq S, Khuram S, Shah NS. Lignin and lignin based materials for the removal of heavy metals from waste water – An overview. Z Phys Chem 2019, 233, 315–345.

[31] Castro D, Rosas-Laverde NM, Aldás MB, Almeida-Naranjo C, Guerrero VH, Pruna AI. Chemical modification of agro-industrial waste-based bioadsorbents for enhanced removal of Zn(II) ions from aqueous solutions. Materials Submitted for publication 2021.

[32] Upadhyay AK, Singh R, Singh DPP. Restoration of wetland ecosystem: A trajectory towards a sustainable environment, restoration of wetland ecosystem: A trajectory towards a sustainable environment. Springer, Singapore, 2020.
[33] Vázquez-Núñez E, Molina-Guerrero CE, Peña Castro JM, Fernández-Luqueño F, de la Rosa-álvarez MG. Use of nanotechnology for the bioremediation of contaminants: A review. Processes 2020, 8, 1–17.
[34] Wołowiec M, Komorowska-Kaufman M, Pruss A, Rzepa G, Bajda T. Removal of heavy metals and metalloids from water using drinking water treatment residuals as adsorbents: A review. Minerals 2019, 9, 487.
[35] Rambabu K, Bharath G, Banat F, Show PL. Green synthesis of zinc oxide nanoparticles using *Phoenix dactylifera* waste as bioreductant for effective dye degradation and antibacterial performance in wastewater treatment. J Hazard Mater 2021, 402, 123560.
[36] Sheet I, Kabbani A, Holail H. Removal of heavy metals using nanostructured graphite oxide, silica nanoparticles and silica / graphite oxide composite. Energy Proc 2014, 50, 130–138.
[37] Kołodyńska D, Araucz K. New titanium oxide sorbent for As (V) and Cr (VI) removal as well as La (III) and Nd (III) recovery. J Mol Liq 2020, 315, 113720.
[38] Hasnain MS, Nayak AK. Background: Carbon nanotubes for targeted drug delivery. Briefs Appl Sci Technol Springer, 2019.
[39] Narvaez-Muñoz CP, Carrion-Matamoros LM, Vizuete K, Debut A, Arroyo CR, Guerrero V, Almeida-Naranjo CE, Morales-Flórez V, Mowbray DJ, Zamora-Ledezma C. Tailoring organic-organic poly (vinylpyrrolidone) microparticles and fibers with multiwalled carbon nanotubes for reinforced composites. ACS Appl Nano Mater 2019, 2, 4302–4312.
[40] He K, Zeng Z, Chen A, Zeng G, Xiao R, Xu P, Huang Z, Shi J, Hu L, Chen G. Advancement of Ag–Graphene based nanocomposites: An overview of synthesis and its applications. Small 2018, 14, 1–13.
[41] Liu X, Ma R, Wang X, Ma Y, Yang Y, Zhuang L, Zhang S, Jehan R, Chen J, Wang X. Graphene oxide-based materials for efficient removal of heavy metal ions from aqueous solution: A review. Environ Pollut 2019, 252, 62–73.
[42] Suppan G, Briones-Macías M, Pazmiño Arias E, Zamora-Ledezma C. Fabrication and characterization of metal-free composite electrodes based on few-layer-graphene nanoplatelets for oxygen reduction reaction applications. Phys Status Solidi 2021, 2000515.
[43] Chen Z, Mitchell S, Vorobyeva E, Leary RK, Hauert R, Furnival T, Ramasse QM, Thomas JM, Midgley PA, Dontsova D, Antonietti M, Pogodin S, Lopez N, Perez-Ramírez J. Stabilization of single metal atoms on graphitic carbon nitride. Adv Func Mater 2017, 27, 1605785. 12.
[44] Wang L, Xu H, Gao J, Yao J, Zhang Q. Recent progress in metal-organic frameworks-based hydrogels and aerogels and their applications. Coord Chem Rev 2019, 398, 213016.
[45] Kobielska PA, Howarth AJ, Farha OK, Nayak S. Metaleorganic frameworks for heavy metal removal from water. Coord Chem Rev 2018, 358, 92e107.
[46] Li CP, Zhou H, Wang S, Chen J, Wang ZL, Du M. Highly efficient Cr2O72-removal of a 3D metal-organic framework fabricated by tandem single-crystal to single-crystal transformations from a 1D coordination array. Chem Commun 2017, 53, 9206e9209.
[47] Sikora E, Hajdu V, Muranszky G, Katona KK, Kocserha I, Kanazawa T, Fiser B, Viskolcz B, Vanyorek L. Application of ion-exchange resin beads to produce magnetic adsorbents. Chem Pap 2020, 1, 3.
[48] Jian M, Liu B, Zhang G, Liu R, Zhang X. Adsorptive removal of arsenic from aqueous solution by zeolitic imidazolate framework-8 (ZIF-8) nanoparticles. Colloids Surf A Physicochem Eng Asp 2015, 465, 67e76.
[49] Luo F, Chen JL, Dang LL, Zhou WN, Lin HL, Li JQ, Liu SJ, Luo MB. High-performance Hg2þ removal from ultra-low-concentration aqueous solution using both acylamide- and hydroxyl-functionalized metal-organic framework. J Mater Chem 2015, 3, 9616e9620.

[50] Li J, Wang X, Zhao G, Chen C, Chai Z, Alsaedi A, Hayat T, Wang X. Metaleorganic framework-based materials: Superior adsorbents for the capture of toxic and radioactive metal ions. Chem Soc Rev 2018, 47, 2322e2356.

[51] Wang H, Mi X, Li Y, Zhan S. 3D Graphene-based macrostructures for water treatment. Adv Mater 2020, 1806843 1–11.

[52] Deb VK, Rabbani A, Upadhyay S, Bharti P, Sharma H, Rawat DS, Saxena G. Microbe-assisted phytoremediation in reinstating heavy metal-contaminated sites: Concepts, mechanisms, challenges, and future perspectives. In: Arora P (ed.). Microbial technology for health and environment. Microorganisms for sustainability 2020, 161–189.

[53] Devi P, Kumar P. Concept and application of phytoremediation in the fight of heavy metal toxicity. J Pharm Sci Res 2020, 12.

[54] Abdel Maksoud MIA, Elgarahy AM, Farrell C, Al-Muhtaseb AH, Rooney DW, Osman AI. Insight on water remediation application using magnetic nanomaterials and biosorbents. Coord Chem Rev 2020, 403, 213096.

[55] Haldar S, Ghosh A. Microbial and plant – assisted heavy metal remediation in aquatic ecosystems: A comprehensive review. Biotech 2020, 3, 1–13.

[56] Gholami S, Dimde M, Braatz D, Müller J, Haag R, Wagner O. Reusable biopolymer based heavy metal filter as plant protection for phytoremediation. Environ Technol Innov 2020, 19, 101005.

[57] Khan AG. Promises and potential of in situ nano-phytoremediation strategy to mycorrhizo-remediate heavy metal contaminated soils using non-food bioenergy crops (vetiver zizinoides & cannabis sativa). Int J Phytoremed 2020, 22, 900–915.

[58] Joseph L, Jun BM, Flora JRV, Park CM, Yoon Y. Removal of heavy metals from water sources in the developing world using low-cost materials: A review. Chemosphere 2019, 229, 142–159.

Index

https://doi.org/10.1515/9783111016825-017

Bei Fragen zur Produktsicherheit wenden Sie sich bitte an:
If you have any questions regarding product safety,
please contact:

Walter de Gruyter GmbH
Genthiner Straße 13
10785 Berlin
productsafety@degruyterbrill.com